PRINCIPLES AND PRACTICES OF LIGHT CONSTRUCTION

PRINCIPLES AND PRACTICES OF LIGHT CONSTRUCTION

Fourth Edition

Ronald C. Smith

*Structures Department,
Southern Alberta Institute
of Technology (retired)*

Ted L. Honkala

*Structures Department,
Southern Alberta Institute
of Technology*

PRENTICE-HALL, *Englewood Cliffs, NJ 07632*

Library of Congress Cataloging in Publication Data

SMITH, RONALD C.
Principles and practices of light construction.

Includes index.
1. Building. 2. Carpentry. I. Honkala, T. L.
II. Title.
TH145.S58 1986   690   85-13840
ISBN 0-13-702085-6

Editorial/production supervision and
interior design: *Jane Zalenski and Tom Aloisi*
Manufacturing buyer: *John Hall*

Printed in the United States of America

10   9   8   7   6   5   4   3   2   1

ISBN: 0-13-702085-6   025

Prentice-Hall International (UK) Limited, *London*
Prentice-Hall of Australia Pty. Limited, *Sydney*
Prentice-Hall Canada Inc., *Toronto*
Prentice-Hall Hispanoamericana, S.A., *Mexico*
Prentice-Hall of India Private Limited, *New Delhi*
Prentice-Hall of Japan, Inc., *Tokyo*
Prentice-Hall of Southeast Asia Pte. Ltd., *Singapore*
Editora Prentice-Hall do Brasil, Ltda., *Rio de Janeiro*
Whitehall Books Limited, *Wellington, New Zealand*

CONTENTS

PREFACE

The demand for new buildings, large and small, continues to be heard throughout the world. Increases in population, a continuing rise in the standard of living in many areas, and urban renewal, and rural development plans all contribute to the demand.

New ideas and theories of construction are continually arising, while improved methods and techniques in building are being developed. The concept of prefabrication, for example, has become widespread and applies to almost every facet of the construction industry.

In recent years, the concern with energy conservation has brought forth the evolution of energy-efficient buildings. This book has been written with the hope that it will assist in developing competence in both the conventional and contemporary building arts for those who are, or will be, associated with the light construction industry.

The methods of achieving that purpose are basically threefold. The authors have endeavored to give an accurate, up-to-date account of conventional methods used in light construction. The second method is to elaborate on some of the new ideas in design and construction which have been put into practice. The third method is to help the student of construction realize the importance of construction planning.

We wish to acknowledge my debt and to express my gratitude to all those with whom I have worked through the years and to those who have so generously contributed illustrations and other material. Without their help and cooperation, this book would not have been possible.

Ronald C. Smith
Ted L. Honkala

LIST OF TABLES

PRINCIPLES AND PRACTICES OF LIGHT CONSTRUCTION

Tools are an essential part of the woodworking trades, and, until recent years, carpenters, cabinetmakers, millworkers, and others who are involved in the craft of shaping wood have relied on hand tools to get the job done. Now some of these tools have been adapted to be run by electric power, which makes their operation faster and easier, while, with others, compressed air has been utilized to provide the power.

Over the years tools have been improved in many ways. They have been made lighter, stronger, and of better material. Special tools have been developed to do specific jobs, while others have been redesigned to perform a number of operations.

Probably in no other trade may the workman be called upon to perform so many different operations in the course of his or her day's work as in the woodworking trades. Therefore he/she must have at his/her disposal and be familiar with a wide variety of tools, each with a specific use. It is the purpose of this chapter to describe and explain the use of the common tools—*hand, electric,* and *pneumatic*—which are available for use in modern light construction.

TOOLS

HAND TOOLS

Hand tools may be divided into the following groups, based on the type of work done with them:

1. Assembling tools
2. Boring tools
3. Cutting tools
4. Holding tools
5. Layout and marking tools
6. Leveling and plumbing tools
7. Measuring tools
8. Sharpening tools
9. Smoothing tools
10. Wrecking tools

On occasion we may find a tool which has a place in more than one of these groups. Such cases will be pointed out.

(a) Claw hammer.

(b) Ball peen hammer.

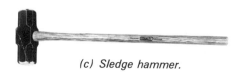

(c) Sledge hammer.

FIGURE 1–1:   *Hammers.*

*(a) Nailing hammer,
steel handle.*

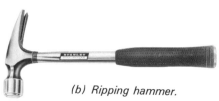

(b) Ripping hammer.

FIGURE 1–2:   *Hammer claws.*

FIGURE 1–4:   *Block to assist pulling.*

Assembling Tools

This group includes such tools as *hammers,* *screwdrivers,* *wrenches,* *nail sets,* and *staplers.*

Hammers. Hammers are classified first of all according to the type of work done with them. Three of the most common classifications are claw hammers, peen hammers, and sledgehammers. The claw hammer is used for driving nails, the peen hammer is for metalwork, and the sledgehammer is for driving stakes [see Fig. 1–1(a), (b), and (c)].

Claw hammers are made with either curved claws for nail pulling or straight claws for ripping and wrecking [see Fig. 1–2(a) and (b)]. The face of a nail hammer may be belled, plain, or knurled (see Fig. 1–3). The weight of a hammer is usually 7–13 oz (200–370 g) for light work, 16–20 oz (455–570 g) for general work, and 22–28 oz (625–800 g) for framing work. You should choose the hammer for the job you have in mind—for framing work the heaviest hammer you can find for driving large nails. On the other hand, for finishing work a very light hammer is best.

(a) Belled.   *(b) Knurled.*   *(c) Plain.*

FIGURE 1–3:   *Hammer faces.*

Hammer handles may be made of wood, steel, or fiberglass. The wooden-handled hammer is preferred by some workmen; they feel it has more give or spring. Others prefer the steel- or fiberglass handled hammers, which are very strong and less likely to break (see Figs. 1–1 and 1–2). Steel-shanked hammers may be furnished with leather or plastic grips, and fiberglass hammers have plastic grips.

Claw hammers should never be used for purposes other than driving or pulling regular nails. Hardened nails, such as those used for concrete, should be driven with a heavier peen hammer, because the metal in the nail may be harder than the hammer face and may damage it. Also, extensions such as sections of pipe should never be slipped over the handle to give more pulling power. If the nail is very hard to pull, use a block, a bar, or a nail puller (see Fig. 1–4).

When a hammer is being used, the handle should be gripped near the end so the entire length of the handle provides leverage.

You should not hold a hammer high on the handle near the head. That way, there is no power behind the swing, and you stand a good chance of missing the nail and putting a dent in the wood. Hold the handle firmly, use short strokes to start the nail, and then drive it home with longer strokes. When driving small nails with short strokes, use only your wrist and forearm in your swing. Spikes need the force of your entire arm to drive them easily (see Fig. 1–5).

FIGURE 1–5: *Hammer grip.*

Screwdrivers. Screwdrivers are made in three types: *flat blade,* for use with slotted screws; *Robertson,* for use with Roberson head screws, which have a square pocket in the head; and *Phillips,* for use with Phillips head screws, which have an indented cross in the head (see Fig. 1–6).

Flat blade screwdrivers are made with a plain handle (Fig. 1–7), with a ratchet handle (Fig. 1–8), or with no handle but with a tapered end (see wood bit) to fit in a bit brace to give the workman extra leverage for driving heavy screws.

Slotted head

Robertson head

Phillips head

FIGURE 1–6: *Screw head types.*

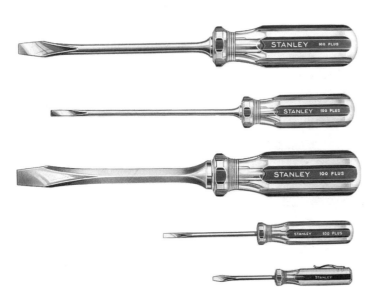

FIGURE 1–7: *Plain-handled screwdrivers.*

FIGURE 1–8: *Ratchet screwdriver.*

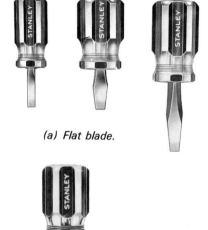

(a) Flat blade.

(b) Phillips.

FIGURE 1-9: *"Stubby" screwdrivers.*

FIGURE 1-10: *Robertson screwdriver.*

FIGURE 1-11: *Phillips screwdrivers.*

FIGURE 1-12: *Nail set.*

The blades of screwdrivers are made in a number of widths and thicknesses to accommodate various sizes of screws (see Fig. 1-7). Some—called *stubbies*—are made with very short blades and handles for use in hard-to-get-at places (see Fig. 1-9). One of these is a handy addition to a tool kit.

Robertson screwdrivers are made in a variety of lengths with four tip sizes—Nos. 0, 1, 2, and 3—to fit the standard sizes of screw sockets (see Fig. 1-10).

Phillips screwdrivers are also made in a variety of lengths and four tip sizes—Nos. 0, 1, 2, and 3—to fit a number of screw sizes (see Fig. 1-11).

Nail sets. Nail sets are designed to drive nail heads below the surface of the wood. The diameter of the tip is the size, ranging from $\frac{1}{16}$ in. (1 mm) to $\frac{3}{16}$ in. (5 mm). The shank is usually knurled for better grip (see Fig. 1-12).

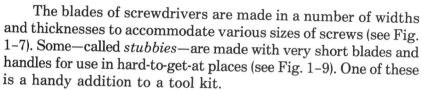

(a) Adjustable.   (b) Open end.

(c) Box end wrench.   (d) Pipe wrench.

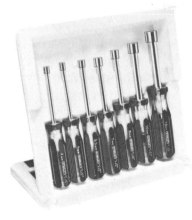

(e) Allen wrenches.   (f) Sockets.

FIGURE 1-13: *Wrenches.*

Wrenches. Wrenches are not ordinarily considered to be carpenter's tools, but one or two can be very useful in a tool box. Wrenches are classified as Allen, socket, open end, box, adjustable, and pipe (see Fig. 1–13). The size of the adjustable and pipe wrenches is designated by length, varying from 6 in. (150 mm) to 24 in. (600 mm). An 8-in. (200-mm) adjustable wrench is considered to be a good general-purpose size. The other wrenches may be purchased in sets, and a set of small sizes will be a wise addition to the tool box.

Staplers. Staplers perform a variety of operations formerly done by hand nailing (see Fig. 1–14). They provide an efficient method of attaching building paper, vapor barrier, ceiling tile, and roofing materials. Staplers are very convenient to use as they leave one hand free to hold the material in place.

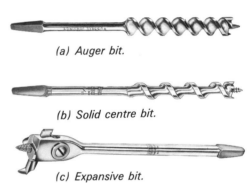

(a) *Hand tacker.*

(b) *Hammer type stapler.*

FIGURE 1-14:

Boring Tools

Included in this group are the tools which actually cut holes in wood or metal—*bits* and *drills*—as well as the tools which hold and turn them—*bit braces, hand drills,* and *push drills.* In addition there are *countersink bits, bit extensions, bit depth gauges,* and *expansive bits.*

Wood bits. Wood bits are made in several styles; two of the more common ones are illustrated in Fig. 1–15, an *auger* bit and a *solid center,* both of which have the same basic parts.

The small, threaded tip at the cutting end is the *feed screw,* which pulls the bit into the wood when it is turned. At the base of the feed screw are two *cutting lips,* which cut the wood from the bottom of the hole. At their outer ends are the two *spurs,* which cut the circumference of the hole. Above these is the *twist,* which carries the wood chips up out of the hole. At the end of the twist is the *shank* and, on the end of the shank, the *tang,* by which the bit is held in the brace.

Wood bits are commonly made in sizes ranging from ¼ to 1¼ in. (6–31 mm) in diameter, increasing in size by increments of ¹⁄₁₆ in. (1.5 mm). A common set, suitable for most tool boxes, will include bits from ¼ to 1 in. (6 to 25 mm). Holes over 1 in. (25 mm) in diameter are commonly bored with an *expansive bit* (see Fig. 1–15), in which an adjustable blade can be set to drill a hole of any size up to its maximum capacity. The bit is equipped with two blades, one boring up to 1½ in. (38 mm) and the other from 1½ to 3 in. (38–75 mm).

Another type of wood bit, the *Forstner* bit, should be mentioned. It is unique in that it has no feed screw and no twist. It is made in the form of a shallow, straight-sided cup with the cutting lips across the open end (see Fig. 1–16). Since it bores a flat-bottomed hole, it is very useful when it is required to bore only part way through a board, as the feed screw of an ordinary bit

(a) *Auger bit.*

(b) *Solid centre bit.*

(c) *Expansive bit.*

FIGURE 1-15: *Wood bits.*

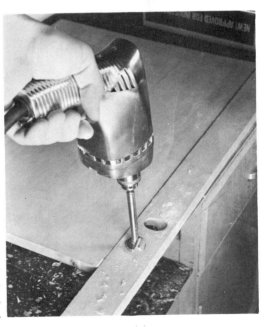

FIGURE 1-16: *Forstner bit in electric drill.*

FIGURE 1-17: *Countersinks.*

FIGURE 1-18: *Countersink bits.*

FIGURE 1-20: *Push drill.*

FIGURE 1-21: *Bit depth gauge.*

might mar the underside of the surface. The type illustrated is for use in an electric drill, but Forstner bits are also made for use in hand braces.

Countersink bits. Countersink bits are used to widen the top of a screw hole so the head of a flat head screw may be set flush with or slightly below the surface. The two illustrated in Fig. 1–17 do the same job, but one is intended for use in a hand brace and the other in a drill press. The bits illustrated in Fig. 1–18 drill a pilot hole at the same time as they countersink the screw head.

Bit extension. Wood bits vary in length from 7 to 10 in. (175–250 mm). If a very deep hole is required, such a bit may not reach far enough; then a bit extension may be used. It is a tool which will hold a wood bit in one end while the other end is held in a hand brace (see Fig. 1–19). It is made up to 24 in. (600 mm) long, so the possible depth of hole bored can be greatly increased.

FIGURE 1-19: *Bit extension.*

Push drill. When holes smaller than ¼ in. (6 mm) are required in wood, some tool other than a wood bit must be used. One such tool is a push drill. This is a spring-loaded tool, made so that when pressure is applied, the bit turns in one direction; when the pressure is released, it turns in the opposite direction, cutting both ways. Bits are interchangeable, usually sold in sets of eight from ¹⁄₁₆ to ³⁄₁₆ in. (1.5–4.5 mm). The top is removable so that a set of bits may be stored inside (see Fig. 1–20).

Bit depth guage. Sometimes it is necessary to bore holes of the same depth or to bore a specified depth. To ensure the proper depth, a bit depth gauge is used. It is clamped to a bit, measuring from the cutting lips to the flat lower end to get the required depth. It will fit any wood bit up to 1 in. (25 mm) (see Fig. 1–21).

Bit brace. Figure 1–22 illustrates a standard bit brace. It is used to hold and turn wood bits. This is a *ratchet* brace, which means that it can be set to drive in one direction only. This feature is very helpful when one is boring in a position where a complete turn of the handle cannot be made.

FIGURE 1-22: *Hand brace with box ratchet. (Courtesy Stanley Tools).*

A specialized type of bit brace is the one shown in Fig. 1-23. It is called a short brace. It has two major advantages over an ordinary brace in some places. First, the distance from the head to the chuck is less, so it can fit into smaller spaces. Second, it is operated by a straight ratchet handle, so it may be used close to a wall or other flat surface.

Twist bits. Another method of drilling small holes is to use a *twist bit.* It is the same diameter from one end to the other, has no tang, and so requires a special chuck to hold it. The cutting end is tapered to a point, the cutting lips each forming an angle of about 60° with the center line of the bit. This type of bit may be used for either wood or metal drilling (see Fig. 1-24).

Drill. The tool used for holding and turning twist bits is illustrated in Fig. 1-25. It is called a *hand drill* and is effective for drilling small holes.

Cutting Tools

Cutting tools include *saws, chisels, axes, snips,* and *knives.* In each of these major categories there are a number of styles or varieties, each adapted for a specific purpose.

Saws. Included in this category are *handsaws, backsaws, compass* saws, *coping* saws, *utility* saws, and *hacksaws.*

Handsaws. Two types of handsaws are made: one designed to cut across the fibers of wood—a *crosscut* saw—and the other designed to cut along the fibers— a *ripsaw.* The main difference between them is the method of shaping and sharpening the teeth.

Crosscut saws have teeth with the cutting edge sloped forward and filed at an angle, so that each tooth, as it is drawn across the wood, severs the fibers like a knife. Ripsaws have the front of the teeth at right angles, or very nearly so, to the blade. As a result the top of each tooth is a cutting edge and acts like a chisel (see Fig. 1-26).

FIGURE 1-23: *Short brace.*

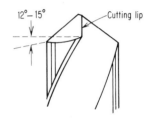

FIGURE 1-24: *Twist bit cutting lips.*

FIGURE 1-25: *Hand drill.*

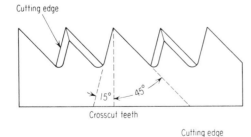

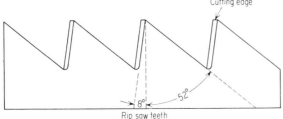

FIGURE 1-26: *Handsaw teeth.*

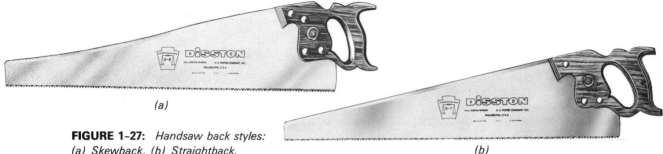

FIGURE 1-27: *Handsaw back styles:*
(a) Skewback, (b) Straightback.

(a)

(b)

Both crosscut saws and ripsaws are made with curved—*skew*—backs and *straight* backs (see Fig. 1-27), and in some saws the blade is tapered; that is, it is made thicker at the cutting edge than it is at the back so that the saw will run more smoothly in the saw kerf (see Fig. 1-28).

FIGURE 1-28: *Full tapered sawblade (Courtesy Disston Inc.).*

Crosscut saws are made in two sizes: *standard,* usually 26 in. (650 mm) long, and *panel,* about 20–22 in. (500–550 mm) long, with finer teeth for finish work.

All handsaws are designated by the number of teeth or *points* they have per unit length—1 in. (25 mm). This number will vary from 5 to 16, depending on the type of saw. The more teeth per unit length, the smaller they will be and consequently the finer the cut that should be obtained (see Fig. 1-29).

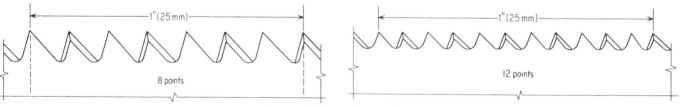

1"(25 mm)

8 points

1"(25 mm)

12 points

FIGURE 1-29: *Fine and coarse saw teeth.*

Backsaws. Backsaws are made with a stiff rib along the back and are intended for fine cutting and small work. Several styles are available, including the *standard* backsaw, 12–14 in. (300–350 mm) with 12–15 teeth per inch (TPI) (see Fig. 1-30). The *dovetail* saw, 10 in. (250 mm) long with 16 TPI, is a smaller version of the backsaw, with a lathe-turned handle in line with the spine.

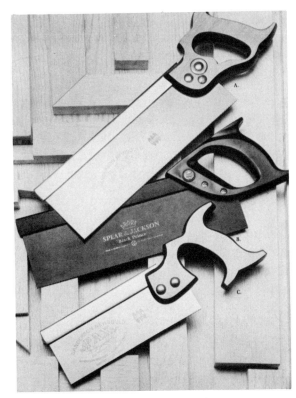

FIGURE 1–30: *Standard backsaws.*

FIGURE 1–31: *Dovetail saw.*

This saw is preferred by cabinetmakers for fine joint work (see Fig. 1–31). A *miter* saw is a long saw made to fit in a frame with guides to hold it in place (see Fig. 1–32). Its purpose is to cut angles or miters, and so the saw and its guides are on an adjustable arm which can be swung through an arc of 90°, up to 45° on either side of the right-angle position. It may be locked in any position to ensure an accurate miter cut. A miter box is also available that will utilize a panel saw in place of a backsaw (see Fig. 1–33).

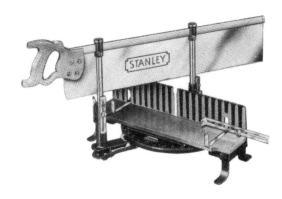

FIGURE 1–32: *Miter saw and frame. (Courtesy Stanley Tool Co.).*

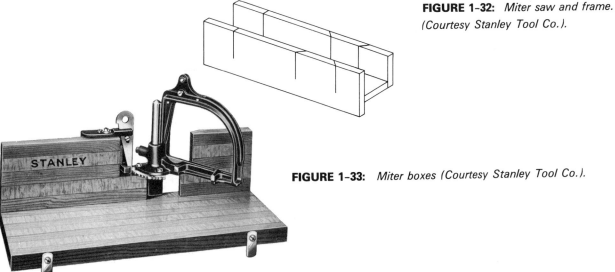

FIGURE 1–33: *Miter boxes (Courtesy Stanley Tool Co.).*

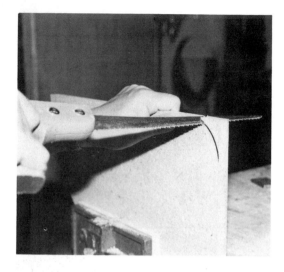

(a) Compass saw in use.

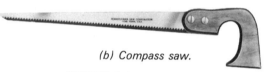

(b) Compass saw.

FIGURE 1-34:   *Compass saw.*

Compass saws. Compass saws are made with narrow, tapered blades and teeth designed to cut either along or across the grain. They are used for cutting holes or sawing along curved or irregular lines (see Fig. 1-34).

Coping saw. A coping saw consists of a bow frame fitted with a very fine blade and is used for cutting thin, curved work. It is very useful in interior finishing work for making *coped* joints in small moldings, casing, baseboard, etc. (see Fig. 1-35).

Utility saw. A utility saw consists of a thin, hard blade with small teeth, which is useful for cutting hard materials such as metal or plastic laminates or gypsum board or similar materials which tend to dull normal saw teeth. The one shown in Fig. 1-36 has a detachable handle which can be rotated into any position.

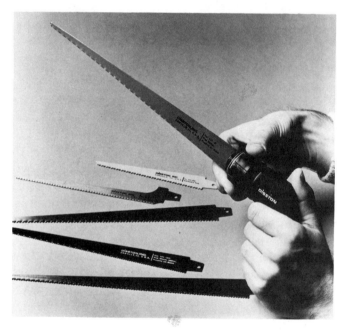

FIGURE 1-36:   *Utility saw with detachable handle.*

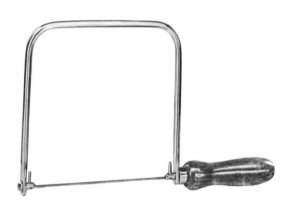

FIGURE 1-35:   *Coping saw. (Courtesy Stanley Tools).*

Hacksaw. A hacksaw is an indispensable metal-cutting saw in your tool box. Most saws will adjust to take 10-in. (250-mm) to 12-in. (300-mm) blades, which will vary from 18 to 32 TPI, depending on the thickness of the metal being cut. The blade should be fine enough so that at least two teeth are in contact with the metal at all times (see Fig. 1-37).

FIGURE 1-37:   *Hacksaw (Courtesy Stanley Tools).*

FIGURE 1-38:
Using a handsaw.

(a) Guiding blade with thumb.

(b) Cross-cut saw angle (45°).

(c) Ripping saw angle (60°).

How to use a handsaw. To begin a cut with a handsaw, rest the blade on the edge of the work, on the waste side of the cutting line. Steady the blade with your thumb [see Fig. 1-38(a)]. Draw the saw toward you slowly and carefully several times until a slight groove is formed. After the cut is started, use long, easy strokes with light pressure on the forward stroke. You will cut most easily with a crosscut if the saw is held at an angle of 45° to the work [see Fig. 1-38(b)]. After cutting has been started, be sure to place your body in a position that will enable you to see the cutting line. The saw, the forearm, and the shoulder should form a straight line at right angles to the work. Be sure to support the work properly [see Fig. 1-38(d)]. Resting it on sawhorses is preferable, but if they are not available, the work may be held in a vise [see Fig. 1-38(e)]. Be sure to support the piece that is being cut off so that it does not break away and splinter the underside of the stock.

When ripping is done, the saw should be held at an angle of 60° to the work [see Fig. 1-38(c)]. If the saw binds when you are ripping long stock, insert a wedge into the cut some distance from the blade.

When it is not being used, a saw should be hung up or placed in a tool box. Make sure other tools are not piled on top of the saw; there should be a place for all tools when they are not in use. A good saw is an important tool and deserves the best of care.

(d) Supporting work with sawhorses.

Chisels. Chisels are divided into two main classifications, depending on whether they are designed to cut wood or metal. Those intended for cutting wood are called *wood* chisels, whereas those used for cutting metal are called *cold* chisels (see Fig. 1-39).

(a) Wood chisel.

(b) Cold chisel.

FIGURE 1-39: *Chisel types.*

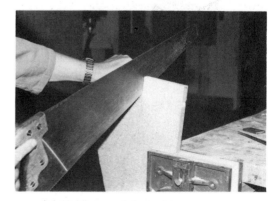

(e) Holding work in a vise.

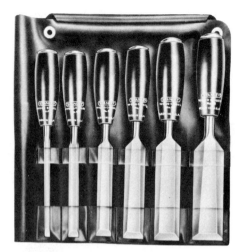

FIGURE 1-40: *Chisel set. (Courtesy Stanley Tools).*

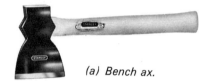

(a) Bench ax.

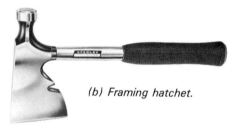

(b) Framing hatchet.

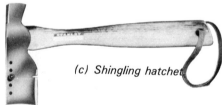

(c) Shingling hatchet.

FIGURE 1-42: *Hatchets. (Courtesy Stanley Tools).*

Chisel widths range from $\frac{1}{8}$ to 2 in. (3–50 mm), but $\frac{1}{4}$-, $\frac{3}{8}$-, $\frac{1}{2}$-, $\frac{3}{4}$-, and $1\frac{1}{4}$-in. sizes are most commonly chosen. Figure 1–40 illustrates a very satisfactory set of chisels for the tool box.

The companion tool for a chisel is a *mallet*—never a hammer. The softer head of a wooden or plastic mallet will not damage the head of your chisel as a hammer will (see Fig. 1–41).

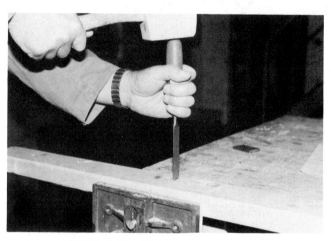

FIGURE 1-41: *Using mallet on head of chisel.*

Axes. In more primitive types of construction the ax was a very useful tool. It was used for felling timber, hewing surfaces flat, cutting notches in logs, and many similar jobs. Today, however, with some changes in size, shape, and weight, the ax is being put to quite different uses. One type, short-handled and broad-bladed, is known as a *bench ax* or *hatchet.* It is used for rough cutting, stake sharpening, etc. [see Fig. 1–42(a)]. Another similar style with a hammer head and a nail slot in the back of the blade is a *framing hatchet* [see Fig. 1–42(b)]. An ax of different style, which has a narrow blade and slim head, is used for laying shingles. The distance between the head and the adjustable pin is a standard exposure distance for shingles [see Fig. 1–42(c)].

Snips. Snips are specifically designed for cutting thin sheet metal, though they can be useful in cutting other materials such as aluminum or vinyl [see Fig. 1–43(a)]. The snips in Fig. 1–43(b) are excellent for cutting smaller circles and can be purchased for left- and right-hand curves as well as straight cuts.

(a)    *(b)*

FIGURE 1-43:   *(a) Tin snips (Courtesy Stanley Tools)*   *(b) Metal masters.*

Knives. Utility knives are an excellent tool for cutting thin veneers, sheathing paper, polyethylene, and gypsum board (see Fig. 1–44). A special hooked blade is designed for cutting asphalt shingles.

FIGURE 1–44: *Utility knife. (Courtesy Stanley Tools).*

Holding Tools

Included in this group are *pliers, C-clamps, bar* or *pipe clamps, adjustable hand screws,* and *vises.*

Pliers. Pliers are made in many sizes, shapes, and styles. Some common types are *adjustable, side-cutting,* and *locking.* The adjustable pliers may be put to any one of a large number of holding uses; the side-cutters are particularly useful for cutting wire, and the vise grip is excellent for holding material on which adjustable pliers may slip (see Fig. 1–45).

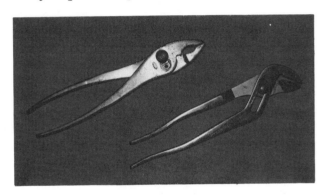

(a) Adjustable pliers.

FIGURE 1–46: *C-clamp (Courtesy Stanley Tools).*

(b) Sidecutting pliers.

FIGURE 1–45: *Pliers.*

C-clamps. C-clamps are useful for holding small sections of lumber together, perhaps for gluing. They are made in a number of common sizes from 3 to 16 in. (75–400 mm), the size indicating the depth of the throat opening (see Fig. 1–46).

Adjustable hand screws. Adjustable hand screws are made of two parallel wooden jaws operated by two long screws. They too are made in a number of sizes, the size indicating the length of the jaw (see Fig. 1–47).

FIGURE 1–47: *Adjustable hand screws.*

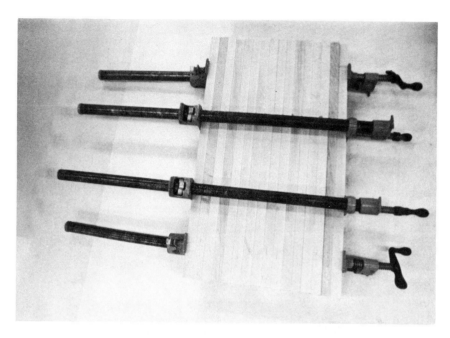

FIGURE 1-48: *Pipe clamps.*

Bar and pipe clamps. These consist of a screw head and a movable tail block mounted on a bar or pipe. They are most useful for gluing pieces together, edge to edge. If pipe is used, almost unlimited lengths may be obtained by joining pieces of pipe together with couplings. A good set of tools should include at least two 6 in. (150 mm) C-clamps and a pair of 3 ft (1 m) bar or pipe clamps (see Fig. 1-48).

Woodworkers' vise. This is a small, sturdy vise that may be clamped to a bench or sawhorse to hold work. It is particularly useful on a sawhorse to hold doors or window sash while they are being dressed to size (see Fig. 1-49).

FIGURE 1-49: *Woodworker's vise in use.*

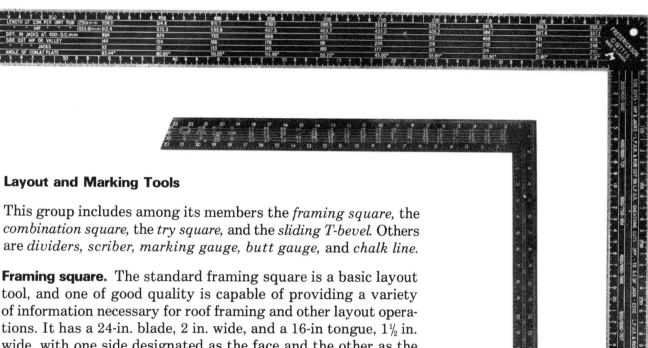

Layout and Marking Tools

This group includes among its members the *framing square*, the *combination square*, the *try square*, and the *sliding T-bevel*. Others are *dividers*, *scriber*, *marking gauge*, *butt gauge*, and *chalk line*.

Framing square. The standard framing square is a basic layout tool, and one of good quality is capable of providing a variety of information necessary for roof framing and other layout operations. It has a 24-in. blade, 2 in. wide, and a 16-in tongue, 1½ in. wide, with one side designated as the face and the other as the back (see Fig. 1–50). The Frederickson square, shown in Fig. 1–51, has a 600-mm blade, 50 mm wide, and a 400-mm tongue, 38 mm wide. Both squares contain an *octagon* table, a *rafter* table, and a *brace* table. Tables dealing with rafters are found in Chapter 7.

The framing square is especially useful for laying out wall plates and roof angles. Good squares are available in steel, aluminum, copper-clad metal, and painted surfaces. Aluminum is light and easy to handle but less likely to withstand rough usage than are some of the others. Steel is strong and sturdy, but may rust, in wet weather unless great care is taken. Copper-clad squares are weather resistant, but are not so easy to read. Painted squares are easy to read, but are susceptible to scratching.

Combination square. This tool is a versatile one and, as its name implies, does the work of several simple tools. The blade slides in a slot in the head and may be locked in any position (see Fig. 1–52). One edge of the head is at right angles to the blade; another is at an angle of 45°. Some manufacturers include in the head two spirit bubbles so that the tool may be used for plumbing and leveling.

FIGURE 1–50: *Imperial framing square.*

FIGURE 1–51: *Metric framing square.*

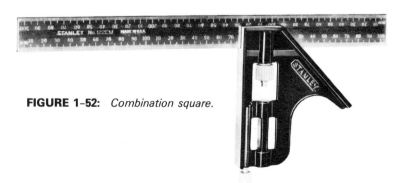

FIGURE 1–52: *Combination square.*

Two more heads are available that will take the same blade. They are a *protractor* head and a *centering* head. With the protractor head on, the tool is used to measure and lay off angles by degrees. With the centering head, the tool is used to locate the center of a circular surface.

Try square. The basic purpose of a try square is to check inside or outside angles for squareness. They are small and sturdily built, with blades ranging from 6 to 12 in. (150–300 mm) in length and steel or wooden handles from 3 to 8 in. (75–200 mm) long (see Fig. 1-53).

FIGURE 1-53: *Try square. (Courtesy Stanley Tools).*

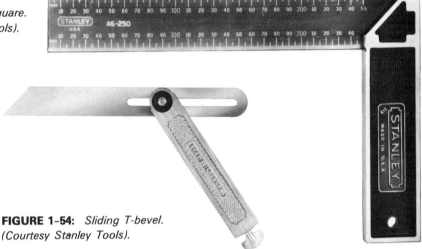

FIGURE 1-54: *Sliding T-bevel. (Courtesy Stanley Tools).*

Sliding T-bevel. The T-bevel, or bevel square, has an adjustable blade which makes it possible to set the tool to any angle desired. Once set, the angle may be duplicated as many times as necessary. The T-bevel is particularly useful in rafter cutting. Once a pattern is laid out and cut, the bevel is set to the exact angle and used to mark all the other rafters (see Fig. 1-54).

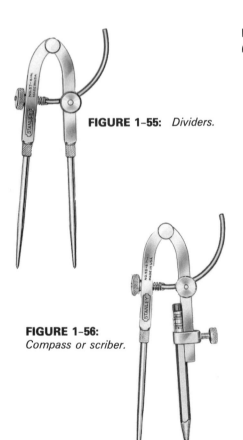

FIGURE 1-55: *Dividers.*

FIGURE 1-56: *Compass or scriber.*

Dividers. A pair of dividers is a very useful tool when it is necessary to lay out a number of equal spaces, as in stair layout work (see Fig. 1-55). One leg has a metal guide bar in the shape of an arc passing through it. This leg may be clamped at any position along this guide bar when the desired step size has been determined. Notice that the end of the guide bar is threaded, passes through the other leg, and has a knurled nut at the end. This arrangement enables the workman to make very fine adjustments in the length of step he/she has set on the dividers.

Scriber. Very similar in design to dividers, a scriber has one metal leg, and a pencil clamped in place serves as the other (see Fig. 1-56). Its main purpose is to reproduce the curves or irregularities of a wall or other such surface on a board, which must be made to fit tightly against it. It is used by holding the piece which must be made to fit in the position which it is supposed to take.

Then, by following the face of the wall with the fixed leg of the scriber, you draw out the shape to be formed. A scriber may also be used as a compass.

Marking gauge. This tool is used to lay out lines parallel to the edges of stock. It consists of a stem marked off in inches (millimeters) and an adjustable head. The face of the head is reinforced with brass for better wear (see Fig. 1–57). When it is being used, the gauge should be pushed away from you and held so that the marking pin trails rather than points straight down. A combination square may be used to do the same job, a pencil being held at the end of the blade to do the marking.

FIGURE 1–57: *Marking gauge.*

Butt gauge. Here is an example of a tool that has been designed for one specific purpose—to lay out the gains for butt hinges. It consists of a metal block containing two sliding pins (see Fig. 1–58). One pin has one marker attached, the other two. The single marker is set to gauge the depth of the hinge gain. The others mark the distance from the back edge of the door that the hinge will reach and the distance from the inner face of the door rabbet that the jamb leaf of the hinge will reach.

A more commonly used tool is a *butt template*, which not only marks the butt location but also cuts around the edge (see Fig. 1–59).

FIGURE 1–58: *Butt gauge.*

Chalk line. A chalk line is an easy way to mark long lines. The chalk-covered line is stretched and held close to the surface to be marked and is snapped. This action leaves a distinct line on the surface. A special reel rechalks the line each time it is wound into the case (see Fig. 1–60).

FIGURE 1–59: *Butt gauge template.*

(a)

(b)

FIGURE 1–60: *(a) Chalk line.*
(b) Chalk line in use.

Leveling and Plumbing Tools

These are tools that are used to check the levelness or plumbness of structures being erected. In this category are the *spirit level, line level, plumb bob,* and *builder's level.* The last is an expensive instrument and is not a regular part of a carpenter's tool kit.

Spirit level. This is really both a level and a plumb. The body is made from wood, aluminum, or magnesium, machined on top and bottom faces for accuracy. It contains three or more vial units, the bubbles in which indicate levelness or plumbness, depending on how the tool is held. Levels should be handled with great care to preserve their accuracy (see Fig. 1-61). For leveling or plumbing long lines, a straightedge should be used in conjunction with the spirit level.

FIGURE 1-61:   *Spirit level (Courtesy Stanley Tools).*

Line level. This is really a very small model of the spirit level, containing only a level bubble. It is used to hang on a building line to check its levelness (see Fig. 1-62).

FIGURE 1-62:   *Line level (Courtesy Stanley Tools).*

Plumb bob. The plumb bob is an ancient but very useful tool. It consists merely of a cone-shaped piece of steel or brass which hangs point down from a cord running from the center of the top (see Fig. 1-63). It is suspended from a point above and, when it comes to rest, will indicate a point on the earth directly below. It has many applications in construction when it is necessary to plumb down.

Builder's level. See Chapter 2 for a description of the builder's level.

Measuring Tools

Tools used for finding, laying out, and checking distances are included among the measuring tools. The more common ones include the *fourfold boxwood rule* (see Fig. 1-64), the spring-loaded pocket tape in lengths from 6 to 16 ft (2-5 m) [see Fig. 1-65(a)], and the rolled steel tape in lengths of 25, 50, and 100 ft (7.5, 15, and 30 m) [see Fig. 1-65(b)].

FIGURE 1-63:   *Plumb bobs. (Courtesy Stanley Tools).*

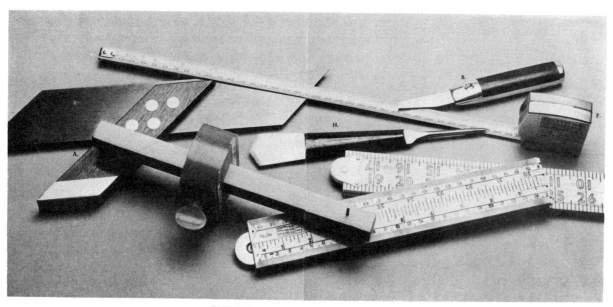

FIGURE 1-64: *Four-fold boxwood rule.*

Sharpening Tools

This group of tools is not used in performing woodworking operations. Their function is to help keep the woodworking tools sharp. Included in this group are *triangular files, mill files, saw sets, bit files, oilstones, slip stones, emery stones,* and *burnishers.*

Files. Triangular files are made specially for filing handsaw teeth. They are designated by length and by their cross-sectional size or taper. Lengths range from 5 to 10 in. (125–250 mm) and taper from *regular* to *double extra slim.*

Mill files are flat, made in single and double cut, in three sizes of teeth: *bastard, second cut,* and *smooth cut.* They are used for jointing handsaws and for general filing. They are available in lengths from 4 to 16 in. (100–400 mm) (see Fig. 1–66).

(a) Spring-loaded tape.

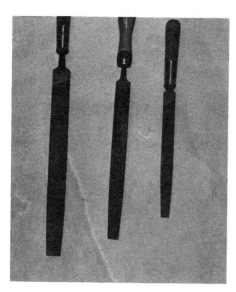

(b) Rolled steel tape.

FIGURE 1-65: *Steel tapes.*

FIGURE 1-66: *Mill files.*

FIGURE 1-67: *Bit file in use.*

Bit files are made especially for filing the cutting lips and spurs of wood-boring bits. One end has teeth on the flat faces but none on the edges, and the other end has teeth on the edges but none on the flat surfaces (see Fig. 1-67).

Sharpening stones. Of the three stones mentioned above, the slip stone is the finest, the oilstone is next, and the emery stone is the coarsest. A slip stone is used for sharpening gouges or other very fine edges and is sometimes used dry. The oilstone is used for sharpening plane blades and chisels and uses light oil as a lubricant. The emery stone is usually turned mechanically, either by hand or by electricity, and is used for grinding down plane blades, sharpening axes, and similar work. A small one which may be clamped to a bench or sawhorse is a useful addition to a tool kit (see Fig. 1-68).

FIGURE 1-68: *Sharpening stones.*

FIGURE 1-69: *Saw set. (Courtesy Stanley Tools).*

Saw sets. A saw set is needed to set the teeth of a handsaw so that it will have clearance in the saw kerf. The set can be adjusted to set more or less, depending on the size of the saw teeth and the type of lumber being cut. Green lumber requires more set in the teeth than dry lumber (see Fig. 1-69).

Burnisher. A burnisher is simply a very hard, tapered steel shaft with a wooden handle. It is used to "turn the edge" of scraper blades (see Fig. 1-70).

Smoothing Tools

There are five main types of tools in this group: *planes, routers, scrapers, rasps,* and *sandpaper blocks.* The purpose of each is to produce smooth surfaces, and each has some specific uses. Planes and scrapers work on exposed, flat surfaces with the grain, while routers reach down into recesses below the surface and may work across the grain. Rasps are useful for curved surfaces, and sandpaper performs the final smoothing operation.

FIGURE 1-70: *Burnisher. (Courtesy Stanley Tools).*

Planes. Planes are of two main kinds, surfacing and special. The latter are among the tools, mentioned at the beginning of the chapter, that have been devised to do specific jobs.

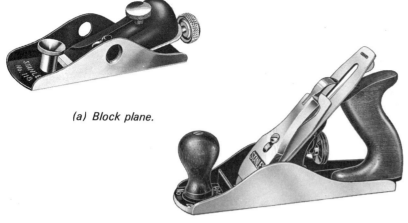

(a) Block plane.

(b) Smooth plane.

(c) Jack plane.

(d) Fore plane.

(e) Jointer.

Surfacing planes are made in five standard sizes, the smallest being a *block* plane, a small, light, one-hand tool usually 6 in. (150 mm) long [see Fig. 1-71(a)]. Its main uses are for surfacing small areas or for planing end grain. The next size is the *smooth* plane, a common style about 9 in. (225 mm) long, used for smoothing and cleaning up surfaces [see Fig. 1-71(b)]. The third is the *jack* plane, perhaps the most widely used of all. About 14 in. (350 mm) long, with a 2 in. (50 mm) blade, it serves most general planing purposes [see Fig. 1-71(c)]. Next is the *fore* plane, really an over-sized jack plane. It varies from 18 to 22 in. (450–550 mm) in length and is useful for straightening edges of lumber as well as for truing surfaces [see Fig. 1-71(d)]. The largest surfacing plane is the *jointer,* which is 22 to 24 in. (550–600 mm) long, usually with a $2\frac{3}{8}''$ (60-mm) blade. It is particulary useful for straightening the edges of doors, window sashes, or long boards. It is also used on large, flat surfaces [see Fig. 1-71(e)].

Special planes, as the name implies, are used to perform operations not possible with regular surfacing planes. There is a *rabbet* plane (see Fig. 1-72) and a *bullnose* plane (see Fig. 1-73).

FIGURE 1-71: *Surfacing planes.*

FIGURE 1-72: *Rabbet plane.*
(Courtesy Stanley Tools).

FIGURE 1-73: *Bullnose plane.*
(Courtesy Stanley Tools).

Rabbet planes are designed to cut a rabbet or groove extending to the edge of a board, along its length. The bullnose plane has its blade placed close to the front end of the body so it can reach close to the obstructed end of a surface.

Routers. Routers are designed to remove wood in order to form a groove or to smooth the bottom of grooves and dadoes made by other means. The side cuts of grooves made with a router must be cut beforehand with a saw. Hand routers are usually supplied with three cutters of varying widths (see Fig. 1–74).

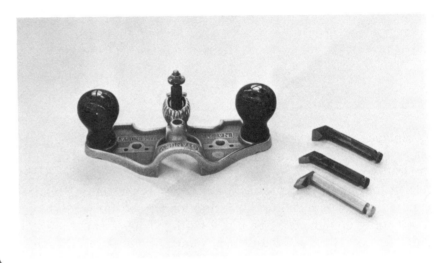

FIGURE 1-74: *Hand router.*

Scrapers. A scraper is used for the same general purpose as a plane—to make a surface smooth—but it is used in a different way. Whereas a plane blade has a sharp edge which cuts its way through wood, a scraper works with the action of a claw. The edge is first sharpened in the same manner as a plane blade. Then that sharp edge is turned over into the shape of a hook. This is done with the aid of a hard, steel instrument known as a *burnisher* (see Fig. 1–70). When the scraper is then pushed or pulled across the surface of the wood, it scrapes off very thin shavings. Its work is limited to the final stages of smoothing a surface.

The simplest type of scraper is a thin piece of alloy steel with one sharpened edge. It may be held in one or both hands [see Fig. 1–75(a)]. Another type has a frame in which the steel blade is held [see Fig. 1–75(b)]. It is used with both hands and is normally pushed away from the operator. Still another type consists of a wooden handle in which a narrow hooked blade is held [see Fig. 1–75(c)]. The paint scraper illustrated in Fig. 1–75(d) is very useful in removing excess paint off the edge of a window pane.

A good plane is a very important tool and the student should, first of all, be acquainted with the names of the parts, which are illustrated in Fig. 1–76. He/she should also know how to sharpen and adjust the plane in order to do a good job with it.

The plane iron and cap iron assembly must be removed first; then the cap iron can be removed from the plane iron. After sharpening, it is ready for assembly and adjusting (for sharpening procedures, read the section later in this chapter on sharpen-

(a) Hand scraper.

(b) Cabinet scraper.

(c) Scraper.

(d) Paint scraper.

FIGURE 1-75: *Scrapers. (Courtesy Stanley Tools).*

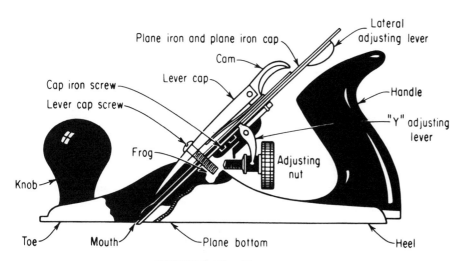

FIGURE 1-76: *Plane parts.*

ing of tools). Replace the cap iron. See that the edge of the cap iron is back about $\frac{1}{16}$ in. (1.5 mm) from the cutting edge of the plane iron. Tighten the cap iron screw and replace the assembly in the plane, beveled edge down. Set the lever cap in place and snap down the cam. If the cam does not snap readily into place, adjust the lever cap screw until it does. Now the blade can be adjusted in the plane. Turn the adjusting nut until the cutting edge projects from the mouth very slightly. If the edge is not projecting evenly, move the lateral adjusting lever to the right or left until both corners of the blade project the same amount. The plane is now ready for use.

Rasps. A wood rasp is a file-like tool used to shape and smooth surfaces which are irregular or difficult to reach with a plane or scraper (see Fig. 1–77).

FIGURE 1-77: *Wood rasp.*

A modern version of the rasp is known as a *Surform* tool. It has a blade with a large number of small, razor-sharp cutting edges. Each cutting edge has its individual throat which allows the shavings to pass through the blade, thus eliminating clogging. These tools are made in several styles, each suited to a particular purpose [see Fig. 1–78(a)]. A couple of the uses to which these tools may be put are illustrated in Fig. 1–78(b).

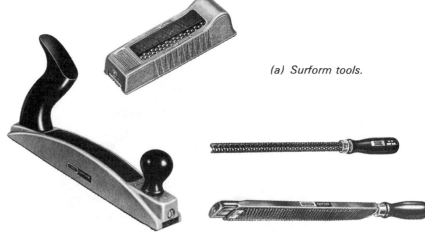

(a) Surform tools.

FIGURE 1-78:

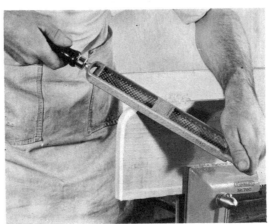

(b) Using Surform tools.

Another new version of a rasp is known as an *abrader*. It operates much like sandpaper but is made from a stainless steel sheet in five different shapes (see Fig. 1-79). Figure 1-80 illustrates operations which may be performed with abraders of different shapes.

FIGURE 1-79: *Abraders.*
(Courtesy Disston Inc.).

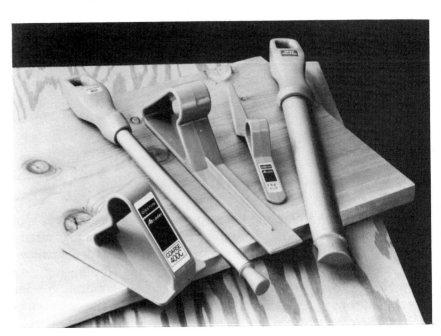

FIGURE 1-80:

(a) Abraders in use
(Courtesy Disston Inc.).

(b) Abraders in use
(Courtesy Disston Inc.).

(c) Abraders in use
(Courtesy Disston Inc.).

Wrecking Tools

Tools belonging to this group are used to dismantle structures that have served their purpose. For example, concrete forms that have been used are taken apart with wrecking tools. The group includes the *rip chisel, ripping bar, nail puller, axe,* and *side-cutting pliers.*

Rip claw hammer. The claws on this style of hammer are made for ripping or prying (see Fig. 1-2). It may also be used, of course, for the same purpose as other hammers.

Rip chisel. This tool is sometimes called a "wonder bar" and is similar to a regular ripping bar. The bar is much thinner, and the ends are sharpened like a chisel to make it much easier to remove baseboards, casings, or other material without damaging them [see Fig. 1-81(b)].

Ripping bar. This tool, sometimes called a "goose neck" bar, is made both for pulling nails and for prying. It is made in several lengths, ranging from 12 to 36 in. (300–900 mm) [see Fig. 1-81(a) and (c)].

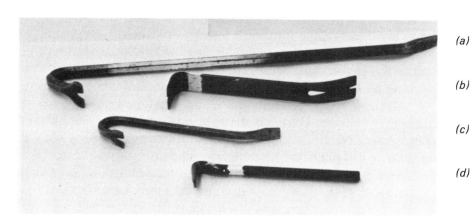

(a)

(b)

(c)

(d)

FIGURE 1-81: *Wrecking bars.*

FIGURE 1-82: *Nail puller.*

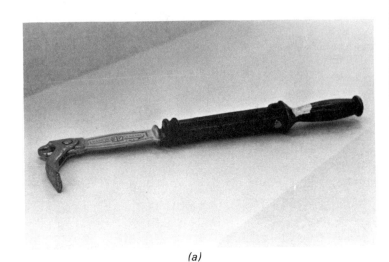

(a)

(b)

(a)

(b)

FIGURE 1-83: *Hand electric saws. (Courtesy Skil Canada Ltd.).*

Nail puller. A nail puller is made for one purpose only—pulling nails. It has a pair of movable jaws that clamp below the nail head and a curved lever arm that gives great pulling advantage. It is also fitted so that the jaws may be driven into wood to reach below the surface for nails whose heads do not protrude above (see Fig. 1-82).

POWER TOOLS

In an age of speed and mass production, more and more power tools are being introduced into construction. There are power tools for every conceivable kind of job today, powered by electricity, compressed air, gasoline, or explosive powder. Some of them will do the job faster, and some will do it more accurately than can be done by hand; all lighten the load of manual labor involved.

All these tools are high-speed, and great care must be taken in their operation. Detailed instructions on how to set up and operate each one are issued by the manufacturer, and they should be followed explicitly.

With a couple of exceptions, the tools described here are held in the hand while being operated, and it cannot be stressed too strongly that *great care and attention are necessary* in their handling and operation.

Electric Power Tools

Hand electric saw. This tool is a quite versatile one. It may be used almost anywhere that an ordinary handsaw may be used. It will both crosscut and rip. The depth of cut is adjustable, and the bed may be tipped to permit angle cutting. It is available in a number of sizes, based on the diameter of the blade—from 6 to 10 in. (150–250 mm) (see Fig. 1–83). The electric hand saw is very convenient as it can be used for horizontal and vertical cutting and is also very transportable—it can be taken up on a roof for cutting. Portable electric saws are usually guided "free-hand" so the accuracy of the cut is determined by the hand guiding the saw (see Fig. 1–84).

The depth of the cut is adjusted by raising or lowering the base of the saw. The proper depth is equal to the thickness of the material plus the full tooth protruding through the work (see Fig. 1–85). The base can be set to cut at a bevel as shown in Fig. 1–86, allowing the operator to make compound cuts. Figure 1–87 shows the proper method for making a plunge cut—note that the

FIGURE 1-84: *Saw in use.*

FIGURE 1-85: *Depth of saw blade.*

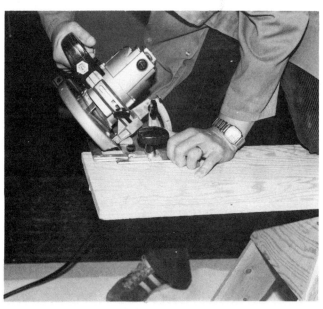

FIGURE 1-86: *Bevel cut.*

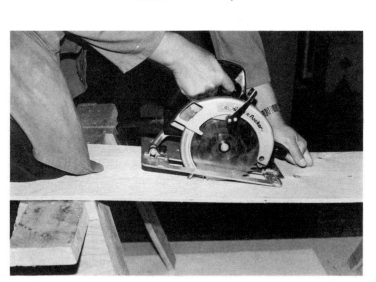

FIGURE 1-87: *Plunge cut.*

front edge of the base is in contact with the work prior to starting the cut. Figure 1–88 shows standard types of blades. The chisel combination is a popular blade as it can be used for ripping as well as crosscutting and is also easy to resharpen. Carbide-tip blades are preferred by some as they stay sharp longer than a standard blade.

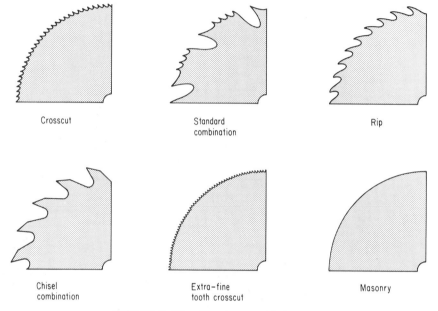

Crosscut   Standard   Rip
                  combination

Chisel   Extra–fine   Masonry
combination   tooth crosscut

FIGURE 1–88: *Circular saw blade types.*

Electric sanders. Illustrated here are two of the many types of sanders on the market today. Figure 1–89(a) illustrates a belt sander for which belts of varying degrees of fineness are available. When this type of sander is being used, it must be kept constantly on the move, or it may leave deep depressions on the surface being sanded. The material must be supported so that it doesn't move while sanding.

Figure 1–89(b) is an orbital sander, designed to do fine sanding, although various grades of paper may be used. In the case of these and other types of sanders, they should be allowed to sand by their own weight only. Never press down on them while they are in operation.

Electric drills. Electric drills are usually designed to receive straight-shanked bits and are made in several sizes, depending on the maximum diameter bit they will take. Figure 1–90(a) illustrates a drill that can be handled with one hand and will take bits up to ¼ in. (6 mm) in diameter. Figure 1–90(b) illustrates another light drill with trigger speed control and a reversing switch. This drill is especially designed to drive screws into gypsum dry wall. The drill will stop driving the screw when the head is slightly below the surface. Figure 1–90(c) shows a rechargeable drill which has the advantage of mobility without the use of a cord. Figure 1–90(d) shows a larger drill which is operated with both hands.

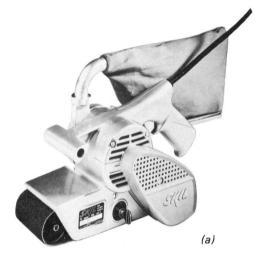

(a)

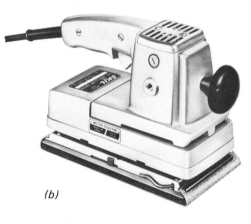

(b)

FIGURE 1–89: *Electric sanders. (Courtesy Skil Canada Ltd.).*

FIGURE 1-90:

(a) Electric drill (¼″ (6 mm) size).
(Courtesy Stanley Tools).

(b) Electric screwdriver.
(Courtesy Skil Tools).

(c) Cordless electric drill (Courtesy Skil Tools).

(d) Electric drill (½″ (12 mm).
size) Courtesy Skil Tools).

It takes bits up to ½ in. (12 mm) in diameter, and extra pressure may be exerted by pushing against the drill with the chest.

Electric drills utilize a variety of drill bits. The twist bit is the most common bit used, and a large range of sizes is available. When larger holes are required, either an auger bit or a speed bore bit is used (see Fig. 1-91). Numerous specialty bits are also available—screwdriver, masonry, spur-edged (see Fig. 1-92), and countersink bits (see Fig. 1-18).

(a) Auger bit.

(b) Speed bore bit.

FIGURE 1-91: Electric drill bits.

FIGURE 1-92: Speciality bits.

(a)

FIGURE 1-93: *Electric routers.*

(b)

FIGURE 1-94: *Router bits. (Courtesy Stanley Tools).*

Routers. Electric routers are versatile tools because they will do many operations: *dadoing, rabbetting, bullnosing, fluting,* to mention only a few. Such a variety of operations is possible because a great many different bits may be used in the machine. Routers are made in a number of sizes, ranging from $\frac{1}{2}$ to $2\frac{1}{2}$ hp (0.37–1.85 kW) (see Fig. 1-93). A variety of router bits is available, producing numerous shapes and sizes of cuts (see Fig. 1-94).

FIGURE 1-95: *Saber saw (Courtesy Skill Tools).*

Saber saw. This is an electric saw that can do all manner of jobs. It can do the work of a crosscut or ripsaw, band saw, keyhole saw, hacksaw, or jigsaw. It is particularly suited to jobs that are hard to get at with any other type of saw (see Fig. 1-95). Although it is quite light, it will cut lumber up to $1\frac{1}{2}$ in. (38 mm) thick. Figure 1-96 illustrates a variety of blades available for a saber saw.

POINTS PER 1 IN. (25 mm)		BLADE DESCRIPTION AND RECOMMENDED USE
	10	Taper ground, clean cuts in paneling, formica, plastic and related materials.
	10	Scroll cutting for wood, plywood, masonite and plastic, up to 5 mm thick.
	12	Taper ground for smooth, clean finish cuts in wood, plywood, plastic and hardboard.
	6	Coarse tooth for extra fast cuts in wood, plywood and hardboard. Extra long.
	8	Flush cutting for wood and plastic.
	18	Metal cutting for non-ferrous metals up to 1/4" (6.5 mm) thick.
	24	Metal cutting for non-ferrous metals up to 1/8" (3 mm) thick.
	36	Metal cutting for non-ferrous metals up to 1/16" (1.5 mm) thick.
	8, 12, 18	General purpose assortment for wood: coarse, medium and fine.
	8	Coarse tooth wood cutting for heavy, fast cuts.
	18	Fine tooth wood cutting for smooth cuts in wood, plywood, plastic and counter top materials.

FIGURE 1-96   *Saber saw blades (Courtesy Stanley Tools)*

The saber saw can be used to make straight or bevel cuts. Circles are easily cut with or without the use of a special guide (see Fig. 1-97). Plunge cuts are used in cutting internal openings. This is achieved by tipping the tool forward with the base resting on the material and the end of the blade just clearing the surface. Turn on the saw and lower it into the work until the blade comes through the other side (see Fig. 1-98).

FIGURE 1-97: *Cutting a circle with a saber saw.*

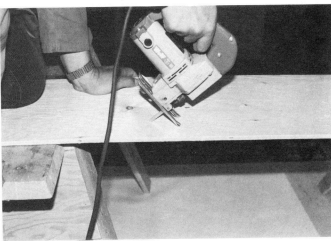

FIGURE 1-98: *Plunge cut.*

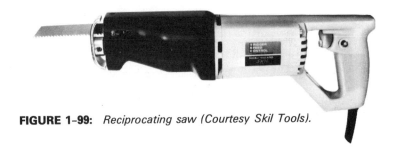

FIGURE 1-99: *Reciprocating saw (Courtesy Skil Tools).*

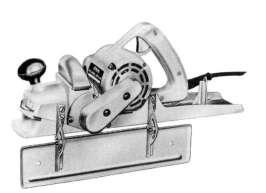

FIGURE 1-100: *Electric plane. (Courtesy Skil Canada Ltd.).*

Reciprocating saw. Another type of all-purpose saw similar to a saber saw is the reciprocating saw. The saw has the blade extending out from the end, allowing it to get into places a saber saw cannot. The blade is longer as well, so cutting depth is increased (see Fig. 1-99).

Electric plane. The electric plane is intended for jointing purposes, straightening edges, etc. It is particularly useful for working on doors or large window units. The fence may be tilted so that a bevel can be planed on an edge, as it is necessary on the closing edge of a door (see Fig. 1-100).

Hammer drill. A hammer drill is used to penetrate hard material such as concrete or stone and operates in such a way that it hammers at the same time as it turns the bit. Figure 1-101 illustrates three styles of drills and two different kinds of drilling operations.

Power screwdriver. Various types of power screwdrivers are in use, some of them as attachments to an electric drill. The one shown in Fig. 1-102 is operated by rechargeable batteries.

Radial arm saw. The radial arm saw is not a hand electric tool, but is operated from a table or bench and gets its name from the fact that the saw is suspended on a horizontal arm, which allows a great variety of movement. The saw can travel along the arm, pivot on the arm, and turn into a horizontal position; the arm itself will turn (see Fig. 1-103). The versatility of the tool is illustrated in Fig. 1-104, which shows it being used for ripping, and mitering as well as standard crosscutting. When crosscutting, mitering, or dadoing, the work is held firmly on the table and against the fence, and the saw is pulled through the work. Care should be taken with heavier materials so that the saw does not attempt to cut through the work too quickly. For ripping, the saw head is turned parallel with the table and locked into position. Lumber is then fed into the blade with care taken to feed the material in the opposite direction to the blade rotation.

(a)

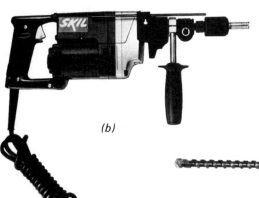

(b)

(c)

FIGURE 1-101: *Hammer drills. (Courtesy Skil Tools).*

FIGURE 1-102: *Battery-powered screwdriver. (Courtesy Disston Inc.).*

FIGURE 1-103: *Radial arm saw.*

(a) Miter cutting.

(b) Ripping.

(c) Compound miter.

(d) Crosscutting.

FIGURE 1-104:

FIGURE 1–105: *Miter saw.*

FIGURE 1–106: *Table saw.*

Miter saw. The miter saw is specially designed for cutting miters and is very useful for cutting moldings. The saw can be locked at 45° either left or right, or in a 90° position. A variety of materials can be cut—wood, plastic, and soft metals. Material being cut must be firmly held in place and carefully positioned prior to pulling the saw down through the material (see Fig. 1–105).

Table saw. This saw is primarily used for cabinetmaking and is sometimes used on projects where there are on-site-built cabinets. Carpenters will also use this saw for other finish work on a project. The saw is very useful for crosscutting, ripping, and dadoing. The size of the saw is determined by the diameter of the saw blade recommended for its use. Typical sizes are 8–10 in. (200–250 mm). The smaller sizes are often chosen for on-site work as they are easier to move through doorways (see Fig. 1–106). There are a number of kinds of saw blades available for the table saw. Make sure you select one with a blade of the correct diameter and arbor hole. Figure 1–107 illustrates the common types of blades.

(a) Rip and cross-cut blades.

(b) Plywood and combination blades.

(c) Hardened tip and carbide tip blades.

FIGURE 1–107:

Jointers. The jointer is one of the most commonly used tools in a shop. Typical uses are for surfacing a board and jointing the edge. Jointers can also be used for cutting rabbets, bevels, or a taper. The size of the jointer is indicated by the length of the knives on the cutterhead. Care should be taken when using a jointer to keep the hand away from the area of the stock over the cutterhead. Use a push stick for smaller pieces (see Fig. 1–108).

FIGURE 1–108:   *(a) Jointer.*   *(b) Combination jointer-thickness planer.*

Bench grinder. This is another tool which is not a hand electric one but is fastened to a bench. The one illustrated in Fig. 1–109 is equipped with a device which holds plane blades at the proper angle while they are being sharpened.

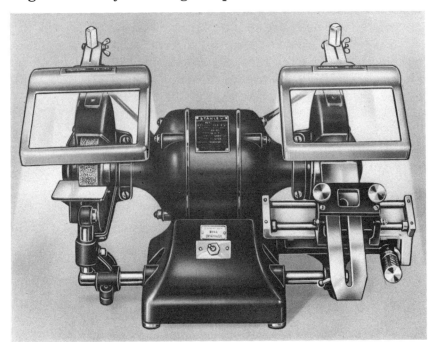

FIGURE 1–109: *Bench grinder.*

FIGURE 1-110:   *Gas powered chain saw (Courtesy Skil Power tools).*

Gasoline-Powered Tools

A number of tools are run by small gasoline engines, including *paint sprayers, soil compactors,* and *chain saws.* The chainsaws are made in a variety of sizes, the size being determined by the length of the bar on which the saw chain runs. The one illustrated in Fig. 1-110 is a 14-in. (350-mm) saw. Such saws are useful for cutting heavy timbers, posts, etc. They have become quite common on framing jobs as they can be used anywhere without needing a power source. The saw cut is quite large, and the blade tends to tear at the wood, so care should be exercised in its use.

Pneumatic Tools

The most common pneumatic tools—those operated by compressed air—are *staplers, nailers,* and *drills.* Staplers are designed to drive staples and are used for fastening paneling, building paper, siding, and shingles. Fasteners used in staplers may be up to $\frac{19}{16}$ in. (40 mm) with a maximum crown of 1 in. (25 mm) [see Fig. 1-111(a)].

Nailers are designed to drive various sizes and styles of nails up to a maximum of $3\frac{13}{16}$ in. (95 mm). They may be used almost anywhere that nails are required [see Fig. 1-111(b)].

Pneumatic drills are designed to duplicate the operations of electric drills. They are generally smaller in size and so can reach places electric drills are too cumbersome to fit. The common bits can be used in these drills [see Fig. 1-111(c)].

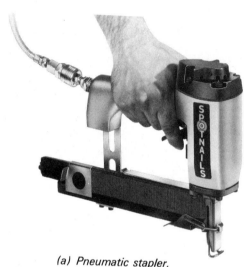

(a) Pneumatic stapler.
(Courtesy Spotnails Inc.).

(b) Pneumatic nailer.
(Courtesy Spotnails Inc.).

FIGURE 1-111:

(c) Pneumatic drill.

Power-Actuated Tools

In many situations, it becomes necessary to attach some object or material to a concrete, masonry, or steel surface. Not only is a special fastener necessary—one that is hard enough to penetrate the surface without bending—but special power is required to drive such a fastener into a hard material.

Tools designed to do such a job use an explosive charge contained in a small cartridge similar to a 22-caliber rifle cartridge. By this means, several types of pins and threaded fasteners (see Fig. 1–112) are driven into steel and concrete or masonry materials.

Cartridges of various energy levels are produced for use with fasteners of various lengths and materials of various densities. The energy level is designated by color: gray, brown, green, yellow, red, and purple denoting energy levels from low to high. Two types of power-actuated tools are available—low velocity and high velocity (see Fig. 1–113). Safety equipment should be used when using power-actuated tools—hard hat, safety glasses, and earmuffs (see Fig. 1–114).

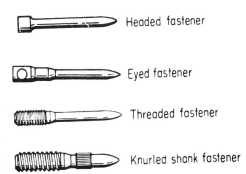

Headed fastener

Eyed fastener

Threaded fastener

Knurled shank fastener

FIGURE 1–112: *Power-actuated fasteners.*

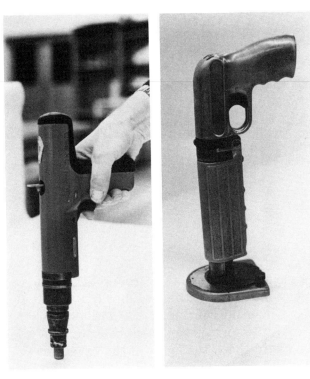

(a) Low velocity tool.   (b) High velocity tool.

FIGURE 1–113: *Power-actuated tools.*

(a) Hard hat and muffs.

(b) Shield and safety glasses.

FIGURE 1–114: *Safety equipment.*

CARE OF TOOLS

The success that one achieves with one's tools depends to a great extent on the care which one takes of them. No tool, no matter how expensive, will give good service unless it is given the proper attention. Tools must be kept sharp and in good general repair and must be protected against the weather and against unnecessarily rough usage.

Sharpening Tools

While all cutting tools must be kept sharp, the ones which will probably need the most careful and constant attention are plane blades, chisels, and saws. Therefore, let us consider the sharpening procedures for these rather carefully.

Plane blade. The first step is to shape the cutting edge. It may be square, or there may be a slight crown on it, depending on how it is to be used. If a surface is to be dressed flat, a crowned edge will be best, but for straightening an edge, a square edge is preferable.

Hold the iron at right angles to the grinding wheel, use light pressure, and grind slowly. If the edge is to be crowned, increase the pressure slightly at the ends of the blade. Test with a try square (see Fig. 1-115).

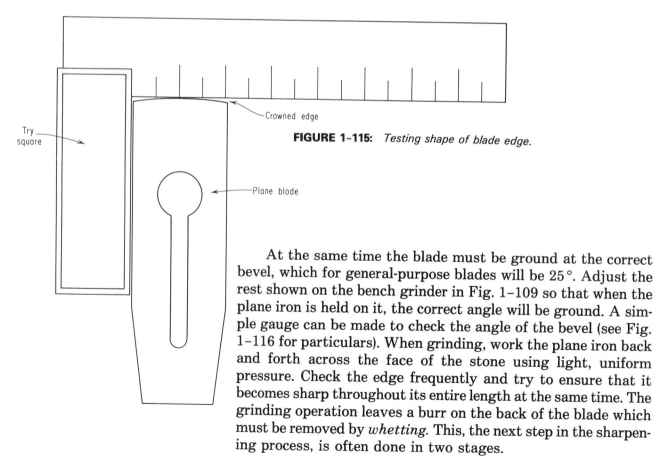

Try square

Crowned edge

Plane blade

FIGURE 1-115: *Testing shape of blade edge.*

At the same time the blade must be ground at the correct bevel, which for general-purpose blades will be 25°. Adjust the rest shown on the bench grinder in Fig. 1-109 so that when the plane iron is held on it, the correct angle will be ground. A simple gauge can be made to check the angle of the bevel (see Fig. 1-116 for particulars). When grinding, work the plane iron back and forth across the face of the stone using light, uniform pressure. Check the edge frequently and try to ensure that it becomes sharp throughout its entire length at the same time. The grinding operation leaves a burr on the back of the blade which must be removed by *whetting*. This, the next step in the sharpening process, is often done in two stages.

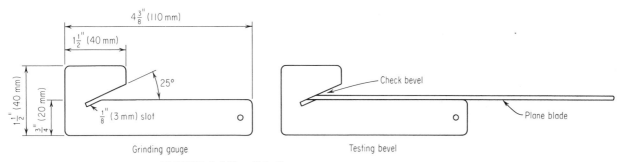

FIGURE 1-116: *Grinding gauge.*

Two stones are generally used for whetting—an oilstone and a finer, Arkansas stone. Use the oilstone first. Apply a few drops of lubricant to start. A mixture of equal parts of light machine oil and solvent is recommended. Now set the bevel of the blade on the stone and raise the heel of the iron about 5°. Only the cutting edge of the beveled surface is in contact with the stone (see Fig. 1-117). Using a circular motion, whet the edge, taking care to hold the iron at a constant angle. After a few strokes, turn the blade over and rub it over the stone a few times, holding it perfectly flat. This action removes the burred edge which forms in whetting. Repeat this procedure until you have produced as sharp an edge as the stone will allow. The edge may be improved with a finer stone, so you repeat the whetting on the Arkansas stone.

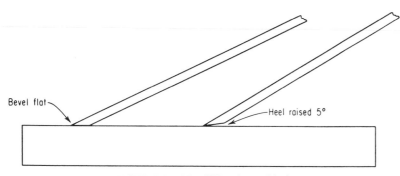

FIGURE 1-117: *Whetting a blade.*

The final step is *stropping* the edge on a piece of leather in much the same way that a barber would do with a razor blade. Glue a piece of heavy leather to a wooden block. Strop the cutting edge back and forth across the leather several times. The last remnants of the burr are removed, leaving an extremely fine, sharp cutting edge.

Chisel. The same procedure is used for sharpening a chisel as was used for a plane blade. The only difference is that the cutting edge of the chisel is always square. Try square and bevel gauge may be used in the same way as for plane blades.

Handsaws. Four operations, namely, *jointing, shaping teeth, setting,* and *finish filing,* are required to make saws cut efficiently.

FIGURE 1-118: *Saw vise.*

Sometimes shaping is not necessary; setting is usually only needed for every three or four finish filings. The first three operations are similar for both crosscut saws and ripsaws, whereas the fourth differs slightly from one to the other.

The first problem is to hold the saw securely. The best way to do this is with a saw vise (see Fig. 1–118). If one is not available, a suitable clamp may be made of wood to be held in a bench vise. The saw must be held rigidly to prevent vibration, with the gullets of the teeth just clearing the top edge of the vise.

The first step is to make sure that the top edges of all the teeth are the same height. This is done by jointing—passing a flat file lightly, lengthwise, along the tops of the teeth. After this is done, it will be seen that some of the teeth have been flattened on top, whereas others may have just barely been touched. Obviously, the former require the most filing.

After jointing, the teeth must be all filed to the same size and shape. Do not bevel the teeth now—beveling is done after setting. To shape the teeth, file in each gullet straight across, at right angles to the saw, using a slim or extra-slim taper, three-cornered file. Use light pressure until the file is well seated in the gullet, and hold the file so as to produce teeth with the proper *rake*. If the teeth are of uneven size, file in each gullet until you reach the center of the flat made by jointing. All gullets must be filed to the same depth and all teeth to the same size.

For setting the teeth, a *saw set* is used (see Fig. 1–69). It bends the top portion of each tooth outward, so that the width of the cut made will be greater than the blade which keeps the saw from "binding" in the saw kerf. Notice that alternate teeth are set in opposite directions on both crosscut saws and ripsaws to about half the thickness of the tooth (see Fig. 1–119).

In filing the teeth of the crosscut, it must be remembered that they cut with both edges and points. Consequently, edges must be bevel-sharpened. There are two methods of bevel filing,

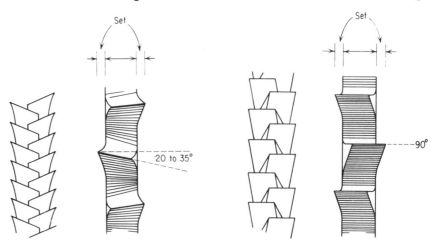

(a) Crosscut teeth set.   *(b) Ripsaw teeth set.*

FIGURE 1-119: *Saw teeth set.*

that of filing toward the point of the saw and that of filing toward the handle. As the first is perhaps easier for the beginner, let us look at that method.

Begin at the point of the saw and work toward the handle. Tilt the saw vise slightly away from you, not more than 10°. Place the saw in the vise with the handle to the right, and, starting at the point of the saw, place the file in the gullet to the left of the first tooth set away from you. Swing the handle of the file toward the handle of the saw, to about the angle shown in Fig. 1-120. Keep the file level throughout the entire stroke—don't "rock" it. To get all the teeth of equal size, you will have to "crowd," or put a little extra pressure toward, the backs of the teeth being filed. Skip the next gullet and every other one thereafter. Then reverse the saw in the vise and file the other half of the teeth, beginning again at the point of the saw. Place the file in the gullet to the right of the first tooth set away from you. Swing the handle of the file toward the handle of the saw as before. Proceed as previously explained, remembering that a little "crowding" to the back is necessary to secure even teeth.

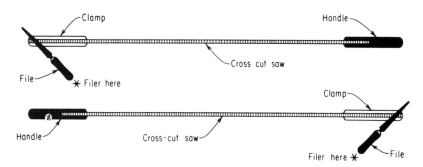

FIGURE 1-120: *Bevel filing.*

A three-cornered file is used for saw filing, its size depending on the size of the saw teeth.

Saw filing is an operation which takes practice to perfect. Do not be discouraged if your first job is not all that you would like it to be. The next one should be better, and you should become progressively more proficient as you practice.

Storing Tools

Proper storage facilities contribute considerably to the life of tools. Sharp-edged tools such as saws or chisels should be so placed in the tool box that the edges do not become damaged from contact with other tools. An excellent way to carry chisels or auger bits, for example, is to have them in a cloth roll with separate compartments for each bit or chisel. A wooden cap may be made for the saw blades, or they may be held separately, teeth down, in the tool box or kit. Take care of all your tools in some such manner, and you will be rewarded with longer use and better results (see Fig. 1-121).

FIGURE 1-121: *Tool storage.*

Handling Tools

Equally important is how you handle and care for your tools *while on the job*. Saws and squares should never be dropped on the floor. Also, they should not be leaned against the leg of a sawhorse or bench unless you are sure that they are not going to slip and fall to the floor. Use some means of holding them securely. Planes should always be laid on the side—never on the bed of the plane. Don't let a great variety of tools get stacked up around you at work on the bench. When you have finished with one tool for a time, put it away where it will not be damaged.

Care in the handling of power tools is very important. This is true not only from the standpoint of damage to the tool but also from the possibility of danger to you. Careless or improper handling of a power tool may result in serious injury.

Develop a pride in your tools—in their appearance and condition—and you will find that you will be able to do a better job, that tool replacement will be less costly, and that you will be looked upon as a workman worthy of the name.

REVIEW QUESTIONS

1-1. Name two *basic* differences between crosscut and ripsaw teeth in handsaws.

1-2. What is the advantage of having a handsaw with a tapered blade?

1-3. How many teeth per inch (25 mm) are there in an *eight-point* saw?

1-4. How does a backsaw differ from most other handsaws?

1-5. Why should the cutting angle of a chisel that is to be used for cutting hardwood be ground to 30°?

1-6. What are the two main differences between a shingling hatchet and a bench ax?

1-7. List four tables found on a framing square.

1-8. What is the main purpose of a sliding T-bevel square?

1-9. List two main differences between a Forstner bit and an auger bit.

1-10. Why should hardened nails not be driven with a nail hammer?

1-11. Matching test: Place the number found beside each item in column 1 in the blank space to the left of each phrase in column 2 with which it matches.

Column 1	*Column 2*
1. Bit extension	_____ Acts as a depth gauge
2. Push drill	_____ For drilling deep holes
3. Bit file	_____ Has 16-in. (400-mm) tongue
4. Butt gauge	_____ Cutting irregular lines
5. Burnisher	_____ Ground straight across
6. Combination square	_____ Has one safe edge
7. Compass saw	_____ For turning scraper edge
8. Rafter square	_____ To lay out hinge gains
	_____ Has an interchangeable blade
	_____ A spring-loaded tool

1-12. Tool manipulation test: See Fig. 1-122 for plan and elevation of project. Carry out operations as follows:

(a) Saw rough stock to 1 in. × 5 in. × 13 in. (25 × 125 × 325 mm).

(b) Plane all surfaces and square ends. Use no sandpaper.

(c) Lay out butt gain and complete.

(d) Lay out and drill holes.

(e) Mark chamfers and do in order A, B, C.

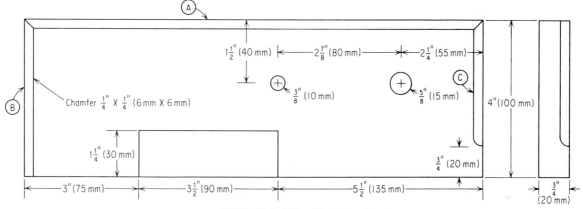

FIGURE 1-122: *Test drawing 1.*

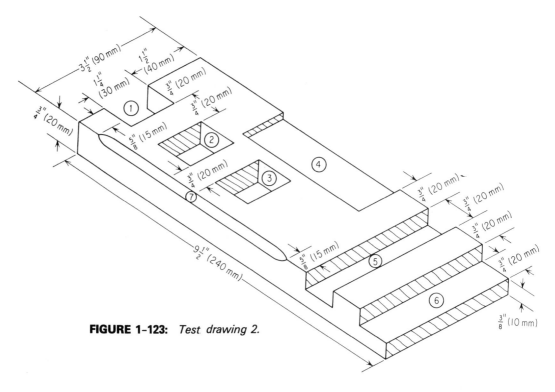

FIGURE 1-123: *Test drawing 2.*

1-13. Chisel test: See Fig. 1–123 for isometric view of project. Carry out operations as follows:

(a) Cut rough stock to 1 in. × 4 in. × 10 in. (25 × 100 × 250 mm).

(b) Plane to ¾ in. × 3½ in. × 9½ in. (20 × 90 × 240 mm).

(c) Lay out and make cuts as indicated:

(1) ¾ in. × ¾ in. × 1¼ in. (20 × 20 × 30 mm) cut

(2) ⅜ in. × 1 in. × 1¼ in. (10 × 25 × 30 mm) blind mortise

(3) ¾ in. × 1 in. × 1¼ in. (20 × 25 × 30 mm) through mortise

(4) Gain for leaf of 3½-in. (89 mm) butt hinge, 1¼ in. (30 mm) wide

(5) ⅜ in. × ¾ in. × 3½ in. (10 × 20 × 90 mm) dado

(6) ⅜ in. × ¾ in. × 3½ in. (10 × 20 × 90 mm) rabbet

(7) ¼-in. (6-mm) stopped chamfer

SITE INVESTIGATION

The erection of a building, large or small, necessitates a good deal of prior study and planning, both in connection with the building itself and with the site on which it will be built.

This study, commonly referred to as *site investigation,* will vary widely as to method and degree, depending on the type, size, and proposed use of the building to be constructed. Planners in the light construction area are likely to be most interested in surface aspects of the site and the subsoil at relatively shallow depths. On the other hand, planners of large, heavy buildings are usually deeply interested in the nature of the subsoil at considerable distance below the surface and are concerned with the surface chiefly from the standpoint of site area and adjacent buildings.

Site Investigation in Light Construction

The surface aspects of the site with which planners in the light construction field will be concerned include

1. Presence or absence of trees and shrubs

2. Contours of the site

3. Elevation of the site in relation to the surrounding area

4. The size, shape, and proximity of surrounding buildings

Particularly in the case of planning for residential construction, the presence of trees and their size often influence the design of the building. The number and kinds of trees and shrubs present will also help to determine the type of equipment necessary to clear the site. In some cases, it may be deemed necessary to remove all the trees, while in others the architect may wish to retain some trees as part of the overall design.

Whether the site is flat or rolling and whether it slopes in one direction and, if so, in what direction in relation to the front of the building are important questions for the designer. For example, a slope or small hill may enable him/her to introduce ground level entrances to the building at different levels of the building. The amount of slope and its direction will also determine the amount and disposition of fill which may have to be brought to the site.

2

THE BUILDING SITE

The level of the area under study, in relation to that of the surrounding area, requires consideration. If it is low in comparison to the surrounding land, the level will probably have to be raised by adding fill. On the other hand, care must be taken that runoff from elevated land does not cause damage to surrounding property.

In the case of residential buildings particularly, it is usually desirable that they conform in a general way to those around them. For example, a tall, narrow building in the midst of the group of low, rambling ones would probably look out of place.

SOIL INVESTIGATION

Soil investigation, carried out in connection with light construction projects, will consist basically of tests to determine the kinds of soil to a depth of twice the building height below the surface, the level of the water table in the soil, where frost is a problem, the depth of frost penetration, and any very unusual soil characteristics.

The type or types of soil present will indicate the bearing strength of the material and will help to determine what sort of excavating equipment is best suited for the job at hand and what measures, if any, must be taken to prevent cave-in of the excavation. Depending on the type of soil, the excavated material will either have to be removed from the site and replaced by other soil, or it must be stored to be used for backfilling and landscaping. Soil type will also be an indication as to whether or not frost penetration is a serious problem. Some very fine clay soils expand substantially when subjected to moisture, and care must be taken when these are encountered at the foundation level of a building.

The level of the water table—the natural water level in the soil—will determine whether particular precautions, e.g., the installation of weeping tile, must be taken to drain water away from around the foundation. It may also influence the waterproofing techniques employed on the exterior of the foundation walls.

Testing for the presence of a number of aggressive chemical substances, particularly sulfates of calcium, sodium, or magnesium, in the groundwater or soil are important considerations in the study of soils on the site. When soil or water containing appreciable amounts of these sulfates comes in contact with concrete, the sulfates react chemically with hydrated lime and hydrated calcium aluminate in the cement paste. This causes considerable expansion in the paste, resulting in corrosion and disintegration of the concrete.

To prevent this deterioration from taking place, concrete which will be in contact with these sulfates should be made using cement which has a low content of calcium aluminate. If the

FIGURE 2-1: *Soil test operation.*

soil contains over 0.10% sulfates or water contains over 150 ppm (parts per million) water-soluble sulfate, sulfate-resistant cement should be used.

Penetration of the soil by frost may have serious consequences under certain conditions. When moisture is present in any of several fine-grained soils, freezing may result in the formation of ice lenses and consequent heaving of the soil. Under these conditions, it is desirable to have the footings for a building below the frost line. Possible alternatives are to eliminate the soil moisture by draining it away or to replace the offending soil with one which is not affected by frost action—gravel or coarse sand.

The method of soil testing commonly used in this type of construction work is done by boring. Machine augers may be used, and soil is brought to the surface for analysis in a *disturbed* condition (see Fig. 2-1). Disturbed samples are generally used for soil grain size analysis, for determining the specific gravity of the soil, and for compaction testing. For determining other properties of soils such as strength and permeability, it is necessary to obtain an *undisturbed* sample, The most common method of obtaining an undisturbed sample is to push a thin *Shelby tube* into the soil, thereby trapping the sample inside the tube (see Fig. 2-2).

FIGURE 2-2: *Shelby tube.*

Another danger which must be considered, although it is sometimes very difficult to detect without special equipment, is the presence of underground watercourses or springs. Such watercourses normally flow to some natural outlet, but when the soil is disturbed and natural outlets dammed off, the result is often trouble. Water begins to collect underground, the hydrostatic pressure builds up, and eventually the water must break out somewhere. Often it will be through the basement floor. If there is the slightest hint of an underground water flow, drainage tile must be installed to take care of it.

AVAILABLE SERVICES

Are services such as electricity, sewer, water, gas, and telephone available? The answer will affect plans and preparations for the building. If a sewer line is already installed, it will be necessary to know its depth in order to get a drop from the building to the sewer line. A trench will have to be dug across the property to the building. If no sewer line is in, it will be necessary to find out what the level of a future line will be or, if none is available, to make plans for a private sewage disposal system.

Is there a water main in the immediate area? If so, what is, or will be, the location of the water connection for the site concerned? This location will determine where the water line will enter the building. If there is no main, plans must be made for a well or other water supply.

The availability of electricity will have an effect on the actual construction work. If there is none, the work will all have to be done by hand or by gasoline engine power. If there is electricity near at hand, a number of power tools may be used which could substantially reduce the time required for building.

If there is gas in the area, it should be ascertained where it is likely to enter the building, so that gas appliances may be located as conveniently as possible. If there is no gas, some other means of supplying heat must be found. It may be that more electrical power must be provided.

It will be necessary to find out from some civic authority the grade level at the site. The level at which the street and sidewalks will be put in must be known before the excavation and landscaping are planned. The grounds in front of the building should be level with or slope toward the sidewalk or street.

ZONING RESTRICTIONS

Most urban areas have zoning laws which restrict specified areas to certain uses. They may be industrial, business, local commercial, multiple-family dwelling area, or single-family dwelling area. In addition, a building in a particular area may be required to have a specified minimum number of square feet of floor space. Regulations stipulate how far back from the front property line the house must be and the minimum distance it must be away from the side property line. Often minimum lot areas are defined and the maximum area of a lot which a building may cover. The setback for garages, whether front-drive or lane-drive, is specified in many cases.

These regulations will vary to some degree from city to city and from community to community. However, the basic reasons for having these restrictions are the same, namely, the preservation of certain standards and the protection of the citizens to whom they apply. Therefore, it is quite essential, when one is contemplating the erection of a building, that he/she be familiar with the regulations which apply to that community and the particular area being considered.

BUILDING LAYOUT

When the design work for a building has been completed, plans are drawn to indicate in detail how the building is to be constructed. In many cases a *site plan* is included to show the exact location of the building on the property. When no site plan is included, more freedom of location may be possible, subject to the local or regional building regulations that will specify the minimum allowable proximity of the building to front and side property lines. In residential construction, for example, the front *setback* is frequently a minimum of 20 ft (5.4 m), while the side clearance might be 10% of the width of the lot. These regulations should be checked before proceeding with any proposed construction.

Armed with a set of plans, the builder must first lay out the position of the building on the site. In some areas it is necessary to employ a registered surveyor to locate the building stakes. If it were to be built on the surface of the ground, it would be a relatively easy matter to set stakes at the corners and other important points in the building by accurate measurements from the property lines, but in most situations some soil at least will be removed from the surface, and stakes would be lost during excavation. It is therefore necessary to set the stakes back from the excavation so they will remain in place after the digging is completed. These stakes are usually set at the lot property line (see Fig. 2–3). The distance to the building corners will be indicated on the stakes. The amount of digging or cut will also show on these stakes (see Fig. 2–4). After the digging is completed, the corners of the building are located in the bottom of the excavation. Footing forms can be set from these corners.

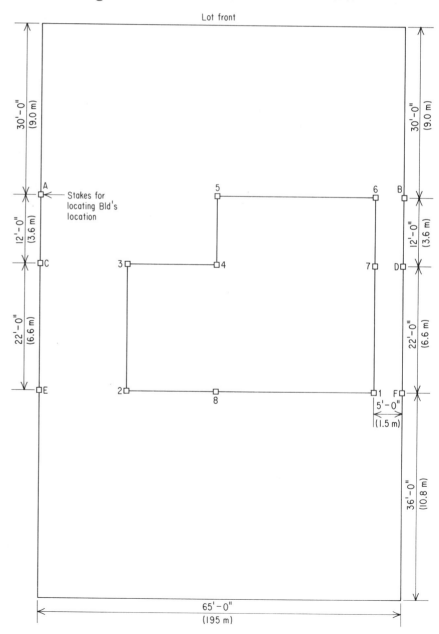

FIGURE 2-3: *Lot layout.*

FIGURE 2-4: *Building stakes.*

For larger buildings an alternate method using batter boards is sometimes utilized (see Fig. 2-5). Batter boards are kept back some distance from the actual building location and so are not disturbed during excavation work. Then, when excavating is complete, building lines are strung from one batter board to another, and each intersection of two building lines represents a point on the building. These points can be established in the excavation by dropping a plumb bob from the line intersections to the ground below.

A batter board is located by driving a stake at the approximate position of each point to be established and then building a batter board to cover each position, placed well back from the area to be excavated. Then by accurate measurements and sometimes with the aid of leveling instruments (see Figs. 2-8 and 2-9), the necessary reference points are placed on the batter boards where they may be marked by a nail or by making a shallow saw kerf on the top edge of the bar.

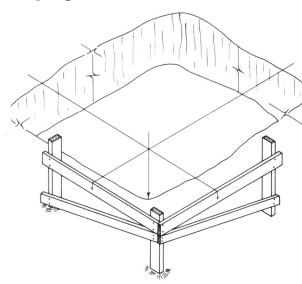

FIGURE 2-5: *Locating building corner in excavation from batter boards.*

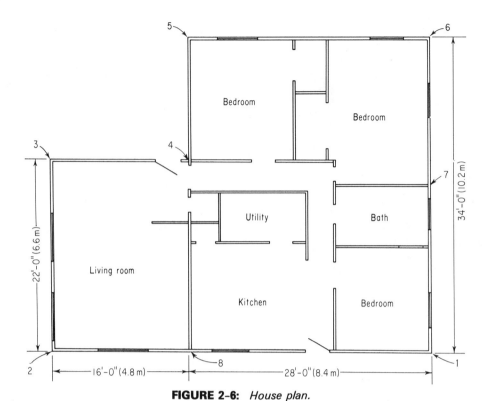

FIGURE 2-6: *House plan.*

A Simple Building Layout

Suppose that a house with a plan as shown in Fig. 2-6 is to be built on a lot measuring 65 ft × 100 ft (19.5 m × 30 m). It is to be set back 30 ft (9 m) from the front property line and in 5 ft (1.5 m) from the east boundary. If the layout is to be done without the aid of a leveling instrument, proceed as follows:

1. Locate the four corners of the lot and stake them.

2. Run tight lines across the front and down both sides of the lot.

3. Measuring from these lines, drive stakes at approximately the positions indicated as 1 to 6 in Fig. 2-3.

4. Erect stakes along the side property lines 30 ft (9 m) back from the front property line at A and B to indicate the front setback of the building. These stakes will have marked on them the distance to the building corner and the amount of cut at that point.

5. Measure back along the side property line another 12 ft (3.6 m) to C and D to mark the location of the front living room wall.

6. Measure back along the side property lines another 22 ft (6.6 m) to the location of the rear building wall at E and F.

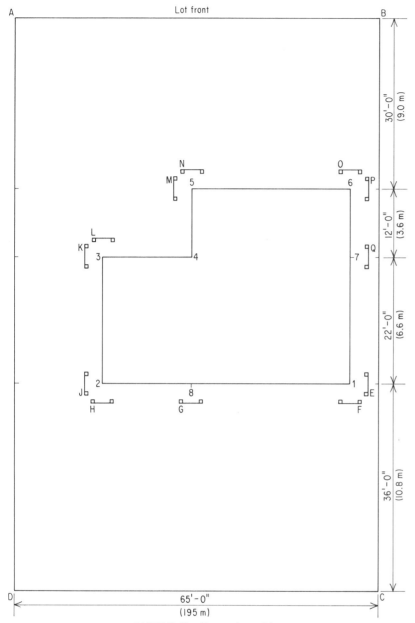

FIGURE 2-7: *Batter board layout.*

7. If batter boards were used for this layout, they would be set
at right angles 4-6 ft (1.2-2 m) back from the corner stakes
at 1, 2, 3, 5, and 6. Single batter boards would be used at
locations 7 and 8 (see Fig. 2-7).

All the points required to lay out the building are now established, and excavating work may proceed.

LEVELING INSTRUMENTS

The alternative to laying out as described above involves the use
of a leveling instrument. There are two basic types commonly
used in construction work, the builder's level (see Fig. 2-8) and
the transit level, illustrated in Fig. 2-9.

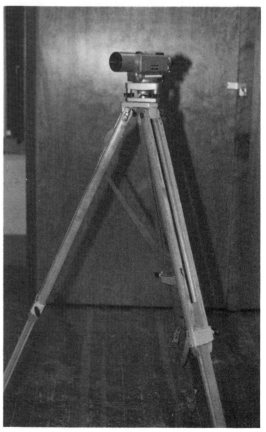

FIGURE 2-8: *Builder's level.*

The builder's level turns in a horizontal plane only, while the transit level moves in both horizontal and vertical planes, making it much more versatile. Some instruments are leveled by means of leveling screws, while others are automatic—they simply require a small bubble to be centered in a circular dial by means of a single adjustment in order to level them.

(a) Transit level.

(b) Transit level equipped with EDM (Electronic distance measuring equipment).

FIGURE 2-9

Builder's Level

The leveling instrument illustrated in Fig. 2-8 consists of a *telescope tube* containing an *objective lens* in front, which can be focused by means of a *focusing knob* on the side. The telescope is mounted in a *frame* and has attached to it a *bubble tube* very similar to a small spirit level. The telescope and bubble tube are leveled by means of three *leveling screws,* turned against a *leveling head.* The screws are adjusted until the bubble is centered in the dial, regardless of the direction in which the instrument points.

At the rear end of the telescope there is an *eyepiece,* containing a small lens and an *eyepiece ring* which can be turned to focus the lens. A pair of *cross hairs* is mounted in front of the eyepiece lens, and they are brought into sharp focus by the adjustment of the eyepiece ring.

At the bottom of the frame there is a *graduated horizontal circle* and a *vernier scale,* used to read horizontal angles accurately. The instrument may be held in any horizontal position by tightening a *horizontal motion clamp screw*. It can then be brought into fine adjustment by a *horizontal motion tangent screw.*

Transit Level

A transit level telescope is pivoted in the frame and held in the horizontal position by a *locking lever.* When it is unlocked, the telescope may be tilted through a vertical arc and held in a tilted position by a *vertical motion clamp screw.* A fine adjustment to the position can then be made by the *vertical motion tangent screw.*

Attached to the telescope is a *graduated vertical arc* which moves as the telescope is tilted. Fixed to the frame is a *vertical vernier scale,* used to read the degree of tilt accurately.

The entire instrument is carried on a *centering head* which sits in a circular opening in the leveling head and allows lateral movement of the instrument within the confines of the opening when the leveling screws are loosened slightly. With some instruments a *plumb bob* is hung from the underside of the centering head and this lateral movement aids in the final centering of the instrument over a pin in the ground. Other instruments use an *optical plummet* (a visual sighting arrangement) to center over a pin.

Setting up a Level

When in use in the field, a level is set on a tripod which has a threaded top plate protected by a cap. To mount the instrument and adjust it in preparation for use, the following steps are suggested:

1. Set the tripod in the desired position and spread the legs so that when the instrument is mounted, the eyepiece will be at a comfortable height for the user. The legs should have a spread of about 3 ft (1 m). Try to set it so that the tripod head is as level as you can judge by eye.

2. Set the leg tips firmly in the ground. On a hard surface, it may be desirable to use a wooden pad under the tips. Some tripods have telescoping legs so that on uneven ground the tripod head can be kept level by lengthening or shortening one of the legs.

3. Tighten the wing nuts at the top of the tripod legs into a firm position.

4. Remove the screw cap from the tripod and, lifting the instrument by the frame, set it on the tripod head. Be sure that the horizontal clamp screw is loose.

5. Hold the level by the frame and screw the leveling head firmly into place.

Leveling an Instrument

The instrument is now ready to be leveled. This operation must be carried out carefully and precisely, not only to obtain accurate results but also to ensure that the screw threads are not damaged by too much pressure. The following procedure may be adopted for this purpose:

1. Check to see that, with a transit level, the locking lever is in the *locked* position and that the horizontal motion clamp screw is loose.

2. Turn the instrument so that the telescope tube is parallel to one pair of leveling screws on an instrument with four leveling screws. On an instrument with three leveling screws the telescope is set parallel to two of the screws.

3. Turn the screws simultaneously with the thumb and forefinger of each hand, one clockwise and the other counterclockwise until the bubble is centered in the tube. Be sure that both screws are exerting light pressure against the leveling head, but *do not* exert excess pressure on them.

4. Turn the instrument until the telescope is at right angles to the first pair of screws and repeat step 3.

5. Turn back over the first pair and recenter the bubble. Return to the second pair and check the levelness.

6. Now turn the instrument 180° and check to see that the bubble is still centered as it should be in this or any other position of the telescope.

7. Check to see whether the cross hairs appear sharp and clear. If they do not, turn the eyepiece ring to bring them into focus for you.

Some instruments utilize a single target bubble for leveling the base and the instrument level itself automatically (see Fig. 2–9).

LEVELING

The instrument is now ready for leveling, and, in order to do so, a *leveling rod* or a thin wooden pole is required. A leveling rod, usually made in two or more sections, is marked off in feet (meters) and subdivided, often by the decimal system into smaller fractions of a foot (meter). A rod with divisions in feet and inches can be useful in construction (see Fig. 2–10).

Some rods are equipped with a sliding *target* (see Fig. 2–11) which is adjusted until the horizontal line coincides with the horizontal cross hair and then locked in position. The reading is then taken *at the horizontal line*. To take a level sight on a rod with target, proceed as follows:

1. Have a colleague hold a leveling rod with target at some convenient distance away [50–100 ft (15–30 m)] from the leveled instrument.

2. Aim the telescope at the rod by sighting along the top.

3. Now sight through the eyepiece and adjust the focus by turning the focusing knob. Adjust the instrument until the object is centered as closely as possible. Tighten the horizontal motion clamp screw and adjust with the horizontal motion tangent screw until the vertical cross hair is centered on the rod.

4. Have the rod man adjust the target up or down until the horizontal cross hair coincides with the horizontal line on the target. He can now take the reading indicated on the rod.

If, for example, the reading taken in the procedure outlined above were 3.625 ft (1.105 m), it would mean that the horizontal cross hair is the distance above the ground on which the rod is standing. This is commonly known as the *height of instrument* (H.I.) for that position.

Now have the rod moved to a second position and again take a reading, e.g., 4.746 ft (1.447 m). Since the *line of sight* of the instrument is level, the bottom end of the rod must be 4.746 − 3.625 = 1.121 ft (0.342 m) lower in the second position than it was in the first. In other words, the *difference in elevation* of the ground between the two positions is 1.121 ft (0.342 m).

Use of a Level to Establish Batter Board Heights

To use a level in building layout, there are a number of basic operations that must be understood. One is the establishment of a common level at a number of different locations. A second is the determination of the elevation of a point or series of points, relative to a point of known elevation—a *benchmark*. A third is the running of a straight line and a fourth the measurement of a given angle or the establishment of an angle of given size.

An example of how to set a common level at a number of positions may be shown by referring to Fig. 2-7 again, where batter boards could be set at a common level—the level of the bar at E. To set the remainder of the bars at the same level with a leveling instrument, proceed as follows:

1. Set up the instrument at some convenient point, preferably within the confines of the batter boards, and level it.

2. Have a wooden pole held vertically on top of the bar at E. Sight on the pole, and have your line of sight on the pole marked with a pencil.

3. Now have the pole held against the side of one of the stakes at F and moved up or down until your line of sight coincides

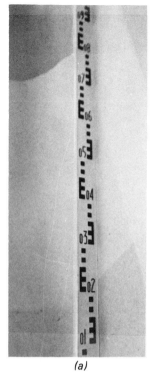

(a)

FIGURE 2–10:
(a) Metric levelling rod
(b) Levelling rod (Divisions in feet and inches).

(b)

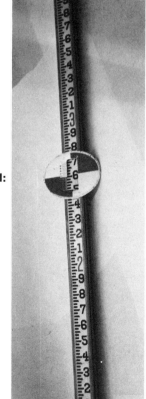

FIGURE 2–11:
Target rod.

with the pencil mark again. Mark the stake at the bottom end of the pole. That point is level with the top of the bar at E.

4. Establish a similar level mark on one of the stakes at each batter board position in the layout.

5. Attach batter board bars to each pair of stakes at the established level. All will be on a common level with E.

Use of a Transit level to Establish Building Lines

A transit is extremely useful for setting a number of points in a straight line—piers for a pier foundation or for resetting the building lines at the bottom of an excavation. The setting of these points can be achieved with a builder's level, but a transit is much more convenient, especially when a difference in elevation is involved.

Set the instrument directly over the reference point and level the instrument. Release the telescope level lock and swing the instrument in the desired direction and align the cross hair on the desired stake. Tighten the horizontal circle clamp so the telescope can only move in the vertical plane. Moving the telescope up or down will locate a number of points in a straight line (see Fig. 2–12). This same process will easily establish a line in the bottom of an excavation from reference points outside of the excavation (see Fig. 2–13).

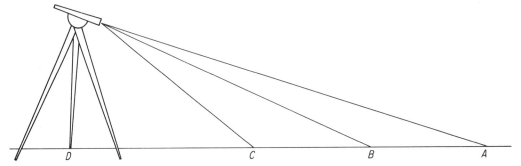

FIGURE 2–12: *Using a transit to establish a straight line.*

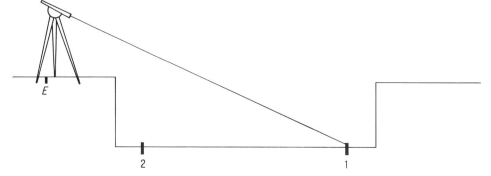

FIGURE 2–13: *Locating corners in a basement.*

Using an Instrument to Turn a Horizontal Angle

To lay out an angle with a builder's level, set the instrument directly over the required point on the ground. This point is commonly marked with a nail driven in the top of a stake. Turn the instrument and sight on station B (see Fig. 2–14) and set the horizontal circle at zero to align with zero on the vernier scale. Swing the instrument to the required angle to locate station C. The line of sight must be dropped with a plumb line to the ground level to establish the exact point (see Fig. 2–15). This dropping is not necessary when a level transit is used as it can be rotated on a vertical plane.

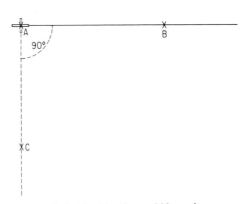

FIGURE 2-14: *Turning a 90° angle with a transit or a level.*

FIGURE 2-15: *Dropping line of sight to ground level when using a builder's level.*

Right angles can be established without the use of an instrument by using the right-angle method illustrated in Fig. 2–16. Once the layout is complete, work can progress on the building foundation.

FIGURE 2-16: *Checking diagonals.*

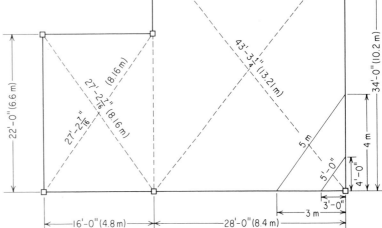

REVIEW QUESTIONS

2-1. Give two reasons for inspecting a residential building site before any plans are drawn.

2-2. Who is responsible for the location of property corner stakes?

2-3. Explain what is meant by *front setback*.

2-4. What do National Housing Standards say about

(a) The retention of existing trees on a lot on which a building is to built?

(b) The handling of topsoil on a lot?

2-5. Explain how the approximate corners of a building are located.

2-6. Give two reasons why batter boards should be set to a common level.

2-7. Explain how you would use a leveling rod and level to determine the difference in elevation of two points.

2-8. Why is it important to keep batter boards back several feet from approximate building corners?

2-9. Explain how a corner is located in the excavation from building lines.

2-10. What is the purpose of each of the following parts of a transit level?

(a) Objective lens

(b) Eyepiece ring

(c) A vernier scale

(d) Horizontal motion tangent screw

(e) Centering head

2-11. A leveling instrument is set up in the center of the building site illustrated in Fig. 2-7. A rod reading taken at position A is 3.156 ft (1.125 m). Readings taken at positions B, C, and D are recorded as 3.287, 5.246, and 5.718 ft (1.187, 1.904, and 1.988 m), respectively.

(a) What is the difference in the elevation between point A and each of the other three points?

(b) In what direction does the lot slope?

The foundation is the supporting base upon which the super-structure of a building is built; it anchors the building to the earth and transmits the loads of that building to the soil beneath. It is therefore of the utmost importance that the foundation be strong, accurately built to size, plumb and level, and of such dimensions that its loads are spread over an area of undisturbed soil large enough to support them safely. Any errors made in the size, shape, or strength of the foundation may lead not only to construction difficulties but may also contribute to instability and future movement.

A building falling into the category of light construction may be constructed on any one of a number of types of foundation, depending on its size, use, location, the prevailing climatic conditions, and the type of soil in the area.

FOUNDATION TYPES

The types of foundation most commonly used include *concrete full basement foundation, concrete surface foundation, slab-on-ground foundation, pier foundation, and preserved wood foundation.*

FOUNDATION EXCAVATION

Regardless of the type of foundation to be used, some earth removal will be necessary, the amount depending on the type of foundation, type of soil, depth of frost penetration, soil drainage conditions, and the proposed use of the building.

For a full concrete basement or preserved wood foundation, the excavation may be of some considerable depth, while for a slab-on-ground, relatively little earth removal may be required. For a surface, pier, or slab foundation, the excavating may be confined to holes or trenches.

In any case, the top soil and vegetable matter must be removed, and, in localities in which termites occur, all stumps, roots, and other wood debris must be removed to a minimum depth of 12 in. (300 mm) in unexcavated areas under a building.

National building codes specify the *minimum depths* of foundations based on the type of soil encountered and on whether or not the foundation will contain an enclosed heated space. In general, for rock or soils with good drainage, there is no depth limit, but for soils with poor drainage and where there will be no

3

THE FOUNDATION

heated space, the minimum depth will normally be 4 ft, 0 in. (1200 mm) or the depth of frost penetration, whichever is greater. The depth of frost penetration in a particular area is established by the local building authority. A stake driven to the proper level, a nearby manhole cover, or other permanent location of known elevation may be used as a reference point from which to measure the depth to be excavated.

The first step in carrying out an excavation is to stake out the area to be excavated. This will include not only the area covered by the building, already established (see Chapter 2), but enough extra space that people may move about outside the foundation forms (see Fig. 3-1). The deeper the excavation, the more necessary it is to allow outside working room. Generally, 2 ft, 0 in. to 3 ft, 0 in. (600–900 mm) on all sides will be sufficient.

FIGURE 3-1: *Excavation with space outside the foundation forms.*

The type of soil involved will also have an influence on the excavation limits. If it is firm and well packed and has good cohesive qualities, it may be possible to excavate and leave perpendicular earth walls standing at the outlined limits. However, if the soil is loose or becomes loose as it dries out, it will be necessary either to *slope the sides* of the excavation, up to about 45°, depending on the type of soil, thus increasing the excavation limits, or to *shore* the sides.

If space does not permit sloping, then shoring becomes necessary. If the excavation walls will hold up, then temporary walls—*cribs*—of plywood or planks may be placed against them and held in place by *cribbing studs*, with one end driven into the bottom of the excavation and the other tied back to a stake driven into solid ground (see Fig. 3–2).

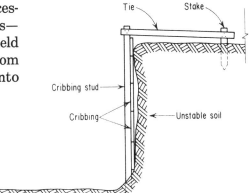

FIGURE 3-2: *Side of excavation cribbed.*

If the soil is very loose, it may be necessary to drive cribbing into the ground around the perimeter of the area and excavate inside it. This type of cribbing may be interlocking sheet piling or wooden or steel piles behind which some type of sheathing is placed as the excavating proceeds.

Another problem involved is the disposal of the earth from the excavation. If the top soil is good loam, it may be required for landscaping on the site after construction, in which case it should be stripped and piled by itself. The remainder of the earth must be disposed of according to circumstances. If some of it is required for backfilling, it should be piled at the site, out of the way of construction. If space does not permit storage, excavation machinery should be used which will allow direct loading onto trucks for removal (see Fig. 3–3).

FIGURE 3-3: *Excavating with bucket on wheeled tractor (Courtesy Hough Machine co.).*

EXCAVATING PROCEDURES

The type of machinery used to do the excavating work will depend on the area and depth of the excavation, the type of soil involved, and the available space outside the excavation.

Shallow excavations may be dug with a *bulldozer blade* provided that there is room around the excavation to deposit the soil or another machine to load it onto trucks. Deeper excavations may be carried out with a front-end loader on a tracked or rubber-tired tractor (see Figs. 3–3 and 3–4).

Another excellent machine for excavating, particularly for light construction, is a *power shovel* (see Fig. 3–5). It can dig either shallow or deep excavations with straight, vertical walls and a level floor.

FIGURE 3–4: *Tracked front-end loader.*

FIGURE 3–5: *Power shovel (back hoe).*

Power trenchers are very useful for the walls for houses with crawl spaces or with slab on grade when the soil is stable enough to resist caving. This is especially suited to shallow foundations.

FOUNDATION CONSTRUCTION

When the excavation has been completed, the work of constructing the foundation can begin. For foundations involving cast-in-place concrete, forms have to be built. For others, gravel must be laid and compacted. In each case, the building lines must be established on the excavation floor with a transit or by dropping a plumb from lines strung on batter boards (see Fig. 2–5).

CONCRETE FULL BASEMENT FOUNDATION

A *full basement* foundation of concrete is one of the most common types of foundation used in residential construction, because of the extra usable space provided. It consists of walls of cast-in-place concrete or concrete block, not less than 8 in. (200 mm) thick and of such a height that there will be at least 7 ft, 10 in. (2350 mm) of headroom. The walls usually encompass an area of the same dimensions as the floor plan of the building and enclose a floor and livable space.

The walls are supported on *continuous footings* wide enough that the building loads are supported safely by the soil beneath them. Interior loads are carried on one or more *girders*, supported on posts resting on individual *post footings* (see Fig. 3–6).

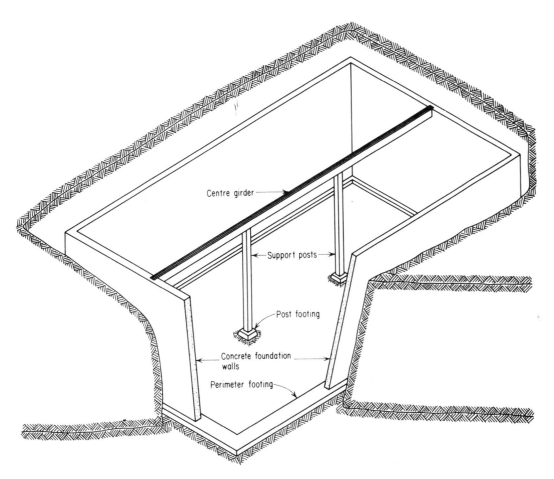

FIGURE 3–6: *Full basement foundation.*

Footing Forms

Forms for the continuous perimeter footings and the interior post footings are the first requirement in the construction of a full basement foundation.

Minimum footing widths for light buildings are given in Table 3–1. The thickness of the footing must not be less than the projection beyond the supported wall or post, except where the footing is suitably reinforced and, in any case, must not be less than 6 in. (150 mm).

TABLE 3-1: *Minimum footing widths for light construction*

Number of floors supported	Minimum widths of strip footings [in. (mm)]		Minimum area of column footings [ft² (m²)]
	Supporting external walls	*Supporting internal walls*	
1	10 (250)	8 (200)	4.5 (0.4)
2	14 (350)	14 (350)	8 (0.75)
3	18 (450)	20 (500)	11 (1.0)

(Courtesy National Research Council.)

NOTES:
1. For each story of masonry veneer over wood-frame construction, width of footings supporting exterior walls are to be increased by 2½ in. (65 mm).
2. For each story of masonry construction other than the foundation walls, width of footings supporting exterior walls must be increased by 5 in. (130 mm).
3. For each story of masonry supported by the footing, the width of footings supporting interior walls must be increased by 4 in. (100 mm).
4. Sizes of column footings shown in the table are based on columns spaced 9 ft, 9 in. (3m) o.c. For other column spacing, the footing areas must be adjusted in proportion to the distance between columns.

The type and thickness of material required for footing forms will depend on their size and on whether they are to be constructed above or below ground level (see Fig. 3–7). If they are to be set above the ground level, they should be built of 1½ in. (38 mm) lumber to withstand the pressure of the freshly placed concrete. Footings below ground level may be made from lighter material.

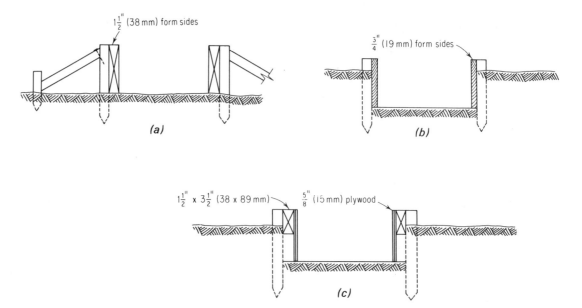

FIGURE 3-7: *Footing forms.*

To lay out and construct footing forms for footings above grade level, proceed as illustrated in Figs. 3–8 through 3–11. These forms can be leveled using a builder's level, a spirit level, or a line level. Forms constructed of 1-in. (25-mm) boards should be held with stakes placed 2–3 ft (600–1000 mm) apart. When 2-in. (38-mm) material is used, the stake spacing may be increased.

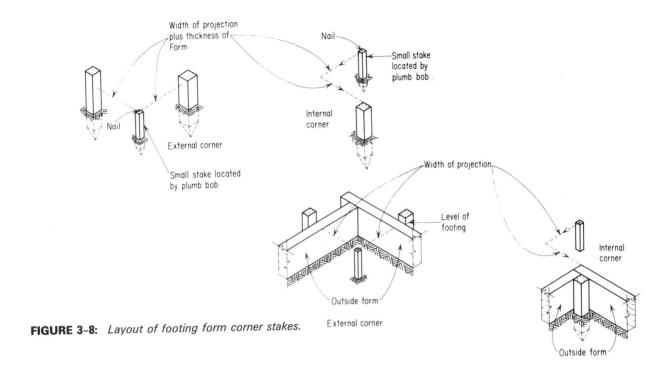

FIGURE 3-8: *Layout of footing form corner stakes.*

FIGURE 3-9: *Outside form for footing.*

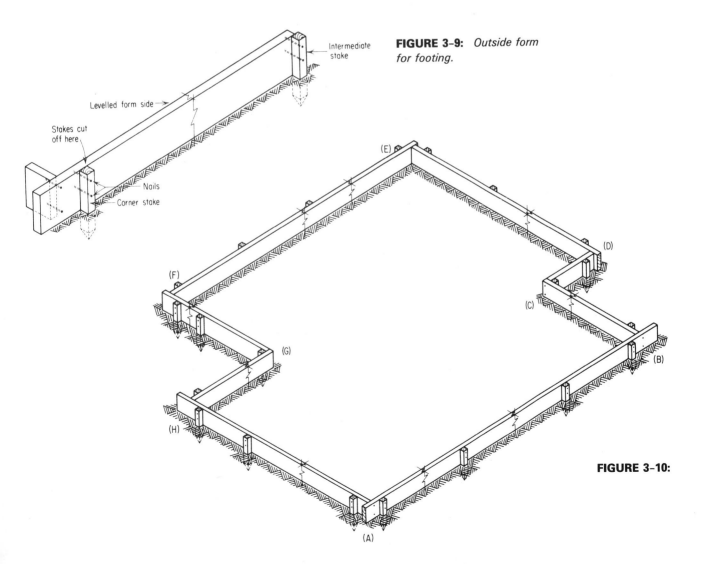

FIGURE 3-10:

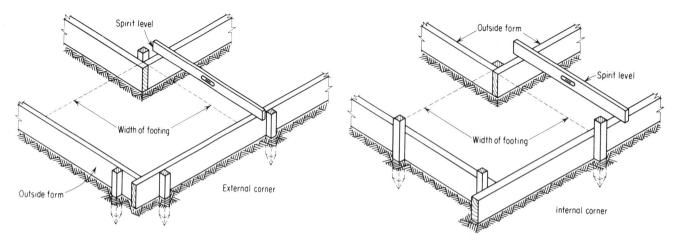

FIGURE 3-11: *Inside form set and levelled.*

Footing-to-Wall Ties

Some means must be provided for tying together the footing and the wall which will rest on it. One method is to insert short *steel dowels* into the footing concrete along the centerline at about 4 ft intervals, when it is first placed. Dowels should project about 3–4 in. (75–100 mm) above the footing. Another method is to set *bricks on edge* into the fresh concrete to about one-half their depth at approximately the same intervals. A third method involves setting an *oiled, tapered wooden keyway* form into the top of the footing, suspended in place by straps nailed across the top of the footing form (see Fig. 3–12).

If the keyway thus formed is coated with a thick coating of asphalt before the wall is cast, it will provide a barrier against penetration of moisture between wall and footing.

FIGURE 3-12: *Footing-to-wall ties.*

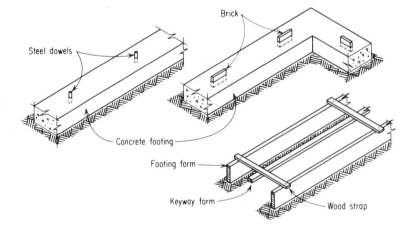

Post Footings

Individual footings, usually square in plan, are required for the posts which will support the center girder (see Fig. 3–6). The minimum area of such footing is given in Table 3–1, and the depth will normally be a minimum of 6 in. (150 mm) for unreinforced footings and not less than the projection. If the column is to be of wood, a layer of polyethylene should be wrapped around the bottom end to protect it from moisture (see Fig. 3–13). Footings for fireplaces and chimneys are normally placed at the same time as other footings. Footings vary in size depending on the soil-bearing capacity and the load they carry.

Bearing Wall Footing

In place of the conventional center girder, supported by posts, it is common practice to use a *bearing wall*, normally made of a 2 × 6 in. (38 × 140 mm) wood frame (see Chapter 4 for framing details).

The wall is supported on a continuous footing, similar in dimensions to the outside wall footings but with a raised center portion to keep the bottom of the bearing wall plate above the level of the basement floor (see Fig. 3–14).

The bottom part of the form is similar to that used for sidewall footings, while the center section is formed by suspending a narrow form the same width as the bearing wall (see Fig. 3–15).

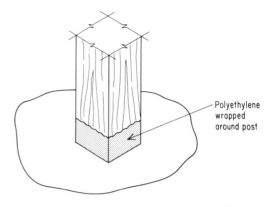

FIGURE 3–13: *Polyethylene around base of wood post.*

FIGURE 3–14: *Bearing wall footing.*

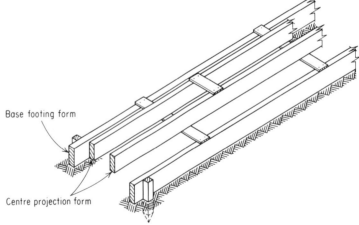

Base footing form

Centre projection form

FIGURE 3–15: *Bearing wall footing form.*

Stepped Footings

Stepped footings are used on steeply sloping lots and where the attached garage or living areas are at basement level. The vertical part of the step should be poured at the same time as the footing. The bottom of the footing should be set level on undisturbed soil below the frost line. The vertical portion of the step should be a minimum of 6 in. (150 mm) thick and the same width as the footing. The vertical portion of the step should not exceed 2 ft (600 mm), and the horizontal portion should not be less than 2 ft (600 mm). In very steep slopes or while building on rock, special footings may be required (see Fig. 3–16).

FIGURE 3-16:   *Stepped footing forms.*

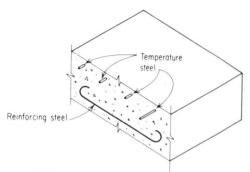

FIGURE 3-17:   *Reinforced footing.*

In residential construction, loads are seldom heavy enough that reinforcement is necessary in the footings. But other types of light construction may require reinforced footings, and, in such cases, the reinforcement should be in the form of *deformed rods* with hooked ends, placed *across the width* of the footing. There should be about 3 in. (75 mm) of concrete below the reinforcement, and the ends of the hooks should not be closer than 1 in. (25 mm) to the side form (see Fig. 3–17). Post footings are reinforced in the same manner, except that in some cases two layers of bars may be placed to run at right angles to one another—*two-way reinforcement.*

Wall Forms

When the footing concrete has hardened and at least partially cured, the forms are removed, and the job is ready for the erection of wall forms. There are many methods of building them, and here only a few of the well-known ones will be discussed.

Wall forms all contain the same basic features, although they may vary as to the details of construction and hardware required.

FIGURE 3-18:   *Wall form section.*

The main component is the *sheathing*, which will give the concrete its desired shape. The sheathing is stiffened and aligned by horizontal *walers*, which, in turn, are supported by *bracing*. The two sides of the form are fastened together by a system of *ties* and are held at their proper spacing by *spreaders* (see Fig. 3–18).

A commonly used forming system involves the use of ¾-in. (19-mm) *plywood panels without any framework, ties* with loops or slots in their ends, and *rods or bars* which are inserted through the tie ends on the outside of the forms. Single *walers* are used to align the forms. The bar and wire ties shown in Fig. 3–19 are commonly used in this type of form.

FIGURE 3–19: *Typical form ties.*

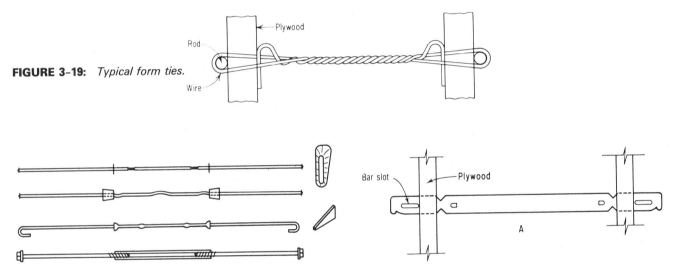

Panels are held together at the corners by a vertical rod running through a series of metal straps with looped ends, which are bolted or screwed along the panel edges, as illustrated in Fig. 3–20.

FIGURE 3-20: *Corner rod.*

The erection of a wall form by this method is quite simple. First, the plywood panels are slotted to receive the tie ends. The number of ties required will depend on the rate of placement and the temperature, but commonly accepted tie spacings for a 4-ft (1200-mm) rate of placement at 70°F (21°C) are as follows:

For a 2 ft × 8 ft (600 × 2400 mm) panel, the *two* end slots are 8 in. (200 mm) from the top and bottom of the panel and 6 in. (150 mm) from the edges. The remainder are in line, 16 in. (400 mm) o.c.

For a 4 ft × 8 ft (1200 × 2400 mm) panel, the *three* end slots are spaced 16 in. (400 mm) o.c., 8 in. (200 mm) from top, bottom, and edges. The remainder are in line, 16 in. (400 mm) o.c. All panels are treated with oil or other form coating.

The panels are erected as follows:

1. Snap a chalk line on the footing ¾ in. (19 mm) outside the foundation wall line.

2. Nail a 2 × 4 plate to the footing on that line with concrete nails (see Fig. 3–21).

3. Hinge the outside corner panels together with a rod and stand them in place (see Fig. 3–22).

FIGURE 3–21: *Plate fastened to footing.*

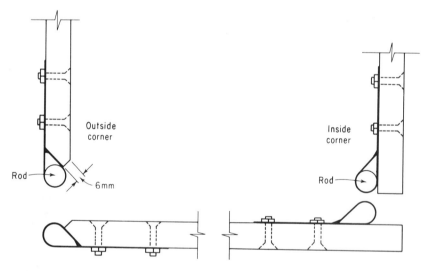

FIGURE 3–22: *Corner straps for plywood forms.*

4. Nail them to the plate with 2-in. (50-mm) nails and plumb and brace them temporarily.

5. Set all the outside panels in place and nail two 2 × 4 (38 × 89 mm) members to them horizontally, one at the top and one at the center (see Fig. 3–23).

6. Place the ties in the slots and insert the rods or bars as shown in Fig. 3–24.

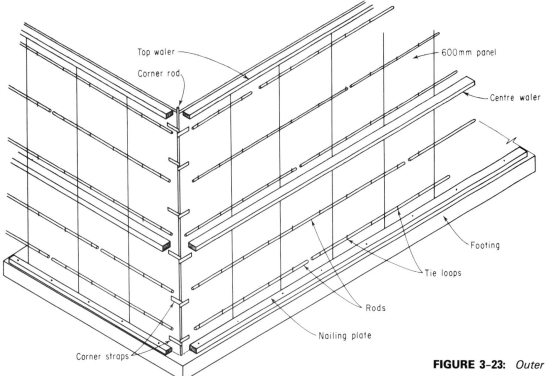

FIGURE 3-23: *Outer panels and walers in place.*

FIGURE 3-24: *Bars placed in tie ends.*

7. Put inside corners together and set them in place, using temporary wooden spreaders to maintain proper spacing.

8. Place the remainder of the inside panels, guiding the tie ends through the slots as each is erected.

9. Insert the inside rods or bars.

10. Finally, align the outside form and brace it as required from the two single walers.

In a cast-in-place floor system the inside forms are set and firmly held in position first (see Fig. 3–25). The outside forms are a little longer to allow fastening to the outside of the floor frame (see Fig. 3–26). When a straighter wall is required, a system of wall forms using 2 × 4 walers is recommended (see Fig. 3–27).

FIGURE 3–25: *Inside forms set in cast-in-place system.*

FIGURE 3–26: *Outside forms in cast-in-place system.*

(a) Single waler system.

(b) Double waler system.

FIGURE 3–27

FIGURE 3-28: *Fastening bottom waler (single waler system).*

(a) Corner support.

This system is more rigid and will not deflect as much due to the concrete pressure. The bottom waler is fastened to the footing as shown in Fig. 3-28. Figure 3-29 illustrates the method used to support the walers and how corners are supported.

Prefabricated Forms

A number of types of prefabricated forms are in occasional use for light construction. One consists of sections or *panels* made with steel frames and sheathed with plywood (see Fig. 3-30). Sections are made in various widths and heights, with special pieces to form corners and pilasters. The units are held together by a system of bar ties and wedges which give the completed form great rigidity. A minimum of walers and bracing is required with this type of form.

(b) Waler bracket and tie.

FIGURE 3-29

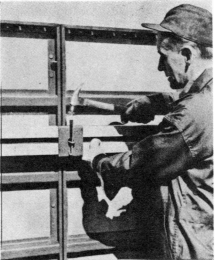

FIGURE 3-30: *Steel frame and wood sheathing form panels.*

FIGURE 3-31: *Steel forms in place (Courtesy Economy Forms Corp.).*

Another prefabricated forming system is made entirely of steel. Again, a variety of pieces is available, including inside and outside corners, angle corners, radius corners, inside fillets, flexible panels for odd-sized spaces, insert angles, spreader ties, spreader tie pins, plate clamps, and aligner clamps. Again, as illustrated in Fig. 3-31, a minimum of walers and bracing is necessary.

Girder Pocket Form

In designs in which the floor frame is to be built on top of the foundation (box sill), it is necessary to provide a pocket in two opposite foundation walls in which the ends of the center girder may rest [see Fig. 3-32(b)]. It is formed by a box, wide enough to allow for an air space around the end of the girder [see Fig. 3-32(a)] and deep enough to allow for a bearing plate under the end of the girder and for the top of the girder to be 1½ in. (38 mm) above the top of the foundation wall.

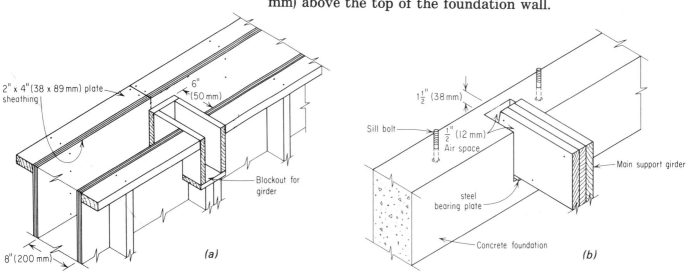

FIGURE 3-32: *Girder pocket in concrete foundation.*

Still Plate Bolts (Wood Floor Frame)

Also, in cases where box sill construction is used, the floor frame must be anchored to the foundation walls. This is done by bolting a 2 × 6 (38 × 140 mm) *sill plate* to the top of the foundation and nailing the floor frame to it. The anchors are ½-in. (12.7-mm) bolts, spaced approximately 4 ft (1200 mm) o.c., set into the concrete [see Fig. 3–32(b)] to hold the sill plate in position. They may be set after the concrete has been placed and before it has hardened, or they may be suspended from wood straps nailed across the top of the form before concrete is placed.

Sill Plates (Steel Floor Frame)

Several different methods are used for setting a sill plate in a foundation wall if the plate is to support a steel floor frame. If the building is to have conventional wood siding or similar finish, a 2 × 4 (38 × 89 mm) sill plate is set into the top *outside* edge of the wall form and held in place by straps, as illustrated in Fig. 3–33. Anchor bolts are suspended from the sill, to be cast into the concrete.

If the foundation wall is to support brick veneer exterior finish, the sill plate is cast into the top *inside* edge of the wall (see Fig. 3–34).

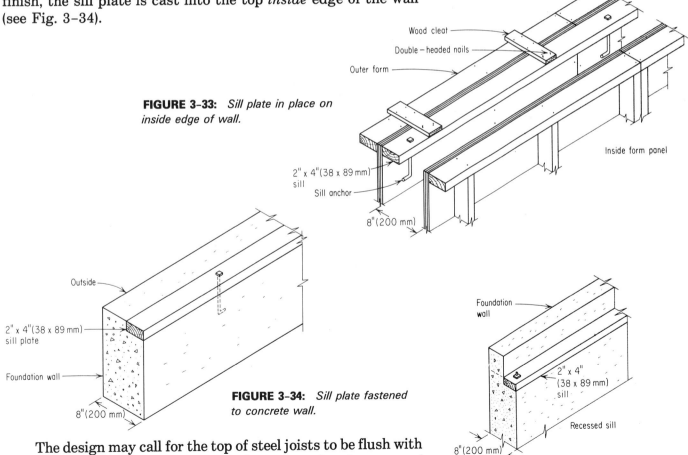

FIGURE 3-33: *Sill plate in place on inside edge of wall.*

FIGURE 3-34: *Sill plate fastened to concrete wall.*

FIGURE 3-35: *Recessed sill plate.*

The design may call for the top of steel joists to be flush with the top of the foundation. The sill plate must then be *recessed* below the top of the wall by the depth of the joists (see Fig. 3–35).

"Cast-in" Wood Floor Frame

One of the most common methods of anchoring the floor frame to the foundation is to use the *cast-in joist* system. Instead of resting on top of the foundation, the ends of the floor joists and the girder ends are embedded in the concrete (see Fig. 3–36).

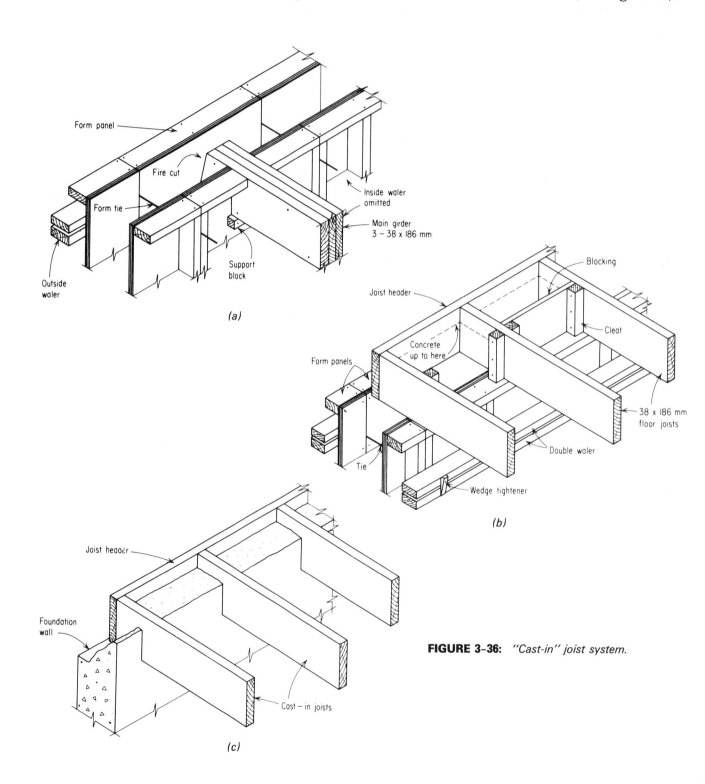

FIGURE 3-36: *"Cast-in" joist system.*

As a result, the girder and floor joists must be set in place before concrete is placed. The procedure is as follows:

1. Treat the ends of the girder and one end of each joist with wood preservative.

2. Cut notches in two opposite inside wall forms, the width and depth of the girder, and nail a support block at the bottom of each notch (see Fig. 3-36).

3. Lay out the joists with one end resting on the center girder and the other on the wall forms.

4. Lay off the header joists according to the specified joist centers and nail the joists to them at these locations (see Fig. 3-37).

5. Position the joist assembly so that the outer face of the header joists is flush with the inner face of the outside form.

6. Nail *blocking* flush with the inner face of the form sheathing.

FIGURE 3-37: *Header joist laid out and joists partially assembled.*

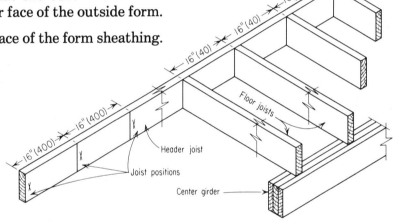

"Cast-in" Steel Floor Frame

Steel joists may be "cast in" in much the same way as wood ones except that steel joist ends are let into the inside form, as shown in Fig. 3-38. Wood or polystyrene blocks are fitted into the ends of the joists, where they protrude through the form, to prevent concrete from flowing out through the opening.

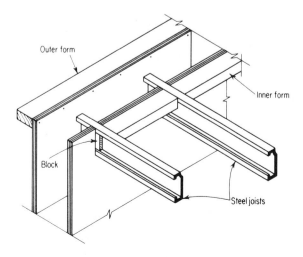

FIGURE 3-38: *Steel joists cast in.*

FIGURE 3-39: *Window frame set into forms.*

Door and Window Openings

Openings for doors and windows may be blocked out in two ways. One method is to secure the door or window *frame* into the form in its correct position. Wood frames must be made to coincide with the wall thickness, and metal frames are produced to fit various wall thicknesses. Wood frames should have *keys* secured to their outside surface so that they will not be able to move in the wall (see Fig. 3–39). Wood frames must also be braced diagonally, horizontally, and vertically so that the pressure of the concrete will not change their shape during placing.

Another method of forming openings is to set *rough bucks* into the form. A rough buck is a frame made from 2-in. (38-mm) material with *outside dimensions* equal to those of the frame to

FIGURE 3-40. *Rough back for door secured in forms.*

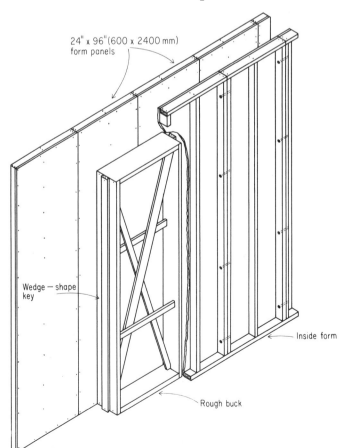

be used. *Wedge-shaped keys* are nailed to the outside of the buck as illustrated in Fig. 3–40. After the concrete is placed and hardened and the forms removed, the rough buck is also removed, leaving the key embedded in the concrete. The door or window frame is then inserted into the opening and held in place by nailing it to the key.

CONCRETE SURFACE FOUNDATION

A surface foundation also consists of concrete or concrete block walls, but they extend into the earth only far enough to reach below the frost line. The earth may or may not be excavated from within the walls, and, in many cases, the space is not great enough to be livable. A concrete floor may or may not be included, depending on the proposed use of the space. The walls rest on footings, similar to those used for full basement foundations.

Forms for a concrete surface foundation will be the same as those for a full basement, except that the wall forms normally will not be as high. The footing forms may be built in trenches, rather than in a full-scale excavation, as illustrated in Fig. 3–41, but otherwise procedures will be very similar to those described for full basement forms.

The only openings in the wall will usually be *crawl space* openings, which will be formed in the same way as those for doors or windows.

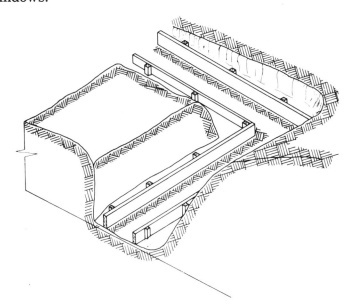

FIGURE 3-41: *Surface foundation footing built in trenches.*

SLAB-ON-GROUND FOUNDATION

Many residential buildings are built without basements. This is very common in southern regions where it is not necessary to excavate to reach a frost line. Two types of concrete floor construction are the *combined slab and foundation* and the *independent concrete slab and foundations walls.*

Combined Slab and Foundation

The combined slab and foundation consists of a shallow perimeter footing or beam that is placed at the same time as the concrete floor slab. This type of foundation is especially useful in southern regions where frost penetration is not a problem and where good predictable ground conditions exist (see Fig. 3–42). The footing is reinforced, and the bottom of the footing should be at least 12 in. (300 mm) below grade line. This thickening of the slab may also occur under interior load-bearing walls.

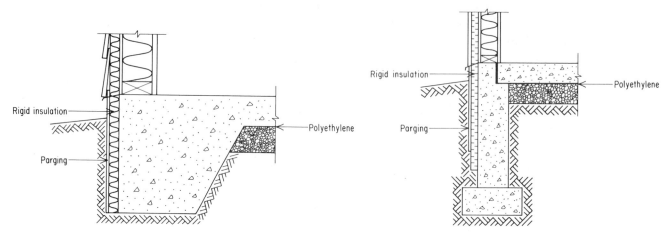

FIGURE 3-42: *Combined slab and foundation.*   **FIGURE 3-43:** *Independent slab and foundation wall.*

Independent Concrete Slab and Foundation Walls

An independent slab-on-ground foundation consists of perimeter footings and stub walls surrounding a reinforced concrete slab cast directly on the ground. The walls will extend down to solid undisturbed soil below frost level (see Fig. 3–43). However, in areas of deep frost penetration, it may be impractical to extend the footings down to below the frost line, and instead the footing is placed on a well-drained gravel pad at least 5 in. (125 mm) in depth.

The slab must not be less than 4 in. (100 mm) in thickness, supported by at least 5 in. (125 mm) of clean, coarse, well-packed gravel or similar granular material.

Forms for the footings for the perimeter bearing wall will be identical to those for a surface foundation and normally will be built in a shallow trench.

Walls may be cast-in-place concrete or concrete blocks and, in many cases, may not exceed 16–32 in. (400–800 mm) in height. If concrete walls are specified, forms for them will be similar to those for other concrete walls except for the height. Cast-in plates at the top of the wall are easily fastened inside the form prior to casting, providing a base connection for framed walls (see Fig. 3–44). A ledge is usually formed in the wall to receive the slab.

FIGURE 3-44: *Connection at top of foundation wall.*

FIGURE 3-45: *Reinforcing inserted into concrete wall.*

Inserting slab reinforcing into the concrete wall will also provide support (see Fig. 3–45). Placing piers inside the perimeter walls will give the slab additional support (see Fig. 3–46). Continue the reinforcing from the piers into the slab, providing a positive tie.

When the walls are complete, the earth within them is leveled and compacted to within approximately 9 in. (225 mm) of the top of the wall. A layer of gravel is added and compacted to a level within 4 in. (100 mm) from the top. Compaction is usually carried out by a power-driven compactor, similar to that shown in Fig. 3–47.

Two very important considerations with this type of foundation are *moisture control* and *insulation*, and Figs. 3–42 and 3–43 illustrate typical slab constructions with these two factors in mind.

FIGURE 3-46: *Pier supports for concrete slab.*

FIGURE 3-47: *Power compactor at work (Courtesy Wacker Co.).*

FIGURE 3-48: *Pier foundation.*

A continuous waterproof membrane is laid over the entire compacted gravel surface to prevent the migration of moisture into the slab from below. Six-mil polyethelene is often used for this membrane. A strip of rigid insulation at least 2 in. (50 mm) thick is applied to the outer exposed wall surface with asphalt adhesive. Then rigid insulation is laid on the entire surface. This insulation may only be placed around the perimeter or may be eliminated in warmer regions. Where it is used on exterior surfaces, it should be covered with a ½ in. (12 mm) of cement parging on wire lath. Finally, welded wire mesh or ⅜ in. (10-mm) steel placed 24 in. (600 mm) on center in both directions is placed over the whole surface, and then concrete is placed over the reinforcing.

PIER FOUNDATION

A pier foundation is one in which the building is constructed on a number of beams, with each supported by several *piers* or posts of pressure-treated wood, masonry, or concrete. Each post rests on an individual concrete footing or is used without a footing when the soil has sufficient bearing capacity (see Fig. 3–48).

Pier footings should be taken down to below the frost line or placed on well-drained gravel pads.

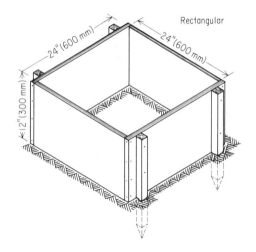

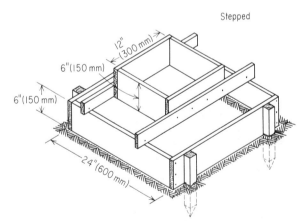

FIGURE 3-49: *Pier footing forms.*

Forms for pier footings are usually built in individual excavations, square in plan, and may be *rectangular* or *stepped*. The footing area is based on the amount of load carried by each one and the type of soil on which it rests. The construction of these footing forms is illustrated in Fig. 3–49.

Footings which are to support wooden piers should have a *steel pin* set in the center of the top surface to anchor the pier in place (see Fig. 3–50). Mortar is sufficient to anchor the masonry pier to the footing.

FIGURE 3-50: *Dowel for wooden pier.*

PRESERVED WOOD FOUNDATION

A preserved wood foundation is a complete *wood-frame foundation* system, built with preservative-treated lumber and intended for buildings falling into the light construction category. In this system, all wood exposed to decay hazard is pressure-treated with chemical preservatives which permanently impregnate the wood cells to the degree that makes the wood resistant to attack by decay organisms and termites.

It can be built as a *full basement* foundation (see Fig. 3–51) with a *concrete slab floor, wood sleeper floor,* or *suspended wood floor* or as a *surface* foundation (see Fig. 3–52) with an excavated crawl space with no floor.

FIGURE 3-51: *Preserved wood foundation with pressure-treated plank footings (Courtesy of Forest Industries of B.C.).*

FIGURE 3-52: *Surface foundation.*

FIGURE 3-53: *Continuous wood footings.*

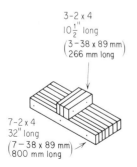

Site Preparation

After the excavation has been completed to the desired level, service and drain lines and a sump, if necessary, are installed and the trenches backfilled and compacted. In some localities, the sump pit may be replaced by a 4 in. (100 mm) perforated, vertical standpipe, at least 24 in. (600 mm) high, surrounded by 16 in. (400 mm) of washed, coarse gravel. The standpipe should extend up through the floor and be capped by a cleanout plug. No drainage system is required for unexcavated crawl spaces if the final grade inside the crawl space is equal to or higher than the grade outside.

Next, 5 in. (125 mm) of clean gravel is laid on undisturbed soil over an area extending 12 in. (300 mm) beyond the dimensions of the building and leveled. The gravel under all *footing plates* is compacted at least 12 in. (300 mm) beyond the edges of the plates to provide good bearing for loads.

Footings

Continuous wood footings, consisting of the *wood footing plates* and the *compacted gravel bed* beneath them, are the most practical and economical for this type of foundation, since they eliminate the building of forms and placement of concrete (see Fig. 3-53).

Wood footing plates are placed directly on the leveled, compacted gravel bed, butted together at end joints and wall intersections. The treated lumber can be ordered in specified lengths to suit the footing layout, and members may extend beyond the line of the wall at corners to avoid the cutting of plates wherever possible. However, if members must be cut, the exposed ends must be thoroughly saturated with wood preservative.

FIGURE 3-54: *Wood post footings.*

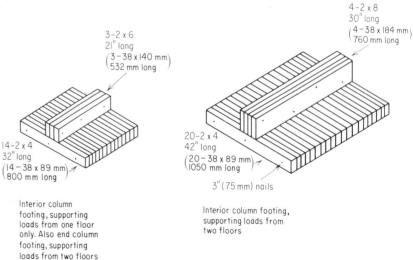

(Sizes apply to columns spaced not more than 8'-0" (2.4 m O.C.))

The sizes of plates required for various positions in the layout and conditions of loading are given in Table 3–2.

TABLE 3-2:   *Wood footing plate sizes (bearing on gravel bed)*

Type of basement floor	Number of floors supported above the basement	Type of exterior siding	Minimum sizes of wood footing plates [in. (mm)]	
			Supporting exterior walls	Supporting interior walls
Slab	1	Conventional	2 × 4 (38 × 89)	2 × 6 (38 × 140)
		Brick veneer	2 × 6 (38 × 140)	2 × 6 (38 × 140)
	2	Conventional	2 × 6 (38 × 140)	2 × 10 (38 × 235)
		Brick veneer	2 × 8 (38 × 184)	2 × 10 (38 × 235)
Treated wood sleeper or suspended wood floor	1	Conventional	2 × 6 (38 × 140)	2 × 8 (38 × 184)
		Brick veneer	2 × 8 (38 × 184)	2 × 8 (38 × 184)
	2	Conventional	2 × 8 (38 × 184)	2 × 12 (38 × 286)
		Brick veneer	2 × 10 (38 × 235)	2 × 12 (38 × 286)

(Courtesy Canadian Wood Council.)

NOTES: 1. Two-story brick veneer houses have brick veneer on only one story.
  2. Width of bottom plate supporting interior walls bearing on slab floors can be reduced 2 in. (50 mm) from values shown if concrete strength is 3000 psi (20 MPa).
  3. Interior woood footing plate supporting only sleeper or suspended floor can be 2 × 4 (38 × 89 mm).
  4. Width of footing plate should be at least equal to the width of the foundation wall stud.

If continuous concrete footings are used under the wood foundation walls, they are placed on top of the gravel bed to allow drainage under the footing. The width of such concrete footings supporting exterior walls can be 3 in. (75 mm) less than the widths specified in Table 3–1, because wood foundation walls are lighter than conventional concrete walls. Concrete footings supporting interior walls should be the same width as those specified in Table 3–1.

When a girder and posts are used to support the interior loads, rather than a bearing wall, the post footings may be either concrete or preservative-treated wood. Concrete footings will be similar to those in conventional foundations, and Fig. 3–54 illustrates the construction of wood post footings for various loading conditions.

Post footings may be set on undisturbed soil below the gravel bed in order to avoid the top of the footings interfering with the basement floor.

Foundation Walls

Wood foundation walls, consisting of a *frame of studs* with *single top and bottom plate, sheathed with plywood* and all treated with preservative, may be prefabricated in sections in a shop or completely assembled on site. Where plywood sheets are applied horizontally, *blocking* is required between studs at the plywood joint. In addition, all plywood joints are caulked with sealant (see Fig. 3-55).

For full basement construction, walls are 8 ft. (2400 mm) high for slab and sleeper floors and 10 ft. (3000 mm) high for suspended wood floors (see section on floors, page 93). The size and spacing of studs depend on the *building loads, species* and *grade of lumber,* and the *height of backfill,* which, in turn, depend on the depth of the excavation. Table 3-3 gives the required stud size and spacing for foundation walls for basements with slab and sleeper floors and Table 3-4 the same information for basements with suspended wood floors.

TABLE 3-3: *Maximum backfill heights for PWFs with slab floors*

			Stud spacing										
			Inches						Meters				
Stud size	Species	Grade	12 in.			16 in.			300 mm			400 mm		
			Vertical load (lb/ft)						Vertical load (kN/m)					
			700	1400	2100	700	1400	2100	10	20	30	10	20	30
2 × 4 in	Douglas fir	No. 1	56	44	32	48	32	—	1.4	1.1	0.8	1.2	0.8	—
(38 × 89 mm)		No. 2	52	40	24	40	28	—	1.3	1.0	0.6	1.0	0.7	—
	Hemlock, fir[a]	No. 1	48	40	24	40	28	—	1.2	1.0	0.6	1.0	0.7	—
		No. 2	44	32	8	36	20	—	1.1	0.8	0.2	0.9	0.5	—
	-P-F[b]	No. 1	44	32	—	36	16	—	1.1	0.8	—	0.9	0.4	—
		No. 2	40	28	—	32	—	—	1.0	0.7	—	0.8	—	—
	Red pine	No. 1	44	28	—	36	—	—	1.1	0.7	—	0.9	—	—
		No. 2	40	24	—	32	—	—	1.0	0.6	—	0.8	—	—
2 × 6 in	Douglas fir	No. 1	92	88	84	80	76	68	2.3	2.2	2.1	2.0	1.9	1.7
(38 × 140 mm)		No. 2	84	80	76	72	68	60	2.1	2.0	1.9	1.8	1.7	1.5
	Hemlock, fir	No. 1	80	76	72	68	64	60	2.0	1.9	1.8	1.7	1.6	1.5
		No. 2	72	68	64	64	56	52	1.8	1.7	1.6	1.6	1.4	1.3
	-P-F	No. 1	80	72	68	68	60	56	2.0	1.8	1.7	1.7	1.5	1.4
		No. 2	72	64	60	60	56	48	1.8	1.6	1.5	1.5	1.4	1.2
	Red pine	No. 1	76	68	64	64	56	52	1.9	1.7	1.6	1.6	1.4	1.3
		No. 2	68	60	56	56	52	44	1.7	1.5	1.4	1.4	1.3	1.1
2 × 8 in	Douglas fir	No. 1	96	96	96	96	96	96	2.4	2.4	2.4	2.4	2.4	2.4
(38 × 184 mm)		No. 2	96	96	96	96	92	84	2.4	2.4	2.4	2.4	2.3	2.1
	Hemlock, fir	No. 1	96	96	96	84	84	80	2.4	2.4	2.4	2.1	2.1	2.0
		No. 2	96	92	88	84	76	72	2.4	2.3	2.2	2.1	1.9	1.8
	-P-F	No. 1	96	96	96	80	80	76	2.4	2.4	2.4	2.0	2.0	1.9
		No. 2	92	88	80	80	72	68	2.3	2.2	2.0	2.0	1.8	1.7
	Red pine	No. 1	96	96	88	84	80	72	2.4	2.4	2.2	2.1	2.0	1.8
		No. 2	88	84	76	76	72	64	2.2	2.1	1.9	1.9	1.8	1.6

[a]Hemlock, fir: western hemlock, amabilis fir (same as Hem-fir).
[b]-P-F: lodgepole pine, jack pine, alpine fir, and balsam fir (same as S-P-F except spruce species are excluded).

(b) Blocking at plywood joints

(c) Sealant at joints.

FIGURE 3-55: (a) Prefabricated foundation walls.

TABLE 3-4: Maximum backfill heights for PWFs with suspended floors

Stud size	Species	Grade	Stud spacing											
			Inches						Meters					
			12 in.			16 in.			300 mm			400 mm		
			Vertical load (lb/ft)						Vertical load (kN/m)					
			700	1400	2100	700	1400	2100	10	20	30	10	20	30
2 × 4 in (38 × 89 mm)	Douglas fir	No. 1	84	84	68	72	56	40	2.1	2.1	1.7	1.8	1.4	1.0
		No. 2	80	64	48	56	40	24	2.0	1.6	1.2	1.4	1.0	0.6
	Hemlock, fir[a]	No. 1	72	60	44	52	40	24	1.8	1.5	1.1	1.3	1.0	0.6
		No. 2	56	40	24	40	24	—	1.4	1.0	0.6	1.0	0.6	—
	-P-F[b]	No. 1	68	52	36	48	32	—	1.7	1.3	0.9	1.2	0.8	—
		No. 2	52	36	—	36	—	—	1.3	0.9	—	0.9	—	—
	Red pine	No. 1	64	44	28	44	28	—	1.6	1.1	0.7	1.1	0.7	—
		No. 2	48	32	—	32	—	—	1.2	0.8	—	0.8	—	—
2 × 6 in (38 × 140 mm)	Douglas fir	No. 1	108	108	108	96	96	96	2.7	2.7	2.7	2.4	2.4	2.4
		No. 2	108	108	108	96	96	96	2.7	2.7	2.7	2.4	2.4	2.4
	Hemlock, fir	No. 1	96	96	96	88	88	88	2.4	2.4	2.4	2.2	2.2	2.2
		No. 2	96	96	96	88	88	80	2.4	2.4	2.4	2.2	2.2	2.0
	-P-F	No. 1	96	96	96	84	84	84	2.4	2.4	2.4	2.1	2.1	2.1
		No. 2	96	96	92	84	84	68	2.4	2.4	2.3	2.1	2.1	1.7
	Red pine	No. 1	100	100	96	88	88	80	2.5	2.5	2.4	2.2	2.2	2.0
		No. 2	100	96	88	88	76	56	2.5	2.4	2.2	2.2	1.9	1.4
2 × 8 in (38 × 184 mm)	Douglas fir	No. 1	120	120	120	112	112	112	3.0	3.0	3.0	2.8	2.8	2.8
		No. 2	120	120	120	112	112	112	3.0	3.0	3.0	2.8	2.8	2.8
	Hemlock, fir	No. 1	116	116	116	100	100	100	2.9	2.9	2.9	2.5	2.5	2.5
		No. 2	116	116	116	100	100	100	2.9	2.9	2.9	2.5	2.5	2.5
	-P-F	No. 1	112	112	112	96	96	96	2.8	2.8	2.8	2.4	2.4	2.4
		No. 2	112	112	112	96	96	96	2.8	2.8	2.8	2.4	2.4	2.4
	Red pine	No. 1	120	120	120	104	104	104	3.0	3.0	3.0	2.6	2.6	2.6
		No. 2	120	116	108	104	104	96	3.0	2.9	2.7	2.6	2.6	2.4

[a]Hemlock, fir: western hemlock, amabilis fir (same as Hem-fir).
[b]-P-F: lodgepole pine, jack pine, alpine fir, and balsam fir (same as S-P-F except spruce species are excluded).

TABLE 3-5: *Foundation wall plywood sheathing thickness (unsanded)*

Face grain direction	Stud spacing [in. (mm)]	Plywood thickness		
		½ in. (12.5 mm)	⅝ in. (15.5 mm)	¾ (18.5 mm)
		Height of backfill [ft–in. (m)]		
Perpendicular to studs	12 in. (300 mm)	9–10 (3.0)	9–10 (3.0)	9–10 (3.0)
	16 in. (400 mm)	6–10 (2.1)	8–10 (2.7)	9–10 (3.0)
Parallel to studs	12 in. (300 mm)	5–7 (1.7)	7–10 (2.4)	9–10 (3.0)
	16 in. (400 mm)	3–7 (1.1)	4–7 (1.4)	6–10 (2.1)

(Courtesy Canadian Wood Council.)

The thickness of plywood used for sheathing depends on the *direction of face grain*—face grain parallel or perpendicular to the studs—the *stud spacing*, and the *height of backfill*. The plywood thickness required, based on these factors, is given in Table 3-5.

Plywood may be fastened to the frame with common nails or with 14 or 16 gauge staples, minimum length 2 in. (50 mm). Nails and 14 gauge staples are spaced 6 in. (150 mm) o.c. along outside edges and 12 in. (300 mm) o.c. along intermediate supports, while 16-gauge staples are spaced at 4 in. (100 mm) o.c. on outside edges and 8 in. (200 mm) o.c. on intermediate supports. Staples should be driven with the crown parallel to the framing.

Openings in foundation walls are framed in the same manner as those for conventional wall frames. Nailing requirements at openings are indicated in Fig. 3-56.

All joints between plywood panels below grade are sealed by bedding the plywood edges in a sealant applied to the face of the framing member or by leaving a ⅛-in. (3-mm) gap between plywood edges and caulking it with a sealant.

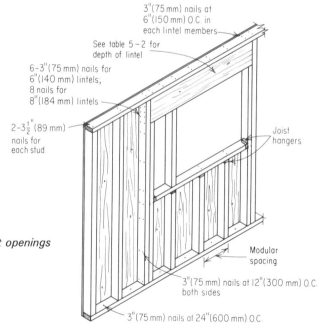

FIGURE 3-56: *Nailing requirements at openings in preserved wood foundation wall.*

The exterior of foundation walls below grade is also covered with a 6-mil polyethylene membrane, extending from a minimum of 3 in. (75 mm) above the finished grade line to the bottom of the footing plate, where it should be cut off. The membrane *should not* extend under the gravel pad or under the footing plate (see Fig. 3–57).

The polyethylene is cemented to the sheathing at the top edge by a 6-in. (150-mm) band of adhesive (see Fig. 3–58). It is also protected at the grade level by a 12 in. wide (300 mm wide) strip of treated plywood, set with its top edge at least 3 in. (75 mm) above the finished grade line. A strip of sealant about 3 in. (75 mm) wide is applied to the top inside face of the plywood before it is nailed to the foundation wall.

FIGURE 3-57: *Polyethylene membrane.*

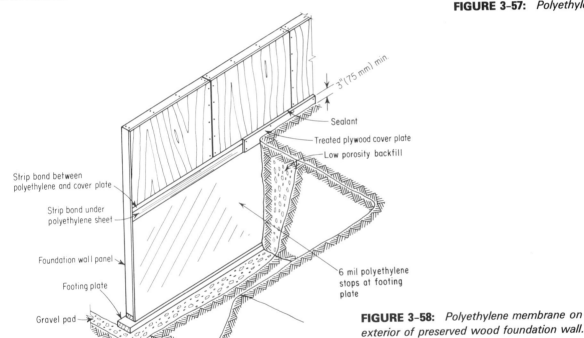

FIGURE 3-58: *Polyethylene membrane on exterior of preserved wood foundation wall.*

Concrete Slab Floors

Concrete slab floors used in conjunction with wood foundations are similar to those used for conventional basement floors. Basically, they consist of a minimum 4-in. (100-mm) concrete slab, placed over a 5-in. (125-mm) gravel bed, with a 6-mil polyethylene moisture barrier between concrete and gravel.

To transmit lateral soil loads from the wall into the slab, the top edge of the slab must butt directly against the bottom ends of the wall studs. This may be done by fastening a continuous, treated wooden strip along the lower edge of the foundation wall (see Fig. 3–59) wide enough that distance "d" will be from 1 to 2 in. (25–45 mm), depending on stud spacing and depth of backfill. This strip can be used as a screed to level the concrete slab and will remain in place after the concrete has hardened.

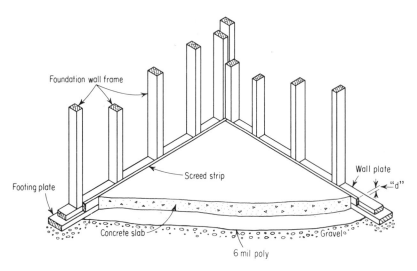

FIGURE 3-59: *Concrete slab floor in wood foundation basement.*

Wood Sleeper Floors

Wood sleeper floors (see Fig. 3-60) are damp-proofed by laying 4-mil, 4-ft-wide (1200-mm-wide) strips of polyethylene, overlapped 4 in. (100 mm) at the edges—not a continuous membrane—over the leveled gravel bed.

The 2 × 4 in. (38 × 89 mm) treated wood *sleepers* are placed on the polyethylene cover at spacings of from 4 to 6 ft (1200–1800 mm), depending on the depth of the floor joists to be used. Then floor joists, 2 × 4 in.(38 × 89 mm) or wider, span between the footing plates and the sleepers with at least 1½ in. (38 mm) bearing on the footing plate. To achieve this, it may be necessary to use wider footing plates than the design requires.

Joists are placed in line with foundation wall studs (see Fig. 3-60) and butt in line over the sleeper supports. They are toe-nailed to the sleepers and to the wall studs with at least two 3-in. (75-mm) nails at each junction.

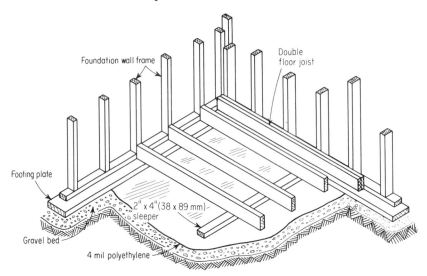

FIGURE 3-60: *Wood floor frame on wood sleepers.*

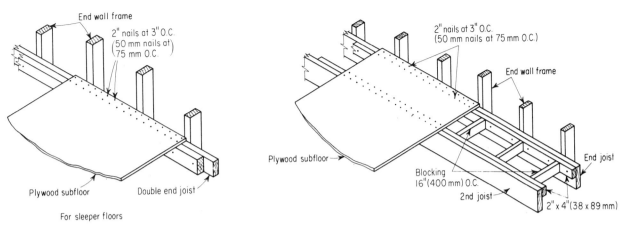

FIGURE 3-61: *Sleeper floors end nailing and extra framing.*

Plywood subflooring is installed over the floor joists and acts as a diaphragm to resist lateral earth loads. For the thickness of plywood required, see Chapter 4, page 134.

To ensure the proper transfer of lateral soil loads from the *end* walls to the floor frame and plywood subfloor, it is necessary to provide additional nailing and, in some cases, additional framing. See Fig. 3-61 to illustrate this additional support.

Suspended Wood Floors

Joists for a suspended wood floor are supported above the gravel base on a continuous 2 × 4 in. (38 × 89 mm) *ledger* at the foundation walls and by a low bearing wall at the inner end (see Fig. 3-62). Joist size and spacing is in accordance with Table 4-4, page 114.

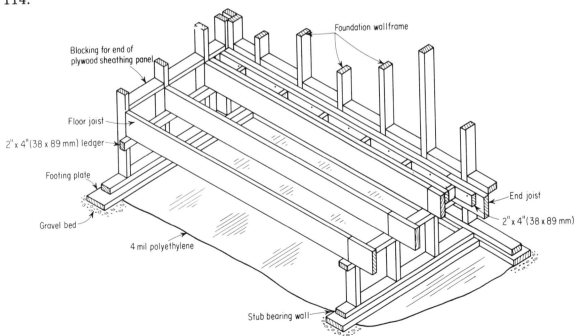

FIGURE 3-62: *Suspended wood floor.*

Joists are placed directly in line with the foundation wall studs and butted in line over the bearing wall (see Fig. 3-62). Plywood subflooring is applied over the joists and nailed as required in the same manner as for sleeper floors. Again, consult, page 134 for plywood thickness.

As was the case with wood sleeper floors, additional framing and nailing may be required to ensure the proper transfer of lateral soil loads to the floor.

CONCRETE PLACING

Regardless of the care spent in building forms, the final test of the strength and durability of the foundation being built will lie in the quality of the concrete used. That quality will depend on a number of factors, all of major importance. They include the following:

1. Clean and well-graded aggregate and properly proportioned fine and coarse aggregate.

2. Clean water; water fit for human consumption is the best test.

3. The amount of water used in the mix per unit of cement—a matter of primary importance. This ratio to a large extent controls the strength of the concrete. The less water used, within limits, the stronger the concrete will be, and, conversely, the more water used, the less strength will be achieved. The ratio, expressed in pounds (kilograms) of water per pound (kilogram) of cement, will ordinarily vary from 0.40 to 0.70 lb/lb (kg/kg).

4. Whether or not *entrained air* is included in the mix. Entrained air consists of thousands of tiny, stable bubbles of air which are introduced into the fresh concrete to improve the *flowability* of the mix. Air entrainment does, however, result in some reduction in compressive stength.

TABLE 3-6: *Compressive strengths of concrete for various water/cement ratios*

Water/cement ratio [lb/lb (kg/kg) of cement]	Probable compressive strength after 28 days [psi (MPa)]	
	Non-air-entrained	*Air-entrained*
0.40	5800 (40)	4800 (33)
0.45	5300 (36.5)	4200 (29)
0.50	4800 (33)	3650 (25)
0.55	4350 (30)	3200 (22)
0.60	3900 (27)	2750 (19)
0.65	3350 (23)	2300 (16)
0.70	3050 (21)	1750 (12)

Table 3-6 indicates the probable compressive strength which will be achieved, using a number of different water/cement ratios, both with and without entrained air.

The concrete may be mixed on the job, delivered ready-mixed from a concrete mixing plant, or delivered from a batching plant, mixed in transit. No matter which method is used, great care must be taken in placing the concrete in the forms:

1. Concrete should not be allowed to drop freely more than 48 in. (1200 mm). If the height of the form is greater than that, some type of chute is required so that the concrete may be conducted to at least within 48 in. (1200 mm) of the bottom.

2. Concrete should be placed in such a way that it will drop straight down, not bounce from one form face to the other.

3. Place in even layers around the form—don't try to place all in one spot and allow the concrete to flow to its final position.

4. Start placing at the corners and work toward the center of the form. All of these precautions will help to prevent *segregation*, the separation of the aggregates from the water-cement paste.

Concrete can best be consolidated in the form by vibration. This may be done either internally or externally. An internal vibrator (see Fig. 3-63) is inserted into the concrete and operated until consolidation has taken place. An external vibrator is operated against the outside of the form. Care must be taken not to over-vibrate, because excess paste will be brought to the top or out to the face of the forms.

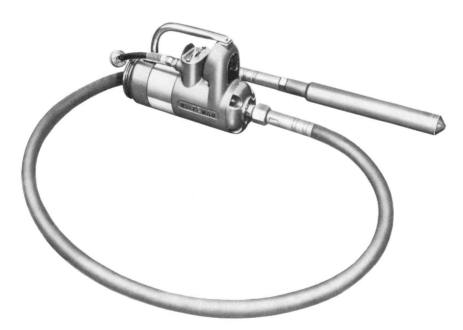

FIGURE 3-63: *Electric internal vibrator.*

Proper curing of the concrete, that is, allowing it to gain its rated strength, is very important. Temperature and moisture conditions control the curing. Concrete cures best at a temperature of about 70°F (21°C) and cures very slowly below 40°F (5°C). Moist conditions are required for good curing. If the concrete is allowed to dry out soon after it is placed, it cannot be expected to gain the strength required of it. Keep the concrete warm and moist for as long as possible.

One means of keeping the concrete moist is to leave the forms in place. However, if it is necessary to remove the forms early in the curing process, care must be taken not to damage the green concrete. Remove the braces, take off the wedges from the rod ends, and remove the walers. Take out all the nails or bolts holding the sections of form together and remove panels carefully.

There are occasions when it is necessary to have the concrete made with other than normal cement. If the land on which the foundation will rest contains alkali salts, concrete made with normal cement will set and cure poorly. In such cases, it is wise to specify alkali-resistant cement, which will produce concrete that will set and cure under alkaline conditions. Sometimes it may be necessary to have the concrete set and cure more quickly than is normally the case. If this is so, a special cement, called high-early-strength cement, should be used in making the concrete. Concrete so made will cure much more rapidly in its early stages than that made with normal cement.

BASEMENT FLOOR

The concrete basement floor may be placed in either one or two layers. In the first case, the complete thickness is placed in one operation; in the second, a base slab is placed first, over which a topping slab, perhaps slightly different in character, is laid later.

In either case, proper preparations for the floor are important. First, there should be a well-packed gravel base provided, at least 6 in. (150 mm) deep.

A 6-mil polyethylene membrane should then be laid over the gravel, initially to prevent the loss of water from the concrete to the base and eventually to prevent the migration of moisture upward through the concrete slab.

Screeds are then laid to the proper level. A screed is simply a *guide strip*, the top of which represents the level of the finished floor. They should be so spaced that a straightedge 8–10 ft (2–3 m) in length, can span from one to the next, to *strike off* the concrete to the right level (see Fig. 3-64). Straight 2×4 or 2×6 (38 × 89 or 38 × 140 mm) members, set on edge on some type of support which rests on the covered gravel base, or 25-mm pipe, carried on adjustable *chairs* (see Fig. 3-64), are commonly used as screeds. In either case, they must be set to provide for a minimum depth of floor of 3½ in. (90 mm).

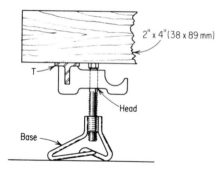

Wood screed

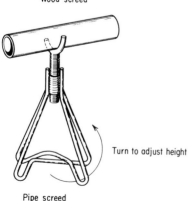

Pipe screed

FIGURE 3-64: *Floor screeds.*

Concrete Joints and Reinforcement

The joint between floor and wall may require some special consideration. Since the floor is probably placed after the wall concrete has hardened, there is little bond between them, and eventually shrinkage will produce a crack around the perimeter. This crack may be a source of trouble if moisture collects under the footings, because the water will find its way up through the joint. Such an occurrence may be prevented by sealing the joint with an asphalt caulking compound.

An oiled, wedge-shaped strip should be placed against the wall, outside the outer screed (see Fig. 3–65) and left there when the screed is removed. After the concrete has hardened, the strip is removed and the space packed with caulking compound (see Fig. 3–65).

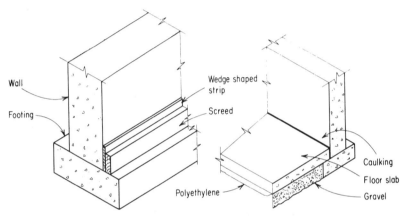

FIGURE 3–65: *Floor to wall joint.*

The need for reinforcement will depend on the size of the floor and its use. Residential floors do not usually require reinforcing, but larger ones may. Reinforcing may be done for two reasons, one being to give the concrete greater strength in bending and the other to control contraction and expansion of the top surface due to temperature changes. In the first case, the reinforcement will be rod or heavy wire mesh placed near the bottom of the slab. In the second, it will be light wire mesh placed close to the top surface.

Check the plans for any slope that may be required in the floor. If a plumbing drain is present, the floor will be sloped to it, and the screeds must be set so that the slope will appear.

The final step in placing the floor is producing its surface finish, which may be done either by hand or with a power trowel. Whichever method is used, it is important that the job be done at just the right time. The concrete should be partially set, hard enough to bear the workman's weight but with the surface still workable. It is first worked smooth with a float and then given a final treatment with a steel trowel.

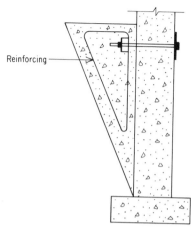

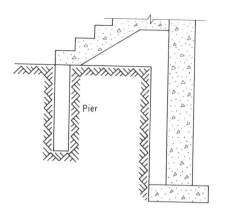

FIGURE 3-67: *Pier to support step.*

FIGURE 3-66: *Bolts used to fasten support for step.*

CONCRETE STEPS

Entrance steps must be attached to the foundation wall by means of reinforcing placed during wall construction or bolts fastened through the wall after construction (see Fig. 3-66). The foundation for the step can be piers resting on a footing when needed (see Fig. 3-67) or brackets fastened to the wall (see Fig. 3-68).

FIGURE 3-68: *Brackets for step support.*

FIGURE 3-69: *Step forms.*

Steps may be included and poured as part of the platform (see Fig. 3–69). When the steps are over 3 ft (1 m) wide, 2-in. (38-mm) material should be used for the risers, and additional support should be added when the width is over 6 ft (2 m), as shown in Fig. 3–69.

The risers of the steps should be set at an angle of about 10° to provide a nosing. The bottom edge of the riser should be beveled to permit troweling of the entire surface of the step (see Fig. 3–70).

Precast platforms and steps are commonly used, particularly where custom work is not required (see Fig. 3–68).

SIDEWALKS AND DRIVEWAYS

Sidewalks and driveways are not usually poured until after construction of the building is complete. In some areas the slab may be poured directly in contact with the soil. If the base is subjected to moisture and frost action, a gravel base is advised. Concrete thickness can also be reduced from 5 in. (125 mm) to 3½ in. (90 mm) when put down on a 5-in. (125-mm) gravel base.

The forms for the slab should be set so that there is a drainage slope of at least 1:60. Reinforcing of deformed steel or welded wire mesh is commonly used to keep the floating slab together and to minimize cracking (see Fig. 3–71). Expansion joints are used when the slab comes in contact with a garage slab, curb, or public sidewalk or every 20 ft (6 m) (see Fig. 3–72). Control joints act as a stress relief to limit uncontrolled cracking. The joint should extend to a depth of one-fourth of the slab thickness (see Fig. 3–73). For sidewalks the distance between joints should be equal

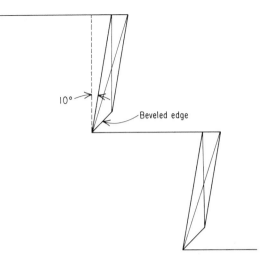

FIGURE 3–70: *Placement of the riser form.*

FIGURE 3–71: *Deformed steel in slab.*

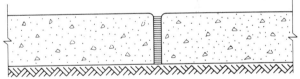

FIGURE 3–72: *Expansion joint.*

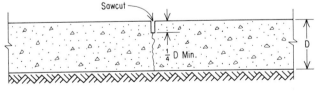

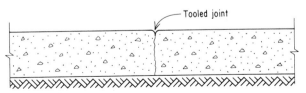

FIGURE 3–73: *Control joints.*

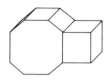

Cobble

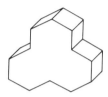

Unistone

Cloverleaf

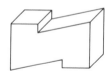

Finetta

FIGURE 3-75: *Paving stone shapes.*

FIGURE 3-74: *Sidewalk form.*

to the width of the slab. Form sides are usually 2 × 4s or 2 × 6s so the edges can be kept straight (see Fig. 3-74).

Edges and expansion joints should be rounded with an edger. A jointing tool is commonly used to establish a control joint as well. The surface of the slab can be troweled, or when extra traction is required, a fine brush finish is used.

Recently, interlocking paving stones have been used for sidewalks and driveways. Numerous different shapes are available in a variety of colors (see Fig. 3-75). The unique geometric shape of the paving stones provides a completely interlocking surface which will transfer individual loads to adjacent stones, allowing the surface to move like a mesh. The flexibility of the system makes repairing the surface an easy procedure.

Interlocking concrete paving consists of a layer of *concrete pavers,* a *sand bedding layer,* and a compacted *sub*grade and is held in place by solid *curbs.* The thickness of the pavers can vary from 2⅜ in. (60 mm) to 4 in. (100 mm) depending on the type of traffic. The herringbone pattern provides the best locking effect, though other patterns are also used (see Fig. 3-76). The subgrade

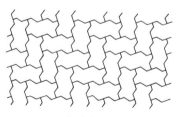

Herringbone

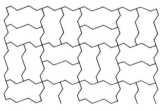

Parquet

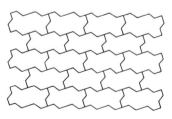

Runner

FIGURE 3-76: *Paving stone patterns.*

TABLE 3-7: *Minimum subgrade thickness*

	Walk or patio	*Driveway*
CBR of greater than 8% (well-compacted rocky round, gravel, sandy loam, or sand with a small amount of silt)	Not required	Not required
CBR of 4–8% (sand, loam, and stiff clay not subject to moisture)	Not required	3 in. (75 mm)
CBR of less than 4% (wet clays or soils which may easily deform when wet or subject to car traffic)	3 in. (75 mm)	5 in. (125 mm)

should be level, and the thickness depends on ground conditions and use of the area. The strength of the soil is measured in terms of the California bearing ratio (CBR). Minimum subgrade thicknesses are given in Table 3-7.

The sand bedding layer should consist of well-graded concrete sand. Uniformity of the layer is important to achieve a good-quality surface. This layer should not be over 1½-in. (40 mm) thick after compaction and should not be used for leveling low spots (see Fig. 3-77). Solid curbs are needed around the perimeter to prevent moving and slipping of the stones. Curbs can be made of concrete, pressure-treated wood, lawn, compacted earth, or an existing building (see Fig. 3-78).

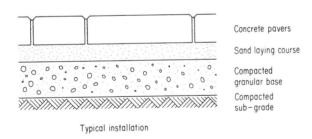

Concrete pavers
Sand laying course
Compacted granular base
Compacted sub-grade

Typical installation

FIGURE 3-77: *Concrete paving stone structure.*

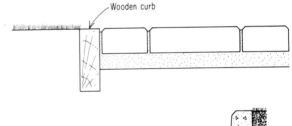

Wooden curb

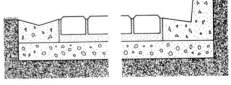

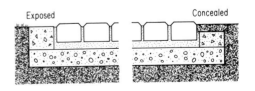

Exposed

Concealed

FIGURE 3-78: *Edge restraints.*

Construction procedures should conform to the following steps:

1. Ground should be cut out and shaped for area needed and then compacted.

2. Subgrade should be spread to the required level and compacted.

3. Edge restraint can be placed before or after placing of stones.

4. Sand is spread and screeded to a level that will result in proper level after compaction.

5. Lay stones starting from a straightedge and a corner directly onto screeded sand in pattern selected. Joints between stones should not exceed ⅛ in. (3 mm), and the surface should have a 2% slope to provide drainage.

6. Vibrate stones with a plate vibrator to final level.

7. Spread dry sand over surface and brush into joints to complete the interlock.

Now that the concrete work and foundation are complete, the job is ready for the erection of the superstructure. If it has been carefully planned and carried out, the completion of the remainder of the building will be easier, with fewer chances of errors as work continues.

REVIEW QUESTIONS

3-1. What is the general purpose of a building foundation?

3-2. Outline the essential difference between a full basement foundation and a surface foundation.

3-3. Explain the reason for placing foundation footings below the level of frost penetration.

3-4. How does a gravel pad under a slab foundation compensate for the fact that the foundation is not below the frost line?

3-5. Why is it necessary, in many cases, to make the excavation larger than the size of the foundation?

3-6. Why is cribbing used in excavating?

3-7. Explain the reason for coating a footing keyway with asphalt before a wall is placed on it.

3-8. What is the reason for placing a layer of mortar under a sill plate before bolting it down?

3-9. Explain what is meant by *cast-in* joists.

3-10. According to Table 3-4, what size and spacing of studs for a preserved wood foundation are required under each of the following sets of conditions?

   (a) One story house with a load of 700 lb/ft (10 kN/m), Douglas, No. 1 lumber. with backfill 6 ft, 6 in. (2 m) deep.

   (b) House with load of 1400 lb/ft (20 kN/m), hemlock-fir, No. 2 lumber, with backfill 5 ft, 8 in. (1.7 m) deep

   (c) House with load of 2100 lb/ft (30 kN/m), -P-F, No. 1 lumber, with backfill 6 ft, 0 in. (1.8 m) deep.

3-11. What is the purpose of using screeds when placing a basement floor?

3-12. Explain the reason for using a bearing wall, rather than a center girder, to support interior floor loads.

COMPONENT PARTS

After the foundation has been completed, the next logical step in the construction of a building is the erection of the floor frame. This, as the name implies, is the part of the structure which carries the floor and interior walls, along with its supporting members. This floor frame consists of *bearing posts,* the *girder* which they support, the *floor joists* carried by the girder and foundation walls, the *bridging* between the joists, and the *subfloor;* sometimes a *bearing wall* will replace posts and girder. When the box sill type of construction is used, a sill plate becomes one of the components of the floor frame.

POSTS AND GIRDERS

Posts and girders are fundamental structural components of the floor frame which support approximately half the total load of the building and transmit it to end foundation walls and footings (see Fig. 3–6).

Posts

The bearing posts may be made of either steel or wood. Wooden ones are sometimes one solid piece of timber but more often are built up of three or four pieces of $1\frac{1}{2}$-in. (38-mm) material laminated together. The cross-sectional area depends on the load to be carried, but usually 6×6 (140×140 mm) will prove ample. One factor governing the size will be the width of the girder. One dimension of the post should be equal to that width in order to provide full bearing.

A steel post, usually round, will be smaller in cross section than a wooden one. It must be capped by a steel plate to provide a suitable bearing area. Steel posts are manufactured which have a thread on the inside of the top end so that a short, heavy stem may be turned into them (see Fig. 4–1). Consequently, the post becomes adjustable in length. This is a decided advantage, because the post can be adjusted to the exact length required on installation, and later if the girder shrinks in its depth, the post can be lengthened to take up the shrinkage.

4

THE FLOOR FRAME

FIGURE 4-1: *Steel adjustable post.*

Girders

Girders may be made of wood and steel. A wooden girder has been most popular for light construction. It is often built up of a number of pieces of 1½-in. (38-mm) material laminated together, although it may be one solid piece of timber. When the girder in laminated, care must be taken in its construction. Pieces may be nailed, bolted, or glued and nailed together, the latter method providing the most rigid unit. A glue must be chosen that will stand up under the atmospheric conditions to which the girder will be subjected. Rarely will it be possible to find pieces long enough to reach from one end of the girder to the other, and consequently pieces must be end-jointed, usually with butt joints. Select or cut pieces of such a length that the joints will come directly over posts [see Fig. 4-2(a)]. When this is not possible, some codes will allow joints within 6 in. (150 mm) of the quarter point in the span [see Fig. 4-2(b)]. If only nails are used in laminating, they should be spaced not over 12 in. (300 mm) apart and staggered, one at the top and the next at the bottom of the girder, as illustrated in Fig. 4-2(b). On girders of extra depth an extra row of nails may be necessary along the center line.

FIGURE 4-2: *(a) Glue laminated girder (made with plywood laminations.) (b) Laminated girder and post.*

(a)

Joint should be within 6"(150 mm) of ¼ point of clear span

Staggered nails

Joint over post

(b)

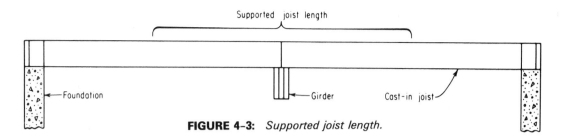

FIGURE 4-3: *Supported joist length.*

The size of the girder depends on the load which it must support, the species and grade of lumber, and the spacing of the posts. The load is determined by the *supported joist length*—one-half the width of the building (see Fig. 4-3). Using these factors, the maximum allowable *free span* of the girder or, in other words, the maximum allowable spacing of posts, is determined. Tables 4-1a and 4-1b give maximum allowable free spans for wooden girders which will support not more than one residential floor and Tables 4-2a and 4-2b the same information for girders which will support not more than two residential floors, both for a number of sizes of built-up girders.

The method for setting the bottom of the bearing post has already been discussed in Chapter 3. At the top end some provision must be made for adjustment, so that the girder will be held exactly level and also so that shrinkage in the depth of the girder can be taken up at a later date. This provision is made by placing two wedge-shaped pieces between the top of the post and the girder (see Fig. 4-4). By driving them both inward, the girder is raised to its correct level; later, further compensation may be made for girder shrinkage in the same way.

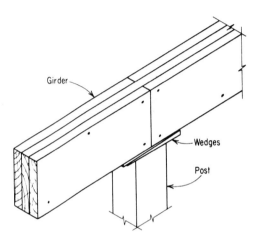

FIGURE 4-4: *Wedged post.*

How to Construct and Set a Wooden Girder

1. Obtain the length of the girder and the position of the bearing posts from the plans.

2. Select and cut the material so that all end joints will come over posts or within 6 in. (150 mm) of the quarter point of the clear span.

3. Lay out the material for one lamination, crown up, and place the second on top, making sure that the joints are staggered. Fasten these two together.

4. Apply the third lamination in the same way, still keeping the joints staggered. Do likewise with any further laminations.

5. Apply two coats of wood preservative to each end.

6. Determine the spacing and position of joists from the plan and lay off the top face of the girder for joists.

7. Set the girder in place and adjust until it is perfectly level.

TABLE 4-1a: *Maximum spans for built-up wood beams supporting not more than one floor in houses (ft-in.)*

Commercial designation	Grade	Supported joist length (ft)	Size of built-up beam (in.)											
			3-2 × 8		4-2 × 8		3-2 × 10		4-2 × 10		3-2 × 12		4-2 × 12	
			ft.	in.	ft.	in.	ft.	in.	ft.	in.	ft.	in.	ft.	in.
Douglas fir-larch	No. 1	8	12	0	13	10	15	4	17	8	18	7	21	6
		10	10	9	12	5	13	8	15	10	16	8	19	3
		12	9	9	11	4	12	6	14	5	15	2	17	7
		14	8	10	10	5	11	4	13	4	13	9	16	3
		16	7	10	9	9	10	1	12	6	12	3	15	2
	No. 2	8	10	10	12	6	13	9	15	11	16	9	19	5
		10	9	8	11	2	12	4	14	3	15	0	17	4
		12	8	10	10	2	11	3	13	0	13	8	15	10
		14	8	2	9	5	10	5	12	0	12	8	14	8
		16	7	8	8	10	9	9	11	3	11	10	13	8
Hem-fir	No. 1	8	10	4	12	0	13	3	15	4	16	1	18	7
		10	9	3	10	9	11	10	13	8	14	5	16	8
		12	8	2	9	9	10	5	12	6	12	8	15	2
		14	7	2	9	1	9	2	11	7	11	1	14	1
		16	6	5	8	2	8	2	10	5	10	0	12	8
	No. 2	8	9	3	10	8	11	9	13	7	14	4	16	7
		10	8	3	9	6	10	6	12	2	12	10	14	10
		12	7	6	8	8	9	7	11	1	11	8	13	6
		14	7	0	8	1	8	11	10	3	10	10	12	6
		16	6	5	7	6	8	2	9	7	10	0	11	8
Eastern hemlock-tamarack	No. 1	8	11	7	13	5	14	10	17	1	18	0	20	10
		10	10	4	12	0	13	3	15	4	16	1	18	7
		12	9	6	10	11	12	1	14	0	14	9	17	0
		14	8	9	10	1	11	2	12	11	13	7	15	9
		16	7	11	9	6	10	1	12	1	12	3	14	9
	No. 2	8	10	4	12	0	13	3	15	4	16	1	18	7
		10	9	3	10	9	11	10	13	8	14	5	16	8
		12	8	6	9	9	10	10	12	6	13	2	15	2
		14	7	10	9	1	10	0	11	7	12	2	14	1
		16	7	4	8	6	9	4	10	10	11	5	13	2
Spruce-pine-fir	No. 1	8	9	11	11	6	12	8	14	8	15	5	17	10
		10	8	11	10	3	11	4	13	1	13	10	15	11
		12	7	8	9	4	9	9	11	11	11	11	14	7
		14	6	9	8	7	8	7	10	11	10	5	13	4
		16	6	0	7	8	7	8	9	9	9	5	11	11
	No. 2	8	9	0	10	4	11	6	13	3	13	11	16	1
		10	8	0	9	3	10	3	11	10	12	6	14	5
		12	7	4	8	6	9	4	10	10	11	5	13	2
		14	6	9	7	10	8	7	10	0	10	5	12	2
		16	6	0	7	4	7	8	9	4	9	5	11	5
Western cedars	No. 1	8	10	2	11	9	13	0	15	0	15	9	18	3
		10	9	1	10	6	11	7	13	5	14	1	16	4
		12	8	2	9	7	10	5	12	3	12	8	14	10
		14	7	2	8	10	9	2	11	4	11	1	13	9
		16	6	5	8	2	8	2	10	5	10	0	12	8
	No. 2	8	9	0	10	4	11	6	13	3	13	11	16	1
		10	8	0	9	3	10	3	11	10	12	6	14	5
		12	7	4	8	6	9	4	10	10	11	5	13	2
		14	6	9	7	10	8	8	10	0	10	6	12	2
		16	6	4	7	4	8	1	9	4	9	10	11	5

NOTE: This table provides the maximum allowable spans for main beams or girders which are built up from nominal 2-in. members in the species, sizes, and grades indicated. Allowable spans for solid wood beams, glued-laminated wood beams, or built-up beams in sizes or grades other than shown must be determined from standard engineering formulas. Supported joist length means one-half the sum of the joist spans on both sides of the beam.

TABLE 4-1b: *Maximum spans for built-up wood beams supporting not more than one floor in houses (meters)*

Lumber species	Grade	Supported joist length (m)	3–38 × 184 (m)	4–38 × 184 (m)	3–38 × 235 (m)	4–38 × 235 (m)	3–38 × 286 (m)	4–38 × 286 (m)
			Size of built-up beam (mm)					
Douglas fir-larch	No. 1	2.4	3.70	4.27	4.72	5.45	5.74	6.63
		3.0	3.31	3.82	4.22	4.87	5.13	5.93
		3.6	3.02	3.49	3.85	4.45	4.69	5.41
		4.2	2.76	3.23	3.53	4.12	4.29	5.01
		4.8	2.46	3.02	3.14	3.85	3.82	4.69
	No. 2	2.4	3.33	3.84	4.24	4.90	5.16	5.96
		3.0	2.97	3.44	3.80	4.38	4.62	5.33
		3.6	2.71	3.14	3.47	4.00	4.22	4.87
		4.2	2.51	2.90	3.21	3.70	3.90	4.51
		4.8	2.35	2.71	3.00	3.46	3.65	4.22
Hem-fir	No. 1	2.4	3.19	3.69	4.07	4.71	4.96	5.72
		3.0	2.85	3.30	3.64	4.21	4.43	5.12
		3.6	2.61	3.01	3.33	3.84	4.05	4.67
		4.2	2.30	2.79	2.93	3.56	3.87	4.33
		4.8	2.06	2.61	2.92	3.33	3.69	4.05
	No. 2	2.4	2.86	3.31	3.65	4.22	4.45	5.13
		3.0	2.56	2.96	3.27	3.77	3.98	4.59
		3.6	2.34	2.70	2.98	3.45	3.63	4.19
		4.2	2.16	2.50	2.76	3.19	3.36	3.88
		4.8	2.02	2.34	2.58	2.98	3.14	3.63
Eastern hemlock-tamarack	No. 1	2.4	3.56	4.11	4.54	5.25	5.53	6.38
		3.0	3.18	3.68	4.06	4.69	4.94	5.71
		3.6	2.91	3.36	3.71	4.28	4.51	5.21
		4.2	2.69	3.11	3.43	3.97	4.18	4.82
		4.8	2.46	2.91	3.14	3.71	3.82	4.51
	No. 2	2.4	3.19	3.69	4.07	4.71	4.96	5.72
		3.0	2.85	3.30	3.64	4.21	4.43	5.12
		3.6	2.61	3.01	3.33	3.84	4.05	4.67
		4.2	2.42	2.79	3.08	3.56	3.75	4.33
		4.8	2.26	2.61	2.88	3.33	3.50	4.05
Spruce-pine-fir	No. 1	2.4	3.09	3.57	3.95	4.56	4.80	5.55
		3.0	2.77	3.19	3.53	4.08	4.30	4.96
		3.6	2.44	2.92	3.11	3.72	3.79	4.53
		4.2	2.14	2.70	2.74	3.45	3.33	4.19
		4.8	1.92	2.44	2.45	3.11	2.98	3.79
	No. 2	2.4	2.78	3.21	3.54	4.09	4.21	4.98
		3.0	2.48	2.87	3.17	3.66	3.85	4.45
		3.6	2.26	2.62	2.89	3.34	3.52	4.06
		4.2	2.10	2.42	2.68	3.09	3.26	3.76
		4.8	1.92	2.26	2.45	2.89	2.98	3.52
Western cedar	No. 1	2.4	3.13	3.62	4.00	4.62	4.86	5.62
		3.0	2.80	3.24	3.58	4.13	4.35	5.02
		3.6	2.56	2.95	3.26	3.77	3.97	4.59
		4.2	2.26	2.73	2.88	3.49	3.51	4.24
		4.8	2.02	2.56	2.58	3.26	3.14	3.97
	No. 2	2.4	2.80	3.23	3.57	4.12	4.34	5.02
		3.0	2.50	2.89	3.19	3.69	3.88	4.49
		3.6	2.28	2.64	2.91	3.37	3.55	4.10
		4.2	2.11	2.44	2.70	3.12	3.28	3.79
		4.8	1.98	2.28	2.52	2.91	3.07	3.55

NOTE: This table provides the maximum allowable spans for main beams or girders which are built up from 38-mm members in the species, sizes, and grades indicated. Allowable spans for solid wood beams, glued-laminated wood beams, or built-up beams in sizes or grades other than shown must be determined from standard engineering formulas. Supported joist length means one-half the sum of the joist spans on both sides of the beam.

TABLE 4–2a: *Maximum spans for built-up wood beams supporting not more than two floors or in houses (ft-in.)*

Commercial designation	Grade	Supported joist length (ft)	3-2 × 8		4-2 × 8		3-2 × 10		4-2 × 10		3-2 × 12		4-2 × 12	
			ft.	in.	ft.	in.	ft.	in.	ft.	in.	ft.	in.	ft.	in.
Douglas fir-larch	No. 1	8	8	10	10	5	11	4	13	4	13	9	16	3
		10	7	4	9	4	9	4	11	11	11	5	14	6
		12	6	4	8	0	8	0	10	3	9	9	12	5
		14	5	7	7	0	7	1	9	0	8	8	10	11
		16	5	0	6	4	6	5	8	0	7	10	9	9
	No. 2	8	8	2	9	5	10	5	12	0	12	8	14	8
		10	7	3	8	5	9	4	10	9	11	4	13	1
		12	6	4	7	8	8	0	9	10	9	9	11	11
		14	5	7	7	0	7	1	9	0	8	8	10	11
		16	5	0	6	4	6	5	8	0	7	10	9	9
Hem-fir	No. 1	8	7	2	9	0	9	2	11	7	11	1	14	1
		10	5	11	7	7	7	7	9	8	9	3	11	9
		12	5	2	6	6	6	7	8	3	8	0	10	1
		14	4	7	5	9	5	10	7	4	7	2	8	11
		16	4	2	5	2	5	4	6	7	6	6	8	0
	No. 2	8	7	0	8	1	8	11	10	3	10	10	12	6
		10	5	11	7	2	7	7	9	2	9	3	11	2
		12	5	2	6	6	6	7	8	3	8	0	10	1
		14	4	7	5	9	5	10	7	4	7	2	8	11
		16	4	2	5	2	5	4	6	7	6	6	8	0
Eastern hemlock-tamarack	No. 1	8	8	9	10	1	11	2	12	11	13	7	15	9
		10	7	4	9	1	9	4	11	7	11	5	14	1
		12	6	4	8	0	8	1	10	3	9	10	12	5
		14	5	7	7	0	7	1	9	0	8	8	10	11
		16	5	0	6	4	6	5	8	1	7	10	9	10
	No. 2	8	7	10	9	1	10	0	11	7	12	2	14	1
		10	7	0	8	1	8	11	10	4	10	11	12	7
		12	6	4	7	5	8	1	9	5	9	10	11	6
		14	5	7	6	10	7	1	8	9	8	8	10	7
		16	5	0	6	4	6	5	8	1	7	10	9	10
Spruce-pine-fir	No. 1	8	6	9	8	7	8	7	10	11	10	5	13	4
		10	5	7	7	1	7	2	9	1	8	9	11	0
		12	4	10	6	1	6	3	7	10	7	7	9	6
		14	4	4	5	5	5	7	6	11	6	9	8	5
		16	3	11	4	10	5	1	6	3	6	2	7	7
	No. 2	8	6	9	7	10	8	7	10	0	10	5	12	2
		10	5	7	7	0	7	2	8	11	8	9	10	11
		12	4	10	6	1	6	3	7	10	7	7	9	6
		14	4	4	5	5	5	7	6	11	6	9	8	5
		16	3	11	4	10	5	1	6	3	6	2	7	7
Western cedars	No. 1	8	7	2	8	10	9	2	11	4	11	1	13	9
		10	5	11	7	7	7	7	9	8	9	3	11	9
		12	5	2	6	6	6	7	8	3	8	0	10	1
		14	4	7	5	9	5	10	7	4	7	2	8	11
		16	4	2	5	2	5	4	6	7	6	6	8	0
	No. 2	8	6	9	7	10	8	8	10	0	10	6	12	2
		10	5	11	7	0	7	7	8	11	9	3	10	11
		12	5	2	6	5	6	7	8	2	8	0	9	11
		14	4	7	5	9	5	10	7	4	7	2	8	11
		16	4	2	5	2	5	4	6	7	6	6	8	0

NOTE: This table provides the maximum allowable spans for main beams or girders which are built up from nominal 2-in. members in the species, sizes, and grades indicated. Allowable spans for solid wood beams, glued-laminated wood beams, or built-up beams in sizes or grades other than shown must be determined from standard engineering formulas. Supported joist length means one-half the sum of the joist spans on both sides of the beam.

TABLE 4-2b: *Maximum spans for built-up wood beams supporting not more than two floors in houses (meters)*

Lumber species	Grade	Supported joist length (m)	3–38 × 184 (m)	4–38 × 184 (m)	3–38 × 235 (m)	4–38 × 235 (m)	3–38 × 286 (m)	4–38 × 286 (m)
					Size of built-up beam (mm)			
Douglas fir-larch	No. 1	2.4	2.78	3.24	3.55	4.13	4.32	5.03
		3.0	2.30	2.90	2.93	3.70	3.57	4.50
		3.6	1.97	2.51	2.52	3.21	3.07	3.90
		4.2	1.74	2.20	2.23	2.81	2.44	3.07
		4.8	1.57	1.97	2.01	2.52	2.44	3.07
	No. 2	2.4	2.52	2.91	3.22	3.72	3.92	4.52
		3.0	2.26	2.61	2.88	3.33	3.50	4.05
		3.6	1.97	2.38	2.52	3.04	3.07	3.69
		4.2	1.74	2.20	2.23	2.81	2.71	3.42
		4.8	1.57	1.97	2.01	2.52	2.44	3.07
Hem-fir	No. 1	2.4	2.31	2.80	2.95	3.57	3.59	4.34
		3.0	1.92	2.44	2.45	3.12	2.99	3.79
		3.6	1.66	2.10	2.12	2.68	2.58	3.25
		4.2	1.48	1.85	1.89	2.36	2.29	2.87
		4.8	1.34	1.66	1.71	2.12	2.08	2.58
	No. 2	2.4	2.17	2.51	2.77	3.20	3.37	3.89
		3.0	1.92	2.24	2.45	2.86	2.99	3.48
		3.6	1.66	2.05	2.12	2.61	2.58	3.18
		4.2	1.48	1.85	1.89	2.36	2.29	2.87
		4.8	1.34	1.66	1.71	2.12	2.08	2.58
Eastern hemlock-tamarck	No. 1	2.4	2.70	3.12	3.45	3.98	4.19	4.84
		3.0	2.30	2.79	2.93	3.56	3.57	4.33
		3.6	1.97	2.51	2.52	3.21	3.07	3.90
		4.2	1.74	2.20	2.23	2.81	2.71	3.42
		4.8	1.57	1.97	2.01	2.52	2.44	3.07
	No. 2	2.4	2.42	2.80	3.09	3.57	3.76	4.34
		3.0	2.16	2.50	2.76	3.19	3.36	3.88
		3.6	1.97	2.28	2.52	2.91	3.07	3.54
		4.2	1.74	2.11	2.23	2.70	2.71	3.28
		4.8	1.57	1.97	2.01	2.52	2.44	3.07
Western cedar	No. 1	2.4	2.27	2.74	2.90	3.50	3.53	4.26
		3.0	1.89	2.40	2.41	3.06	2.94	3.73
		3.6	1.64	2.06	2.09	2.63	2.54	3.20
		4.2	1.45	1.82	1.86	2.32	2.26	2.82
		4.8	1.32	1.64	1.68	2.09	2.05	2.54
	No. 2	2.4	2.12	2.45	2.71	3.13	3.29	3.81
		3.0	1.89	2.19	2.41	2.80	2.94	3.40
		3.6	1.64	2.00	2.09	2.55	2.54	3.11
		4.2	1.45	1.82	1.86	2.32	2.26	2.82
		4.8	1.32	1.64	1.68	2.09	2.05	2.54
Spruce-fine-fir	No. 1	2.4	2.16	2.71	2.75	3.46	3.35	4.21
		3.0	1.80	2.28	2.30	2.91	2.79	3.53
		3.6	1.56	1.96	1.99	2.50	2.42	3.04
		4.2	1.39	1.73	1.77	2.21	2.16	2.69
		4.8	1.26	1.56	1.61	1.99	1.96	2.42
	No. 2	2.4	2.10	2.43	2.69	3.10	3.27	3.77
		3.0	1.80	2.17	2.30	2.77	2.79	3.38
		3.6	1.56	1.96	1.99	2.50	2.43	3.04
		4.2	1.39	1.73	1.77	2.21	2.16	2.69
		4.8	1.26	1.56	1.61	1.99	1.96	2.42

NOTE: This table provides the maximum allowable spans for main beams or girders which are built up from 38-mm members in the species, sizes, and grades indicated. Allowable spans for solid wood beams, glued-laminated wood beams, or built-up beams in sizes or grades other than shown must be determined from standard engineering formulas. Supported joist length means one-half the sum of the joist spans on both sides of the beam.

FIGURE 4-5: (a) Steel girder.

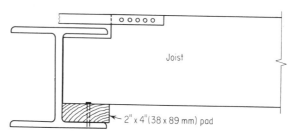

(b) Steel girder supporting joist ends.

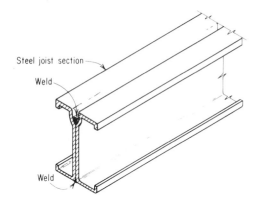

FIGURE 4-6: Steel joist beam.

Steel Girder

A steel girder for light construction may be a *standard* or *wide flange rolled shape* (see Fig. 4–5), or it may be made up of *two cold-formed steel joist sections* welded or bolted together, as shown in Fig. 4–6.

Several depths of rolled shapes are commonly used, the number of supports (posts) required being determined by the supported joist length (see Tables 4–3a and 4–3b). If the joists rest on top of the girder, a wooden pad is generally used to facilitate fastening the joist ends, and the height of the girder must be

TABLE 4-3a: Maximum spans for steel girders in residential construction (ft)

No. of stories	Minimum depth (in)	Minimum weight/foot (lb)	Supported joist length				
			8 ft	10 ft	12 ft	14 ft	16 ft
1	4	7.7	10	9	8.5	8	7.5
	5	10.0	12.5	11.5	11	10.5	10
	6	12.5	15	14	13	12.5	12
	7	15.3	18	17	16	15	14.5
	8	18.4	21	19.5	18.5	17.5	16.5
1½ or 2	4	7.7	8	7.5	7	6.5	6
	5	10.0	10.5	9.5	8.5	8	7.5
	6	12.5	12.5	11.5	10.5	9.5	9
	7	15.3	15	14	13	12	11
	8	18.4	17.5	16	15	14	13

NOTE: For supported joist lengths intermediate to those shown in the table, straight-line interpolation may be used to determine the beam span.

TABLE 4-3b: *Maximum spans for steel girders in residential construction (meters)*

No. of floors supported	Min. depth (mm)	Min. mass (kg/m)	Supported joist length				
			2.4 m	3.0 m	3.6 m	4.2 m	4.8 m
1	101	S 11.46	4.06	3.63	3.33	3.07	2.90
	127	S 14.88	5.11	4.57	4.19	3.89	3.63
	152	S 18.60	6.25	5.61	5.16	4.77	4.47
	152	W 23.07	7.01	6.30	5.77	5.38	5.03
	203	W 25.30	8.28	7.47	6.81	6.33	5.87
	203	S 27.38	8.66	7.80	7.01	6.63	6.20
2	101	S 11.46	3.08	2.74	2.52	2.34	2.18
	127	S 14.88	3.89	3.48	3.18	2.94	2.72
	152	S 18.60	4.77	4.27	3.91	3.61	3.38
	152	W 23.07	5.38	4.80	4.39	4.06	3.81
	203	W 25.30	6.33	5.66	5.18	4.80	4.50
	203	S 27.38	6.63	5.96	5.44	5.03	4.72

NOTES: 1. S: standard rolled section.
2. W: wide flange rolled section.
3. For supported joist lengths intermediate between those shown in the table, straightline interpolation may be used to determine the maximum girder span.

regulated accordingly [see Fig. 4-5(a)]. If the top of joists and girder are to be flush, the joist ends must be carried by the bottom flange of the girder, as illustrated in Fig. 4-5(b). In the latter case, the joists must be spliced over the girder by 2 × 2 (38 × 38) lumber at least 2 ft (600 mm) in length.

Cold-formed joist sections normally have a standard depth of 8 in. (184 mm), and the number of supports required for a girder made from two such sections is also determined from the supported joist length.

BEARING WALLS

In a good many cases, posts and girder are replaced by a *bearing wall* as the primary support for a building. Such a wall will carry the load quite adequately and has the advantage of providing a wall framework if the basement space is to be divided into rooms.

As illustrated in Chapter 3, the bearing wall is supported by a continuous footing having a raised center portion to which the wall is anchored (see Fig. 3-15). The material must be 2 × 6 (38 × 140 mm), minimum, with top and bottom plates and studs spaced not more than 16 in. (400 mm) o.c. (see Fig. 4-7). At the midpoint between top and bottom, blocks must be fitted snugly between the studs, staggered if required, to facilitate nailing, as illustrated in Fig. 4-7. The double top plate will be level with the top of the sill plate or with the bottom edge of the joists, depending on whether box sill or cast-in construction is used.

When an opening occurs in the bearing wall, it must be framed with a lintel across the top (see Fig. 4-7) made with two 2 × 6 (38 × 140 mm) members, if the opening is not more than 4 ft (1200 mm) wide.

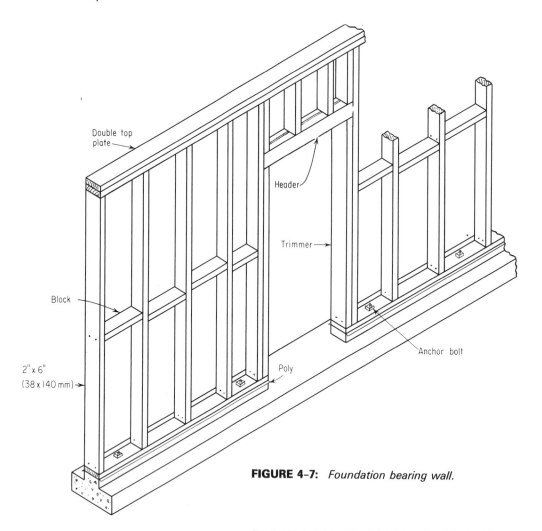

FIGURE 4-7: *Foundation bearing wall.*

SILL PLATE (WOOD FLOOR JOISTS)

The sill plate, sometimes called a mud sill, is set on the foundation wall to provide a connection between the wood floor frame and the concrete foundation wall. If the top of the wall is flat, the plate may sit directly on the wall with the junction caulked.

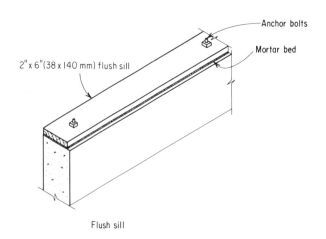

FIGURE 4-8: *Flush sill.*

It may also be laid on a strip of fiberglass sill sealer before fastening down. If the top of the wall is not level, the sill is set in a bed of mortar in order to provide full bearing between the bottom of the plate and the wall on which it rests (see Fig. 4-8). The sill is set in this mortar, with the anchor bolts projecting through previously drilled bolt holes. The sill is tapped down so that it is bedded in the mortar, firm and level. Mortar will squeeze out from under the sill as the anchor bolts are tightened to bring it to its correct position. After the mortar has set, the nuts may be tightened slightly more.

The anchor bolts should be ½ in. (12.7 mm) in diameter spaced no more than 8 ft (2.4 m) apart with at least two bolts in each piece of material. The embedded end of the bolt should have either a head or a bend to prevent withdrawal. The length of the bolt should be sufficient to be embedded at least 4 in. (100 mm) into the concrete.

A more common connection is provided by embedding a 2 × 4 on edge at the top of the concrete wall, as illustrated in Fig. 3-46.

SILL PLATE (STEEL FLOOR JOISTS)

The location of sill plates for a steel joist floor frame has been described in Chapter 3 (see Figs. 3-33, 3-34, and 3-35).

JOISTS

Floor joists, the members which span from foundation wall to girder or from wall to wall, in some cases, and transfer the individual building loads to those members, may be of wood, steel, or reinforced concrete. Wooden joists have been widely used, particularly in residential construction (see Fig. 4-9), but steel joists are gaining in popularity, particularly for long spans (see Fig. 4-10). Reinforced concrete joists are normally reserved for heavier construction.

Wood Joists

Wood joists consist of 2-in. (38-mm) material, in widths varying from 6 to 12 in. (140-286 mm), depending on their *load, length, spacing,* and the *species and grade* of lumber. Common spacings are 12, 16, and 24 in. (300, 400, and 600 mm) o.c. Tables 4-4a and 4-4b gives maximum spans for floor joists for residential construction living quarters and Tables 4-5a and 4-5b the same information for residential construction bedrooms and attics which are accessible by a stairway, based on *lumber species* and *grade, joist size, spacing,* and, in the case of bedrooms and attics, also the *type of ceiling.*

For buildings other than residential, with heavier loads, the size of joists should be calculated by a competent authority.

FIGURE 4-9: *Wood joists in residential construction.*

FIGURE 4-10: *Steel joists in residential construction.*

TABLE 4-4a: *Maximum spans for floor joists—living quarters (live load 40 lb/ft²)*

Lumber species	Grade	Size (in.)	all ceilings, joist spacing							
			12 in.		16 in.		20 in.		24 in.	
			ft	in.	ft	in.	ft	in.	ft	in.
Douglas fir-larch	Select structural	2 × 4	7	1	6	5	5	11	5	7
		2 × 6	11	1	10	1	9	4	8	10
		2 × 8	14	8	13	4	12	4	11	7
		2 × 10	18	9	17	0	15	9	14	10
		2 × 12	22	9	20	8	19	2	18	1
	No. 1	2 × 4	7	1	6	5	5	11	5	7
		2 × 6	11	1	10	1	9	4	8	10
		2 × 8	14	8	13	4	12	4	11	7
		2 × 10	18	9	17	0	15	9	14	10
		2 × 12	22	9	20	8	19	2	18	1
	No. 2	2 × 4	6	10	6	2	5	9	5	5
		2 × 6	10	9	9	9	9	0	8	5
		2 × 8	14	2	12	10	11	11	11	2
		2 × 10	18	1	16	5	15	3	14	3
		2 × 12	22	0	20	0	18	6	17	4
	No. 3	2 × 4	6	2	5	4	4	9	4	4
		2 × 6	9	1	7	10	7	0	6	5
		2 × 8	12	0	10	4	9	3	8	6
		2 × 10	15	4	13	3	11	10	10	10
		2 × 12	18	7	16	1	14	5	13	2
Hem-fir	Select structural	2 × 4	6	10	6	2	5	9	5	5
		2 × 6	10	9	9	9	9	0	8	6
		2 × 8	14	2	12	10	11	11	11	3
		2 × 10	18	1	16	5	15	3	14	4
		2 × 12	22	0	20	0	18	6	17	5
	No. 1	2 × 4	6	10	6	2	5	9	5	5
		2 × 6	10	9	9	9	8	11	8	1
		2 × 8	14	2	12	10	11	9	10	9
		2 × 10	18	1	16	5	15	0	13	8
		2 × 12	22	0	20	0	18	3	16	8
	No. 2	2 × 4	6	7	5	11	5	6	5	0
		2 × 6	10	3	8	10	7	11	7	3
		2 × 8	13	6	11	8	10	5	9	6
		2 × 10	17	3	14	11	13	4	12	2
		2 × 12	20	11	18	2	16	3	14	10
	No. 3	2 × 4	5	2	4	5	4	0	3	8
		2 × 6	7	9	6	9	6	0	5	6
		2 × 8	10	3	8	10	7	11	7	3
		2 × 10	13	1	11	4	10	2	9	3
		2 × 12	15	11	13	10	12	4	11	3
Eastern hemlock-tamarack	Select structural	2 × 4	6	6	5	10	5	5	5	2
		2 × 6	10	2	9	3	8	7	8	1
		2 × 8	13	5	12	3	11	4	10	8
		2 × 10	17	2	15	7	14	6	13	7
		2 × 12	20	11	19	0	17	7	16	7
	No. 1	2 × 4	6	6	5	10	5	5	5	2
		2 × 6	10	2	9	3	8	7	8	1
		2 × 8	13	5	12	3	11	4	10	8
		2 × 10	17	2	15	7	14	6	13	7
		2 × 12	20	11	19	0	17	7	16	7
	No. 2	2 × 4	6	3	5	8	5	3	4	11
		2 × 6	9	10	8	11	8	3	7	10
		2 × 8	13	0	11	9	10	11	10	3
		2 × 10	16	7	15	1	14	0	13	2
		2 × 12	20	2	18	4	17	0	16	0
	No. 3	2 × 4	5	11	5	2	4	7	4	2
		2 × 6	8	9	7	7	6	9	6	2
		2 × 8	11	7	10	0	8	11	8	2
		2 × 10	14	9	12	10	11	5	10	5
		2 × 12	18	0	15	7	13	11	12	8

TABLE 4–4a: *continued*

Lumber species	Grade	Size (in.)	all ceilings, joist spacing							
			12 in.		16 in.		20 in.		24 in.	
			ft	in.	ft	in.	ft	in.	ft	in.
Northern species	Select structural	2 × 4	6	2	5	7	5	2	4	11
		2 × 6	9	9	8	10	8	2	7	9
		2 × 8	12	10	11	8	10	10	10	2
		2 × 10	16	5	14	11	13	10	13	0
		2 × 12	19	11	18	1	16	10	15	10
	No. 1	2 × 4	6	2	5	7	5	2	4	11
		2 × 6	9	9	8	10	8	2	7	7
		2 × 8	12	10	11	8	10	10	10	0
		2 × 10	16	5	14	11	13	10	12	10
		2 × 12	19	11	18	1	16	10	15	7
	No. 2	2 × 4	6	0	5	5	5	0	4	8
		2 × 6	9	5	8	4	7	6	6	10
		2 × 8	12	5	11	1	9	10	9	0
		2 × 10	15	10	14	1	12	7	11	6
		2 × 12	19	3	17	2	15	4	14	0
	No. 3	2 × 4	4	11	4	3	3	10	3	6
		2 × 6	7	5	6	5	5	9	5	3
		2 × 8	9	9	8	6	7	7	6	11
		2 × 10	12	6	10	10	9	8	8	10
		2 × 12	15	2	13	2	11	9	10	9
Western cedars	Select structural	2 × 4	6	2	5	7	5	2	4	11
		2 × 6	9	9	8	10	8	2	7	9
		2 × 8	12	10	11	8	10	10	10	2
		2 × 10	16	5	14	11	13	10	13	0
		2 × 12	19	11	18	1	16	10	15	10
	No. 1	2 × 4	6	2	5	7	5	2	4	11
		2 × 6	9	9	8	10	8	2	7	9
		2 × 8	12	10	11	8	10	10	10	2
		2 × 10	16	5	14	11	13	10	13	0
		2 × 12	19	11	18	1	16	10	15	10
	No. 2	2 × 4	6	0	5	5	5	0	4	9
		2 × 6	9	5	8	6	7	8	7	0
		2 × 8	12	5	11	3	10	2	9	3
		2 × 10	15	10	14	4	13	0	11	10
		2 × 12	19	3	17	6	15	9	14	5
	No. 3	2 × 4	5	2	4	5	4	0	3	8
		2 × 6	7	9	6	9	6	0	5	6
		2 × 8	10	3	8	10	7	11	7	3
		2 × 10	13	1	11	4	10	2	9	3
		2 × 12	15	11	13	10	12	4	11	3
Spruce-pine-fir	Select structural	2 × 4	6	5	5	10	5	5	5	1
		2 × 6	10	1	9	2	8	6	8	0
		2 × 8	13	4	12	1	11	3	10	7
		2 × 10	17	0	15	5	14	4	13	6
		2 × 12	20	8	18	9	17	5	16	5
	No. 1	2 × 4	6	5	5	10	5	5	5	1
		2 × 6	10	1	9	2	8	6	7	9
		2 × 8	13	4	12	1	11	3	10	3
		2 × 10	17	0	15	5	14	4	13	1
		2 × 12	20	8	18	9	17	5	15	11
	No. 2	2 × 4	6	2	5	7	5	3	4	10
		2 × 6	9	9	8	7	7	8	7	0
		2 × 8	12	10	11	4	10	2	9	3
		2 × 10	16	5	14	6	13	0	11	10
		2 × 12	20	0	17	8	15	9	14	5
	No. 3	2 × 4	5	2	4	5	4	0	3	8
		2 × 6	7	5	6	5	5	9	5	3
		2 × 8	9	9	8	6	7	7	6	11
		2 × 10	12	6	10	10	9	8	8	10
		2 × 12	15	2	13	2	11	9	10	9

TABLE 4–4b: *Maximum spans for floor joists—living quarters (meters) (all ceilings) (live load 1.9 kN/m²)*

Lumber species	Grade	Size (mm)	Joist spacing		
			300 mm	400 mm	600 mm
Douglas fir-larch (N)	Select structural	38 × 89	2.17	1.98	1.72
		38 × 140	3.42	3.11	2.71
		38 × 184	4.51	4.10	3.58
		38 × 235	5.76	5.23	4.57
		38 × 286	7.00	6.36	5.56
	No. 1	38 × 89	2.17	1.98	1.72
		38 × 140	3.42	3.11	2.71
		38 × 184	4.51	4.10	3.58
		38 × 235	5.76	5.23	4.57
		38 × 286	7.00	6.36	5.56
	No. 2	38 × 89	2.10	1.91	1.67
		38 × 140	3.31	3.00	2.59
		38 × 184	4.36	3.96	3.42
		38 × 235	5.56	5.05	4.36
		38 × 286	6.77	6.15	5.31
	No. 3	38 × 89	1.88	1.63	1.33
		38 × 140	2.77	2.40	1.96
		38 × 184	3.66	3.17	2.59
		38 × 235	4.67	4.04	3.30
		38 × 286	5.68	4.92	4.01
Hem-fir (N)	Select structural	38 × 89	2.10	1.90	1.66
		38 × 140	3.30	2.99	2.61
		38 × 184	4.35	3.95	3.45
		38 × 235	5.55	5.04	4.40
		38 × 286	6.75	6.13	5.35
	No. 1	38 × 89	2.10	1.90	1.66
		38 × 140	3.30	2.99	2.49
		38 × 184	4.35	3.95	3.28
		38 × 235	5.55	5.04	4.19
		38 × 286	6.75	6.14	5.09
	No. 2	38 × 89	2.02	1.84	1.54
		38 × 140	3.16	2.73	2.23
		38 × 184	4.16	3.60	2.94
		38 × 235	5.31	4.60	3.76
		38 × 286	6.46	5.60	4.57
	No. 3	38 × 89	1.62	1.40	1.14
		38 × 140	2.39	2.07	1.69
		38 × 184	3.16	2.73	2.23
		38 × 235	4.03	4.39	2.85
		38 × 286	4.90	4.24	3.46
Eastern hemlock-tamarack (N)	Select structural	38 × 89	2.00	1.81	1.58
		38 × 140	3.14	2.85	2.49
		38 × 184	4.14	3.76	3.29
		38 × 235	5.28	4.80	4.19
		38 × 286	6.43	5.84	5.10
	No. 1	38 × 89	2.00	1.81	1.58
		38 × 140	3.14	2.85	2.49
		38 × 184	4.14	3.76	3.29
		38 × 235	5.28	4.80	4.19
		38 × 286	6.43	5.84	5.10

TABLE 4-4b: *(Cont.)*

Lumber species	Grade	Size (mm)	Joist spacing		
			300 mm	*400 mm*	*600 mm*
	No. 2	38 × 89	1.92	1.75	1.53
		38 × 140	3.03	2.75	2.40
		38 × 184	3.99	3.63	3.17
		38 × 235	5.09	4.63	4.04
		38 × 286	6.20	5.63	4.92
	No. 3	38 × 89	1.81	1.57	1.28
		38 × 140	2.66	2.31	1.88
		38 × 184	3.51	3.04	2.48
		38 × 235	4.48	3.88	3.17
		38 × 286	5.45	4.72	3.85
Spruce-pine-fir	Select structural	38 × 89	1.98	1.79	1.57
		38 × 140	3.11	2.82	2.46
		38 × 184	4.10	3.72	3.25
		38 × 235	5.23	4.75	4.15
		38 × 286	6.36	5.78	5.05
	No. 1	38 × 89	1.98	1.79	1.57
		38 × 140	3.11	2.82	2.41
		38 × 184	4.10	3.72	3.18
		38 × 235	5.23	4.75	4.06
		38 × 286	6.36	5.78	4.93
	No. 2	38 × 89	1.91	1.73	1.49
		38 × 140	3.00	2.65	2.16
		38 × 184	3.96	3.49	2.85
		38 × 235	5.05	4.46	3.64
		38 × 286	6.15	5.42	4.43
	No. 3	38 × 89	1.58	1.37	1.12
		38 × 140	2.33	2.02	1.65
		38 × 184	3.07	2.66	2.17
		38 × 235	3.92	3.40	2.77
		38 × 286	4.77	4.13	3.37
Western cedar (N)	Select structural	38 × 89	1.90	1.73	1.51
		38 × 140	2.99	2.73	2.37
		38 × 184	3.94	3.58	3.13
		38 × 235	5.02	4.57	3.99
		38 × 286	6.12	5.56	4.86
	No. 1	38 × 89	1.90	1.83	1.51
		38 × 140	2.99	2.72	2.37
		38 × 184	3.94	3.58	3.13
		38 × 235	5.03	4.57	3.99
		38 × 286	6.12	5.56	4.86
	No. 2	38 × 89	1.84	1.67	1.46
		38 × 140	2.89	2.63	2.18
		38 × 184	3.81	3.46	2.87
		38 × 235	4.87	4.42	3.67
		38 × 286	5.92	5.38	4.46
	No. 3	38 × 89	1.58	1.37	1.12
		38 × 140	2.33	2.02	1.65
		38 × 184	3.07	2.66	2.17
		38 × 235	3.92	3.40	2.77
		38 × 286	4.77	4.13	3.37

TABLE 4-5a: *Maximum spans for floor joists—bedrooms and attics accessible by a stairway*

Lumber species	Grade	Size (in.)	Gypsum board or plastered ceiling, joist spacing								Other ceilings, joist spacing							
			12 in.		16 in.		20 in.		24 in.		12 in.		16 in.		20 in.		24 in.	
			ft	in.	ft	in.	ft	in.	ft	in.	ft	in.	ft	in.	ft	in.	ft	in.
Douglas fir-larch	Select structural	2 × 4	7	9	7	1	6	7	6	2	8	11	8	1	7	6	7	1
		2 × 6	12	3	11	1	10	4	9	8	14	0	12	9	11	10	11	1
		2 × 8	16	2	14	8	13	7	12	10	18	6	16	9	15	7	14	8
		2 × 10	20	7	18	9	17	4	16	4	23	7	21	5	19	11	18	9
		2 × 12	25	1	22	9	21	2	19	11	28	8	26	1	24	2	22	9
	No. 1	2 × 4	7	9	7	1	6	7	6	2	8	11	8	1	7	6	7	1
		2 × 6	12	3	11	1	10	4	9	8	14	0	12	9	11	8	10	7
		2 × 8	16	2	14	8	13	7	12	10	18	6	16	9	15	4	14	0
		2 × 10	20	7	18	9	17	4	16	4	23	7	21	5	19	7	17	11
		2 × 12	25	1	22	9	21	2	19	11	28	1	26	1	23	10	21	9
	No. 2	2 × 4	7	6	6	10	6	4	5	11	8	7	7	10	7	3	6	8
		2 × 6	11	10	10	9	9	11	9	4	13	6	11	9	10	6	9	7
		2 × 8	15	7	14	2	13	2	12	4	17	10	15	6	13	10	12	8
		2 × 10	19	11	18	1	16	9	15	9	22	9	19	9	17	8	16	2
		2 × 12	24	2	22	0	20	5	19	2	27	8	24	1	21	6	19	7
	No. 3	2 × 4	6	11	6	0	5	5	4	11	6	11	6	0	5	5	4	11
		2 × 6	10	3	8	11	7	11	7	3	10	3	8	11	7	11	7	3
		2 × 8	13	7	11	9	10	6	9	7	13	7	11	9	10	6	9	7
		2 × 10	17	4	15	0	13	5	12	3	17	4	15	0	13	5	12	3
		2 × 12	21	1	18	3	16	4	14	11	21	1	18	3	16	4	14	11
Hem-fir	Select structural	2 × 4	7	6	6	10	6	4	5	11	8	7	7	10	7	3	6	9
		2 × 6	11	10	10	9	9	11	9	2	13	6	12	2	10	11	9	11
		2 × 8	15	7	14	2	13	2	12	1	17	10	16	1	14	4	13	1
		2 × 10	19	11	18	1	16	9	15	9	22	9	20	6	18	4	16	9
		2 × 12	24	2	22	0	20	5	18	8	27	8	24	10	22	2	20	4
	No. 1	2 × 4	7	5	6	9	6	2	5	9	8	5	7	8	7	1	6	2
		2 × 6	11	0	10	0	9	3	8	8	12	7	11	5	10	2	9	2
		2 × 8	15	0	13	2	11	9	10	10	16	7	15	0	13	5	12	2
		2 × 10	19	2	16	9	15	0	13	9	21	2	19	0	17	0	15	6
		2 × 12	23	4	20	7	18	5	16	9	25	9	23	1	20	8	18	10
	No. 2	2 × 4	7	3	6	7	6	0	5	8	8	3	7	6	6	11	6	2
		2 × 6	11	5	10	3	9	3	8	8	13	0	11	3	10	1	9	2
		2 × 8	15	0	13	11	11	9	10	10	16	9	15	1	13	4	10	10
		2 × 10	19	2	16	9	15	0	13	9	21	7	19	0	17	0	13	9
		2 × 12	23	4	20	7	18	5	16	9	26	3	23	1	20	8	16	2
	No. 3	2 × 4	5	10	5	1	4	6	4	1	5	10	5	1	4	6	4	1
		2 × 6	8	10	7	7	6	10	6	3	8	10	7	7	6	10	6	3
		2 × 8	11	7	10	0	8	11	8	2	11	7	10	0	8	11	8	2
		2 × 10	14	10	12	10	11	6	11	5	14	10	12	10	11	6	10	6
		2 × 12	18	1	15	8	14	0	15	0	18	1	15	8	14	0	12	9
Eastern hemlock-tamarack	Select structural	2 × 4	7	1	6	6	6	0	5	6	8	2	7	5	6	11	6	6
		2 × 6	11	3	10	2	9	5	8	9	12	10	11	8	10	9	9	2
		2 × 8	14	10	13	5	12	6	11	6	16	11	15	4	14	3	12	5
		2 × 10	18	11	17	2	15	11	15	0	21	8	19	8	18	3	16	2
		2 × 12	23	0	20	11	19	5	18	3	26	4	23	11	22	2	20	11

	No. 1	2 × 4	7	1	6	6	6	0	5	8	8	2	7	5	6	11	6	6
		2 × 6	11	3	10	2	9	5	8	11	12	10	11	8	10	10	10	2
		2 × 8	14	10	13	5	12	6	11	9	16	11	15	5	14	3	13	5
		2 × 10	18	11	17	2	15	11	15	0	21	8	19	8	18	3	17	2
		2 × 12	23	0	20	11	19	5	18	3	26	4	23	11	22	2	20	11
	No. 2	2 × 4	6	10	6	3	5	9	5	5	7	11	7	2	6	8	6	3
		2 × 6	10	10	9	10	9	1	8	7	12	5	11	3	10	1	9	2
		2 × 8	14	3	13	0	12	0	11	4	16	4	14	10	13	4	12	2
		2 × 10	18	3	16	7	15	4	14	6	20	11	19	0	17	0	15	6
		2 × 12	22	2	20	2	18	8	17	7	25	5	23	1	20	8	18	10
	No. 3	2 × 4	6	7	5	10	5	3	4	9	6	9	5	10	5	3	4	9
		2 × 6	9	11	8	7	7	8	7	0	9	11	8	7	7	8	7	0
		2 × 8	13	1	11	4	10	2	9	3	13	1	11	4	10	2	9	3
		2 × 10	16	9	14	6	12	11	11	10	16	9	14	6	12	11	11	10
		2 × 12	20	4	17	8	15	9	14	5	20	4	17	8	15	9	14	5
Spruce-pine-fir	Select structural	2 × 4	7	1	6	5	5	11	5	7	8	1	7	4	6	10	6	5
		2 × 6	11	1	10	1	9	4	8	10	12	9	11	7	10	6	9	7
		2 × 8	14	8	13	4	12	4	11	7	16	9	15	3	13	10	12	8
		2 × 10	18	9	17	0	15	9	14	10	21	5	19	6	17	8	16	2
		2 × 12	22	9	20	8	19	2	18	1	26	1	23	8	21	6	19	7
	No. 1	2 × 4	7	1	6	5	5	11	5	7	8	1	7	4	6	8	6	1
		2 × 6	11	1	10	1	9	4	8	10	12	6	10	9	9	8	8	10
		2 × 8	14	8	13	4	12	4	11	7	16	5	14	3	12	9	11	7
		2 × 10	18	9	17	0	15	9	14	10	21	0	18	2	16	3	14	10
		2 × 12	22	9	20	8	19	2	18	1	25	6	22	1	19	9	18	1
	No. 2	2 × 4	6	10	6	2	5	9	5	5	7	9	6	8	6	0	5	5
		2 × 6	10	9	9	9	8	9	7	11	11	3	9	9	8	9	7	11
		2 × 8	14	2	12	10	11	6	10	6	14	10	12	10	11	6	10	6
		2 × 10	18	1	16	5	14	8	13	5	19	0	16	5	14	8	13	5
		2 × 12	22	0	20	0	17	11	16	4	23	1	20	0	17	11	16	4
	No. 3	2 × 4	5	10	5	1	4	6	4	1	5	10	5	1	4	6	4	1
		2 × 6	8	5	7	3	6	6	5	11	8	5	7	3	6	6	5	11
		2 × 8	11	1	9	7	8	7	7	10	11	1	9	7	8	7	7	10
		2 × 10	14	2	12	3	10	11	10	0	14	2	12	3	10	11	10	0
		2 × 12	17	2	14	11	13	4	12	2	17	2	14	11	13	4	12	2
Western cedars	Select structural	2 × 4	6	10	6	2	5	9	5	5	7	9	7	1	6	7	6	2
		2 × 6	10	9	9	9	9	0	8	6	12	3	11	2	10	4	9	7
		2 × 8	14	2	12	10	11	11	11	2	16	2	14	8	13	8	12	8
		2 × 10	18	0	16	5	15	3	14	4	20	8	18	9	17	5	16	2
		2 × 12	21	11	19	11	18	6	17	5	25	2	22	10	21	2	19	7
	No. 1	2 × 4	6	10	6	2	5	9	5	5	7	9	7	1	6	7	6	1
		2 × 6	10	9	9	9	9	0	8	6	12	3	11	0	9	10	9	0
		2 × 8	14	2	12	10	11	11	11	2	16	2	14	7	13	0	11	11
		2 × 10	18	0	16	5	15	3	14	4	20	8	18	7	16	7	15	2
		2 × 12	21	11	19	11	18	6	17	5	25	2	22	7	20	3	18	5
	No. 2	2 × 4	6	7	6	0	5	6	5	2	7	6	6	10	6	1	5	7
		2 × 6	10	4	9	5	8	9	7	11	11	3	9	9	8	9	7	11
		2 × 8	13	8	12	5	11	6	10	6	14	10	12	10	11	6	10	6
		2 × 10	17	5	15	10	14	8	13	5	19	0	16	5	14	8	13	5
		2 × 12	21	2	19	3	17	10	16	4	23	1	20	0	17	11	16	4
	No. 3	2 × 4	5	10	5	1	4	6	4	1	5	10	5	1	4	6	4	1
		2 × 6	8	10	7	7	6	10	6	3	8	10	7	7	6	10	6	3
		2 × 8	11	7	10	1	9	0	8	2	11	7	10	1	9	0	8	2
		2 × 10	14	10	12	10	11	6	10	6	14	10	12	10	11	6	10	6
		2 × 12	18	1	15	7	14	0	12	9	18	1	15	7	14	0	12	9

TABLE 4-5b: *Maximum spans for floor joists—bedrooms and attics accessible by a stairway (meters) (live load 1.4 kN/m²)*

Lumber species	Grade	Size (mm)	Gypsum board or plastered ceiling, joist spacing			Other ceilings, joist spacing		
			300 mm	400 mm	600 mm	300 mm	400 mm	600 mm
Douglas fir-larch (N)	Select structural	38 × 89	2.41	2.19	1.91	2.76	2.50	2.19
		38 × 140	3.79	3.44	3.00	4.34	3.94	3.44
		38 × 184	4.99	4.54	3.96	5.72	5.19	4.54
		38 × 235	6.37	5.79	5.06	7.30	6.63	5.79
		38 × 286	7.75	7.04	6.15	8.87	8.06	7.04
	No. 1	38 × 89	2.41	2.19	1.91	2.76	2.50	2.19
		38 × 140	3.79	3.44	3.00	4.34	3.94	3.28
		38 × 184	4.99	4.54	3.96	5.72	5.19	4.33
		38 × 235	6.37	5.79	5.06	7.30	6.63	5.52
		38 × 286	7.75	7.04	6.15	8.87	8.06	6.72
	No. 2	38 × 89	2.33	2.11	1.85	2.67	2.42	2.05
		38 × 140	3.66	3.33	2.90	4.18	3.62	2.95
		38 × 184	4.83	4.38	3.83	5.51	4.77	3.89
		38 × 235	6.16	5.60	4.89	7.03	6.09	4.97
		38 × 286	7.49	6.81	5.95	8.55	7.40	6.04
	No. 3	38 × 89	2.14	1.86	1.51	2.14	1.86	1.51
		38 × 140	3.16	2.74	2.23	3.16	2.74	2.23
		38 × 184	4.17	3.61	2.95	4.17	3.61	2.95
		38 × 235	5.32	4.61	3.76	5.32	4.61	3.76
		38 × 286	6.47	5.60	4.57	6.47	5.60	4.57
Hem-fir (N)	Select structural	38 × 89	2.32	2.11	1.84	2.66	2.41	2.11
		38 × 140	3.65	3.32	2.90	4.18	3.74	3.05
		38 × 184	4.81	4.37	3.82	5.51	4.93	4.02
		38 × 235	6.14	5.58	4.87	7.03	6.29	5.13
		38 × 286	7.47	6.79	5.93	8.55	7.65	6.24
	No. 1	38 × 89	2.32	2.11	1.84	2.66	2.38	1.94
		38 × 140	3.65	3.32	2.83	5.29	4.58	3.74
		38 × 184	4.81	4.37	3.74	5.29	4.58	3.74
		38 × 235	6.14	5.58	4.77	6.75	5.84	4.77
		38 × 286	7.47	6.79	5.80	8.21	7.11	5.80
	No. 2	38 × 89	2.24	2.04	1.76	2.49	2.15	1.76
		38 × 140	3.52	3.11	2.54	3.60	3.11	2.54
		38 × 184	4.65	4.11	3.35	4.74	4.11	3.35
		38 × 235	5.93	5.24	4.28	6.05	5.24	4.28
		38 × 286	7.21	6.37	5.20	7.36	6.37	5.20
	No. 3	38 × 89	1.84	1.60	1.30	1.84	1.60	1.30
		38 × 140	2.73	2.36	1.93	2.73	2.36	1.93
		38 × 184	3.60	3.11	2.54	3.60	3.11	2.54
		38 × 235	4.59	3.97	3.24	4.59	3.97	3.24
		38 × 286	5.58	4.84	3.95	5.58	4.84	3.95
Eastern hemlock-tamarack (N)	Select structural	38 × 89	2.21	2.01	1.75	2.53	2.30	2.01
		38 × 140	3.48	3.16	2.76	3.98	3.62	3.16
		38 × 184	4.58	4.16	3.64	5.25	4.77	4.16
		38 × 235	5.85	5.32	4.64	6.70	6.09	5.32
		38 × 286	7.12	6.47	5.65	8.15	7.40	6.47
	No. 1	38 × 89	2.21	2.01	1.75	2.53	2.30	2.01
		38 × 140	3.48	3.16	2.76	3.98	3.62	3.16
		38 × 184	4.58	4.16	3.64	5.25	4.77	4.16
		38 × 235	5.85	5.32	4.64	6.70	6.09	5.32
		38 × 286	7.12	6.47	5.65	8.15	7.40	6.47

TABLE 4–5b: *(Cont.)*

Lumber species	Grade	Size (mm)	Gypsum board or plastered ceiling, joist spacing			Other ceilings, joist spacing		
			300 mm	400 mm	600 mm	300 mm	400 mm	600 mm
	No. 2	38 × 89	2.13	1.94	1.69	2.44	2.22	1.94
		38 × 140	3.35	3.04	2.66	3.84	3.47	2.83
		38 × 184	4.42	4.01	3.51	5.05	4.58	3.74
		38 × 235	5.64	5.12	4.48	6.46	5.84	4.77
		38 × 286	6.86	6.23	5.44	7.85	7.11	5.80
	No. 3	38 × 89	2.05	1.79	1.46	2.07	1.79	1.46
		38 × 140	3.03	2.63	2.14	3.03	2.63	2.14
		38 × 184	4.00	3.46	2.83	4.00	3.46	2.83
		38 × 235	5.11	4.42	3.61	5.11	4.42	3.61
		38 × 286	6.21	5.38	4.39	6.21	5.38	4.39
Western cedar (N)	Select structural	38 × 89	2.11	1.91	1.67	2.41	2.19	1.91
		38 × 140	3.31	3.01	2.63	3.79	3.44	2.99
		38 × 184	4.37	3.97	3.47	5.00	4.54	3.94
		38 × 235	5.57	5.06	4.42	6.38	5.80	5.02
		38 × 286	6.78	6.16	5.38	7.76	7.05	6.11
	No. 1	38 × 89	2.11	1.91	1.67	2.41	2.19	1.90
		38 × 140	3.31	3.01	2.63	3.79	3.41	2.78
		38 × 184	4.37	3.97	3.47	5.00	4.49	3.67
		38 × 235	5.57	5.06	4.42	6.38	5.73	4.68
		38 × 286	6.78	6.16	5.18	7.78	6.97	5.69
	No. 2	38 × 89	2.04	1.85	1.61	2.33	2.11	1.72
		38 × 140	3.20	2.91	2.48	3.51	3.04	2.48
		38 × 184	4.22	3.84	3.27	4.63	4.01	3.27
		38 × 235	5.39	4.90	4.18	5.91	5.12	4.18
		38 × 286	6.55	5.95	5.08	7.19	6.23	5.08
	No. 3	38 × 89	1.80	1.56	1.27	1.80	1.56	1.27
		38 × 140	2.65	2.30	1.88	2.65	2.30	1.88
		38 × 184	3.50	3.03	2.47	3.50	3.03	2.47
		38 × 235	4.47	3.87	3.16	4.47	3.87	3.16
		38 × 286	5.44	4.71	3.84	5.44	4.71	3.84
Spruce-pine-fir	Select structural	38 × 89	2.19	1.99	1.74	2.50	2.28	1.99
		38 × 140	3.44	3.13	2.73	3.94	3.58	2.95
		38 × 184	4.54	4.12	3.60	5.19	4.72	1.89
		38 × 235	5.79	5.26	4.59	5.63	6.02	4.97
		38 × 286	7.04	6.40	5.59	8.06	7.32	6.04
	No. 1	38 × 89	2.19	1.99	1.74	2.50	2.28	1.88
		38 × 140	3.44	3.13	2.73	3.88	3.36	2.75
		38 × 184	4.54	4.12	3.60	5.12	4.44	3.62
		38 × 235	5.79	5.26	4.59	6.54	5.66	4.62
		38 × 286	7.04	6.40	5.59	7.95	6.89	5.62
	No. 2	38 × 89	2.11	1.92	1.68	2.41	2.08	1.70
		38 × 140	3.33	3.02	2.46	3.49	3.02	2.46
		38 × 184	4.38	3.98	3.25	4.60	3.98	3.25
		38 × 235	5.60	5.08	4.15	5.87	5.08	4.15
		38 × 286	6.81	6.18	5.04	7.13	6.18	5.04
	No. 3	38 × 89	1.80	1.56	1.27	1.80	1.56	1.27
		38 × 140	2.65	2.30	1.88	2.65	2.30	1.88
		38 × 184	3.50	3.03	2.47	3.50	3.03	2.47
		38 × 235	4.47	3.87	3.16	4.47	3.87	3.16
		38 × 286	5.44	4.71	3.84	5.44	4.71	3.84

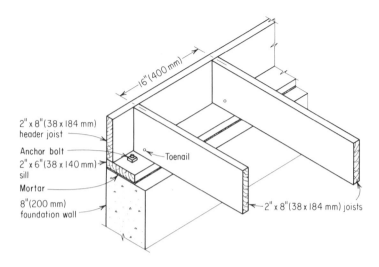

2" x 8"(38 x 184 mm) header joist

Anchor bolt

2" x 6"(38 x 140 mm) sill

Mortar

8"(200 mm) foundation wall

16"(400 mm)

Toenail

2" x 8"(38 x 184 mm) joists

FIGURE 4-11: *Joist framing at foundation wall.*

Joist Framing

Two systems are used for assembling joists—the *cast-in* system, previously described in Chapter 3, and the *box sill* system, in which the whole assembly is mounted on top of the foundation wall, secured to a sill plate.

In the box system, joists are held in position at the foundation wall by the *header joists,* running at right angles to

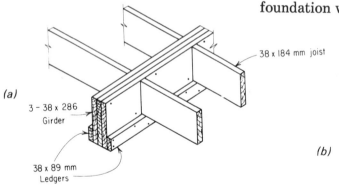

(a)

3 – 38 x 286 Girder

38 x 89 mm Ledgers

38 x 184 mm joist

Figure 4-12: *(a) Butted and scabbed.*
(b) Lapped (c) Ledgers (d) Joist hangers.

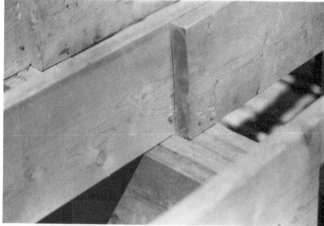

(b)

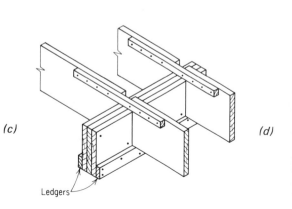

(c)

Ledgers

(d)

the regular joists and bearing on the sill plate (see Fig. 4–11). At the inner end, the joists may be carried *on top* of the girder, or, where more headroom is needed under the girder, set on a ledger nailed to the side (see Fig. 4–12), or supported by joist hangers. Care must be taken to allow for shrinkage of the joist material when they are supported by a ledger or a joist hanger. The top edge of the joists must be above the beam level, or differential shrinkage will cause an uneven floor (see Fig. 4–13).

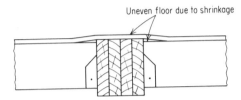

FIGURE 4-13: *Differential shrinkage.*

In locations where an extra load is imposed on the floor frame, as, for example, where a non-load-bearing partition over 6 ft (2 m) in length and containing openings that are not full ceiling height runs parallel to the direction of joist run, joists have to be *doubled*. This may be done either by placing two joists side by side or having two which are separated by blocking (see Fig. 4–14). The blocking should be spaced not more than 4 ft (1.2 m) o.c. When load-bearing partitions run parallel to the joists, the partition is supported by a beam or a load-bearing partition in the basement. Load-bearing partitions at right angles to the floor joists should be located not more than 3 ft (1 m) from the joist support when the wall does not support an upper floor and not more than 2 ft (600 mm) when the wall supports one or more upper floors unless floor joists have been designed to carry the concentrated loads.

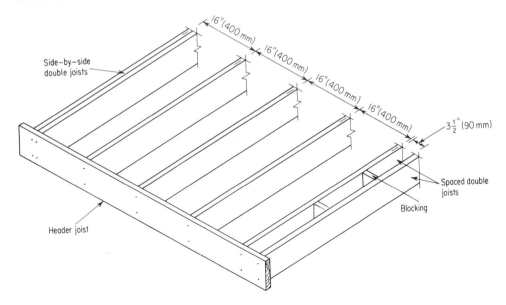

FIGURE 4-14: *Doubled floor joists.*

Openings in the floor frame for a stairwell or where a chimney or fireplace comes through the floor must be framed in the proper way. Trimmer joists are doubled if they support header joists more than 32 in. (800 mm) long. Header joists longer than 48 in. (1.2 m) should also be doubled. Trimmer joists that support header joists more than 6 ft (2 m) long and header joists that are more than 10 ft, 8 in. (3.2 m) long should be designed according to accepted engineering practice. Joist hangers are used to support long joist headers and tail joists (see Fig. 4-15). Most codes require that a chimney opening be at least 2 in. (50 mm) larger on all sides than the dimensions of the chimney and that the framing be kept back 4 in. (100 mm) from the back of a fireplace. See Chapter 9 for stairwell opening calculations.

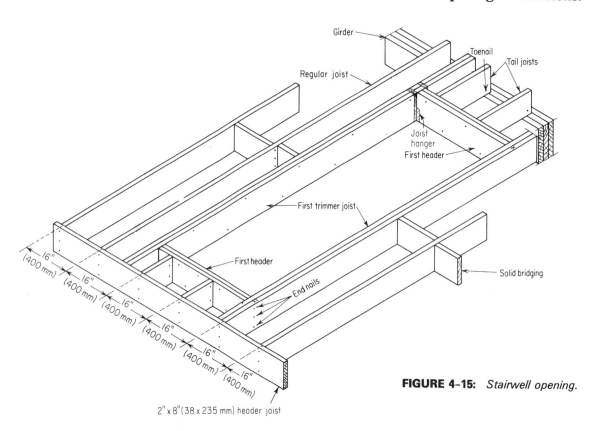

Girder

Toenail

Tail joists

Regular joist

Joist hanger

First header

First trimmer joist

First header

Solid bridging

End nails

16"
(400 mm)
16"
(400 mm)
16"
(400 mm)
16"
(400 mm)
16"
(400 mm)
16"
(400 mm)

2" x 8" (38 x 235 mm) header joist

FIGURE 4-15: *Stairwell opening.*

Bridging

It is important that joists be prevented from twisting after they are in place, and, unless *ceiling furring or plywood cladding* is installed on the underside, restraint must be provided, both at end supports and at intervals not exceeding 7 ft (2100 mm) between supports. At the end supports, the restraint is provided by toenailing the joists to the supports and by endnailing through the header joist.

The intermediate restraint is provided by *bridging*, which may be in the form of *solid blocking* between joists (see Fig. 4-15), *crossbridging*, or *wood or steel strapping*. The solid blocking will

(a)   **FIGURE 4-16:** *(a) Cross bridging (b) Strap bridging.*   *(b)*

be pieces of 1½-in. (38-mm) material the same depth as the joists and may be *offset* one-half their thickness to facilitate endnailing (see Fig. 4–15). The crossbridging is made from 2 × 2 or 1 × 3 (38 × 38 mm or 19 × 64 mm) material, cut to fit as illustrated in Fig. 4–16. The ⅛ × 1 in. (3 × 25 mm) steel strapping must be fastened to the underside of each joist and to the sill or header at each end to prevent overall movement.

The angle cuts on the ends of crossbridging may be obtained by using the *depth of joist* on the tongue of a square and the *distance between joists* on the blade, with the tongue figure on the *upper edge* of the bridging material and the blade figure on the *lower edge,* marking on the tongue, as illustrated in Fig. 4–17.

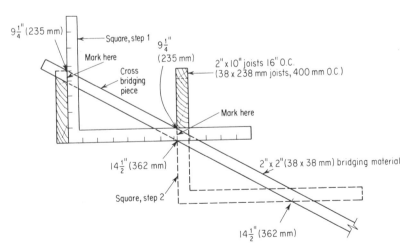

FIGURE 4-17: *Layout of cross bridging.*

Joist Assembly

Setting the joists in their proper location, nailing them securely in place, and providing the extra framing necessary for concentrated loads and frame rigidity are very important steps in the framing of a building. The procedure may be as follows:

1. Study the plans to check the location and centers of the joists. Note where they are to be doubled and if the doubling will be side by side or separated.

2. Select straight header joists and lay them off according to the plans. Be sure that joints in the header joist occur at the center of a regular joist (see Fig. 4-18). Note that if joists are to be *lapped* at the girder, the header joist lay-off on the two opposite sides of the building will have to be offset from one another by the thickness of the joists.

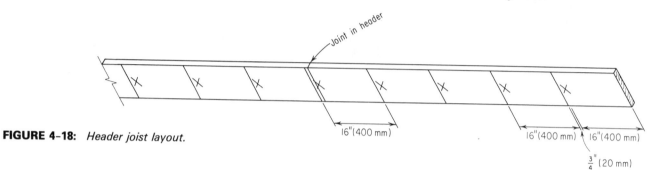

FIGURE 4-18: *Header joist layout.*

3. Check for the number of full-length regular joists required, square one end and cut them to length, unless they are to be lapped.

4. For box sill construction, toenail the header joists in place on the sill plate (see Chapter 3 for procedure in cast-in construction).

5. Where joist hangers are to be used, nail them in place on the girder, in line with the header joist layout.

6. Lay out all the regular joists on one side of the building, set them up in position *with the crown up,* toenail them to the sill and to the girder if they rest on top or on ledgers, and *endnail* into them through the header joist.

7. Repeat with the regular joists on the opposite side and if joists butt at the girder, nail a board 3 ft (1 m) long across the butt joint (see Fig. 4-12). If they lap, nail the two together through the lap joint.

8. Check the plans for the location of floor openings; mark on the joist header the location of the trimmer joists; square the end and endnail in place.

9. Mark the trimmer joists for header positions (see Fig. 4-15); cut headers to exact length and endnail in place.

10. Add the second trimmer if doubling is required.

11. Add the second header if doubling is required.

12. Cut tail joists to exact length and nail them in place *on regular centers.*

13. Locate the position of walls running parallel to floor joists and add extra joists at that location (see Fig. 4–14).

14. Space the top edges of the joists properly over the girder, using a 1 × 4 (19 × 89) nailing strip to hold them in place, and then toenail to girder (see Fig. 4–9).

15. Mark the centerline of the bridging with a chalk line [not more than 7 ft (2.1 m) from end supports] and nail bridging to the line.

(a)

Prefabricated Wood Joists and Truss Systems

In recent years prefabricated wood joists and truss systems have become popular. Prefabricated systems provide greater clear spans and are lighter in weight and straighter than conventional framing. This will provide a flatter floor and straighter ceilings below, and as the cord members are wider than standard joists, the need for strapping is eliminated. Wood prefabricated members are in the form of a wooden I beam or a flat truss (see Fig. 4–19).

Heating ducts, plumbing pipes, and electrical wires are easily incorporated into the joist systems. Trusses have many large openings, allowing room for services. Holes are easily cut into the webs of I beams of plywood or waferboard. Holes 1½ in. (38 mm) can be cut anywhere in the web; larger holes can only be placed in the locations shown in Fig. 4–20. In an area where multiple holes are needed, a distance equal to double the hole's

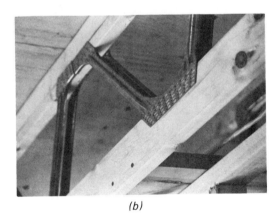

(b)

FIGURE 4-19: *(a) "I" beam floor joists (Courtesy Jager Industries Inc.) (b) Flat truss.*

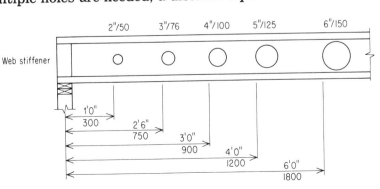

FIGURE 4-20: *Hole chart.*

diameter is required between holes. Square holes must not exceed 75% of the maximum diameter of round holes allowed in the area. Rectangular openings 24 in. (600 mm) × *depth of joist* − 3¼ in. (80 mm) may be cut at midspan only, providing twice the length of the rectangle is left before the next hole.

In a box sill floor frame the joist header can be either 2 × *joist depth* or a ¾-in. *ribbon* [see Fig. 4–21(a)]. Stiffeners are used on both sides of the web [see Fig. 4–21(b)] at intermediate supports. Stairwell opening framing, bridging, and cantilever details are illustrated in Fig. 4–21(c).

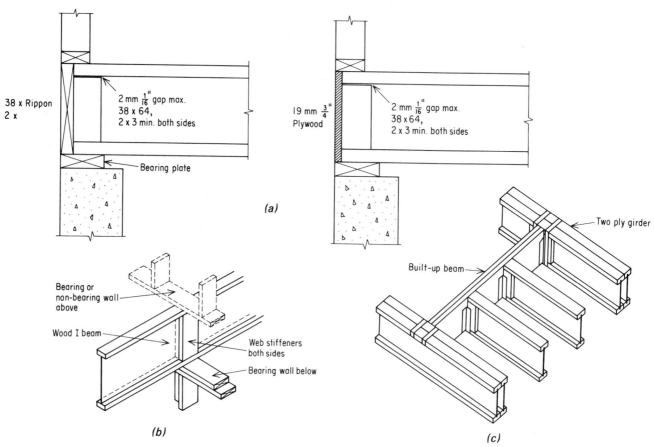

FIGURE 4-21: *(a) Joist header detail. (b) Stiffeners at intermediate support (c) Stairwell opening (Courtesy Jager Industries Inc.).*

Steel Joists

Steel joists of various kinds are available for use in place of wood joists and are particularly useful where relatively long spans are required. They are all prefabricated units, some made in various depths to suit span requirements.

One type, known as *truss joists,* have 2 × 4 (38 × 89 mm) top and bottom chords of wood, with 1½-in. (38-mm) tube webs. A similar type, called *open web joists,* are entirely of steel, as illustrated in Fig. 4–22(a). These require a wooden pad to be attached to the top chord to which the subfloor can be fastened.

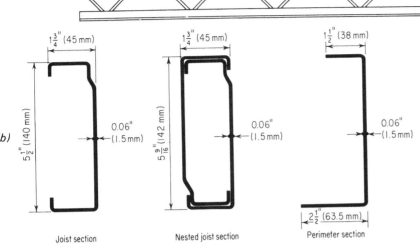

FIGURE 4-22: *a) Open web steel joist. (b) Cold-formed steel joist shapes (Courtesy, Steel Co. of Canada).*

A third type, *cold-formed steel joists,* are made from sheet steel in several standard depths, one of which is shown in Fig. 4–22(b). The single joist, 3½ in. (140 mm) deep, has a greater load-bearing capacity than a 2 × 10 (38 × 235 mm) wood joist, conventionally used. The nested joist section is used as a trimmer joist at floor openings and the perimeter section as a header joist, in an all-steel floor frame.

Cold-formed steel joists may be used with a conventional wooden header joist in box sill construction, as illustrated in Fig. 4–23(a). The joist ends may be fitted with end clips [see Fig. 4–23(b)] and nailed to the header through the clips, or wooden blocks may be fitted into the ends [see Fig. 4–23(b)], in which case the joists are secured by endnailing through the header into the blocks.

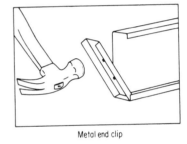

Metal end clip

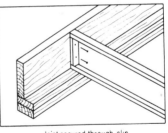

Joist secured through clip

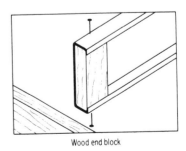

Wood end block

(b)

FIGURE 4-23: *(a) Steel joist with wood header joist (Courtesy U.S. Steel Corp.). (b) End anchors for steel joists.*

(a)

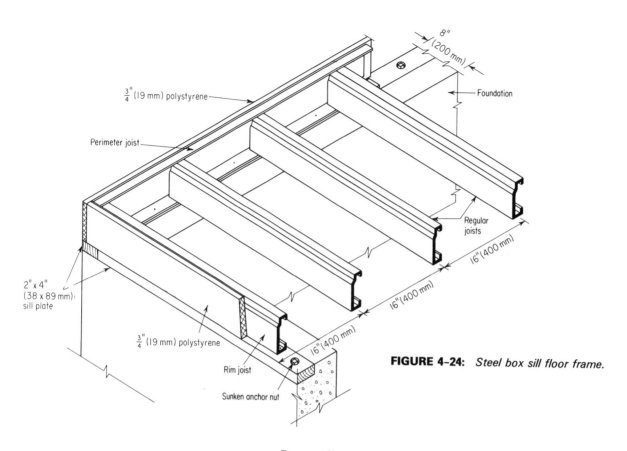

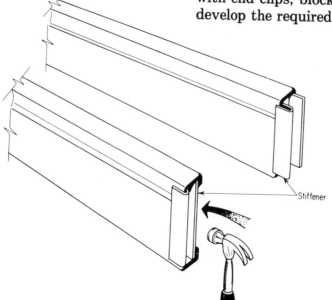

FIGURE 4-24: *Steel box sill floor frame.*

In an all-steel box sill floor frame, a perimeter section [see Fig. 4-22(b)] is used as the header joist, secured to the sill plate with screws or powder-actuated fasteners. It should be set back ¾ in. (20 mm) from the outside edge of the sill plate to allow for rigid insulation to be applied to the outer surface. The rim joist is set back a similar distance, for the same purpose (see Fig. 4-24). Regular joists are normally set at 16 in. (400 mm) o.c. and secured to the perimeter joist by self-tapping screws. The inner ends of joists resting on a girder, ledger, or hanger should be provided with end clips, blocks, or *stiffeners* (see Fig. 4-25) in order to develop the required strength.

FIGURE 4-25: *Joist end stiffener.*

Floor openings in steel frame construction are framed in much the same way as a wood frame. Double trimmers are provided by using the *nested joist section* (see Fig. 4–21) and opening headers, and tail joists are secured by *hangers* (see Fig. 4–26), held in place by self-drilling tapping screws, as illustrated in Fig. 4–27.

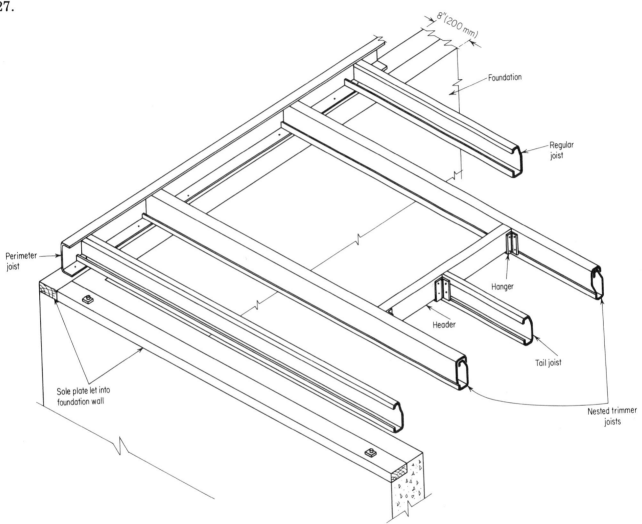

FIGURE 4–26: *Framed opening in steel floor frame.*

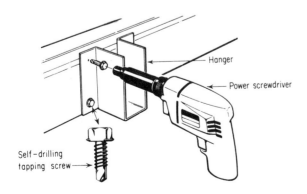

FIGURE 4–27: *Hanger secured with self-drilling tapping screws.*

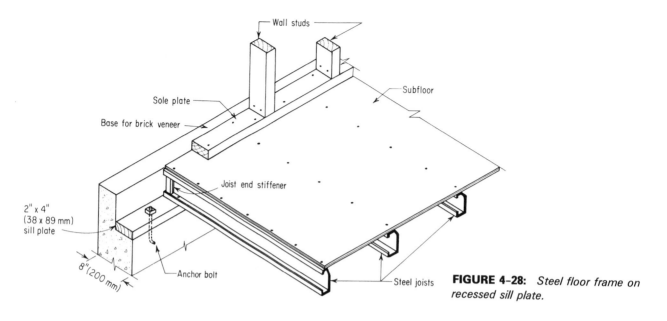

FIGURE 4-28: *Steel floor frame on recessed sill plate.*

Steel joists may be cast in (see Fig. 3–40) or *recessed* onto the top of the foundation to provide a *flush floor.* The recessing may be done by lowering the sill plate into the top, inner edge of the foundation wall (see Fig. 4–28) or by providing joist pockets for ends of the joists, as illustrated in Fig. 4–29. Both methods leave part of the upper surface of the foundation exposed as a base for brick veneer exterior finish.

When the lowered sill plate method is used, joists are provided with end clips or stiffeners and secured to the sill plate by screws through their botton edge. When the ends of joists rest in pockets, the joist ends are blocked and *grouted* into the pockets.

FIGURE 4-29: *Steel joists in joist pockets.*

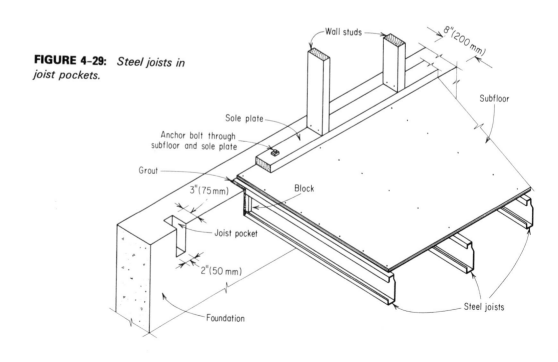

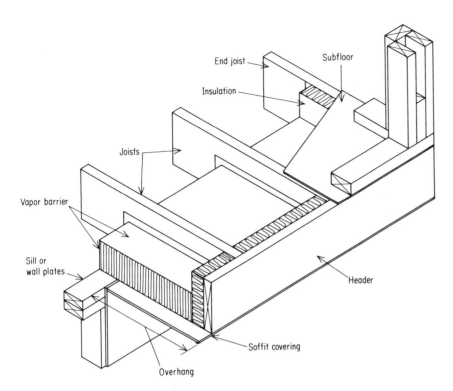

FIGURE 4-30: *Framing floor projections (joists perpendicular to projection).*

FIGURE 4-31: *Framing floor projections (joists parallel to projection).*

Floor Framing at Projections

Floor joists occasionally project beyond the foundation wall to provide support for bay windows or additional floor space in upper rooms. The cantilevered portion of the floor should not exceed 1½ times the joist depth for bay windows or additional interior floor space. The space between the floor joists should be left open to allow for circulation of warm air so even floor temperature occurs throughout the room (see Fig. 4-30). When the floor joists are perpendicular to the projection, the framing is very easy as the joists need only to extend farther to support the projection (see Fig. 4-30). However, if the floor joists are parallel to the projection, cantilevered joists must be at right angles to the regular joists (see Fig. 4-31).

Framing for balconies can be framed in the same method as bay windows, but as less load is supported, the joists can project farther (see Fig. 4-31). Other methods of supporting balconies include building a beam into the foundation wall that extends beyond the outer face of the wall, providing support for the balcony framing (see Fig. 4-32). Bearing posts can also be installed, transferring the load of the balcony to lower floors or foundations. The top level is often cut down for 2-in. decking, which still provides a break in the floor level and does not cause interference with the functioning of the balcony door (see Fig. 4-33).

FIGURE 4-32: *Support for balcony framing.*

FIGURE 4-33: *Balcony framing.*

FIGURE 4–34: *Plywood subfloor.*

FIGURE 4–35: *Fasteners for subfloor-to-steel joists.*

SUBFLOOR

Laying the subfloor is the final step in completing the floor frame. *Center match, common boards, plywood,* or *particle board* may be used for subflooring, and if one of the first two is specified, application should be at an angle of 45° to the joists (see Fig. 4–33).

The minimum thickness of plywood or particle board should be ⅝-in. (15 mm), and the *long dimension* should run *across* the joists, and joints should be broken in successive courses (see Fig. 4–34). To provide a stiffer subfloor, the long dimension of panels should be supported by *blocking* between joists, or the panels used should have tongue-grooved edges. The matching of the tongue with the groove in the plywood is easier to achieve if the tongue is fed into the groove. This allows for hammering the joints together without damaging the tongue. A block of 2 × 4 can be used to hammer against so the edge of the sheet is not damaged. Nails for plywood subflooring should be a minimum of 2 in. (50 mm) long and be spaced 6 in. (150 mm) along the edge of the sheet and 12 in. (300 mm) o.c. on intermediate supports. Annular ringed nails provide extra holding power. Staples of a minimum length of 2 in. (50 mm) can also be used at the same spacing.

Plywood or particle board subflooring is attached to steel joists with *steel joist nails* or *self-drilling* tapping screws (see Fig. 4–35), using a hammer for the nails or a power screwdriver for the screws. Frequently a strip of subfloor adhesive is applied to the edges of joists before subflooring is laid (see Fig. 4–36) in order to provide a more rigid and squeek-free floor.

FIGURE 4–36: *(a) Adhesive for subfloor applied to the edges of steel joists (Courtesy U.S. Steel Corp.). (b) Using subfloor adhesive.*

(a)

(b)

REVIEW QUESTIONS

4–1. A one-story building 32 ft (9.6 m) wide by 40 ft (12 m) long is to have a center supporting girder made up of three 2 × 10 (38 × 235 mm) members of No. 1 Douglas fir. How many evenly spaced posts are required to support it?

4–2. Define the *clear span* of a girder.

4–3. If the floor load in the building mentioned in Question 4–1 is 40 lb/ft² (978 kg)/m², how much load is carried by one post?

4–4. Outline two primary reasons for using bridging.

4–5. What figures would you use on the square to lay out the end cuts for 2 × 2 (38 × 38) cross bridging to fit between 2 × 10 (38 × 235 mm) joists spaced 24 in. (600 mm) o.c.?

4–6. A building 32 ft (9.6 m) wide, with a center supporting girder, is to have hem-fir, No. 1 joists, spaced 16 in. (400 mm) o.c. What size of joist is required?

4–7. An opening 3 ft, 6 in. (1050 mm) wide by 6 ft, 6 in. (1950 mm) long is to be framed in the floor of the building described in Question 4–7, 3 ft (900 mm) from one side and 11 ft (3300 mm) from one end. What length would you cut?

(a) The opening headers?

(b) The trimmer joists?

(c) The tail joists on the outside?

4–8. Carefully draw two sketches to illustrate the differences between a box sill floor frame and a cast-in joist floor frame.

4–9. Fill in the blanks with the word or phrase which makes the sentence correct:

(a) Material for bearing wall studs must be at least _____ in. (mm) in width.

(b) Joists spanning 16 ft (4.8 m) should have _____ rows of bridging.

(c) Strap bridging is usually made from _____ .

(d) Common spacing for anchor bolts in sill plates is _____ .

(e) Joists may be carried on the side of a girder by a _____ or by _____ .

(f) A three-piece, laminated girder, 24 ft (7.2 m) long, supported by end walls and two posts, could be made up of _____ pc. _____ ft (mm) long and _____ pc. ____ ft (mm) long.

(g) _____ bridging is usually used when the space between joists is no greater than the width of joists.

5

THE WALL FRAME

There are two basic systems used for framing walls, namely the *western* or *platform* frame (see Fig. 5–1) and the *balloon* frame, shown in Fig. 5–2. Many of the members are the same in both, and the main differences lie in the method of starting the frame at the ground floor level, the length of the wall framing members, and the method of framing in the floor for the second story.

COMPONENT PARTS, WESTERN FRAME

As illustrated in Fig. 5–1, a western frame begins on the subfloor. The bottom member is the *sole plate* to which are attached *regular studs* one story in length, special arrangements of studs at corners, one of which is called a *western corner*, extra studs to which partitions are attached—*partition junctions*—and other shorter members. On top of these vertical members is a *top plate* and on top of that a *cap plate*. Between studs, halfway up their length, are horizontal blocks called *firestops*. Over door and window openings are *headers* or *lintels* and, supporting the ends of these, *trimmer studs*. The bottom of a window opening is framed by a *rough sill*, and the spaces between rough sill and sole plate and between header and top plate are framed with *cripple studs*. *Wall backing* is placed where sinks, drapes, etc., are to be attached to interior walls, and *sheathing* covers the exterior of the frame. Where platform frame buildings are more than one story high or where the exterior sheathing is to be such material as gypsum board or some type of insulative material (see Fig. 5–1), *wind bracing* is sometimes required.

COMPONENT PARTS, BALLOON FRAME

A balloon frame begins down at the sill plate (see Fig. 5–2). Full-length *regular studs* and *corner posts* rest on the sill and are capped at the top by *top plate* and *cap plate*. At the first floor level, a ¾-in. (19-mm) *ribbon* is let into the studs, and the ends of the *first-floor joists* rest on it. 2 × 4 (19 × 89 mm) *let-in bracing* is nailed to the outside of the studs at an angle to brace the frame. *Partition junctions* are set in only one story high. *Firestops, window and door headers, trimmers, rough sills, cripple studs,* and *wall backing* are similar to those in a platform frame. *Sheathing* completes the frame.

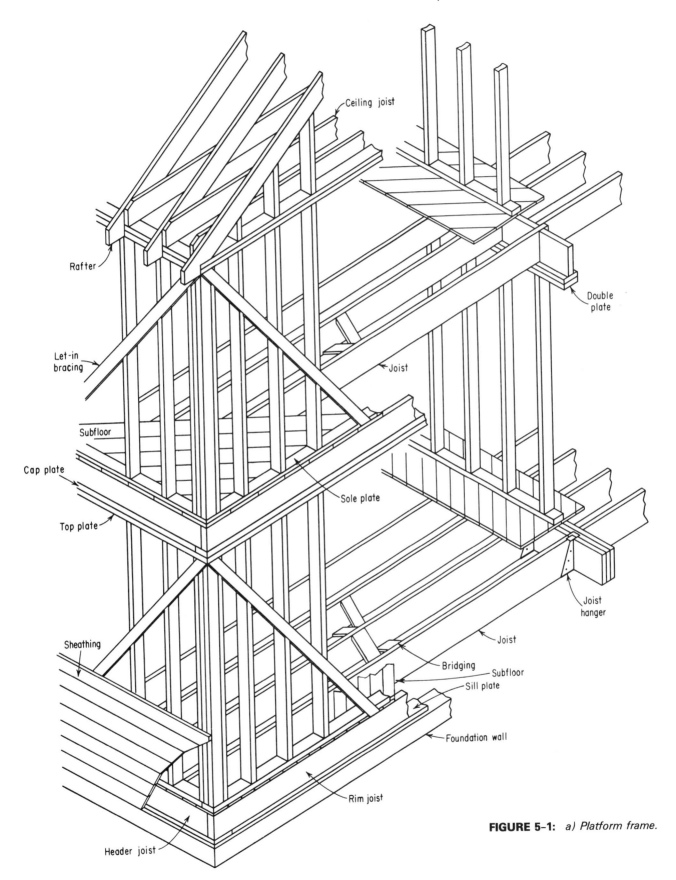

Ceiling joist

Rafter

Let-in bracing

Subfloor

Cap plate

Top plate

Sheathing

Header joist

Rim joist

Joist

Sole plate

Double plate

Joist hanger

Joist

Bridging

Subfloor

Sill plate

Foundation wall

FIGURE 5-1: *a) Platform frame.*

FIGURE 5-1 *Continued*

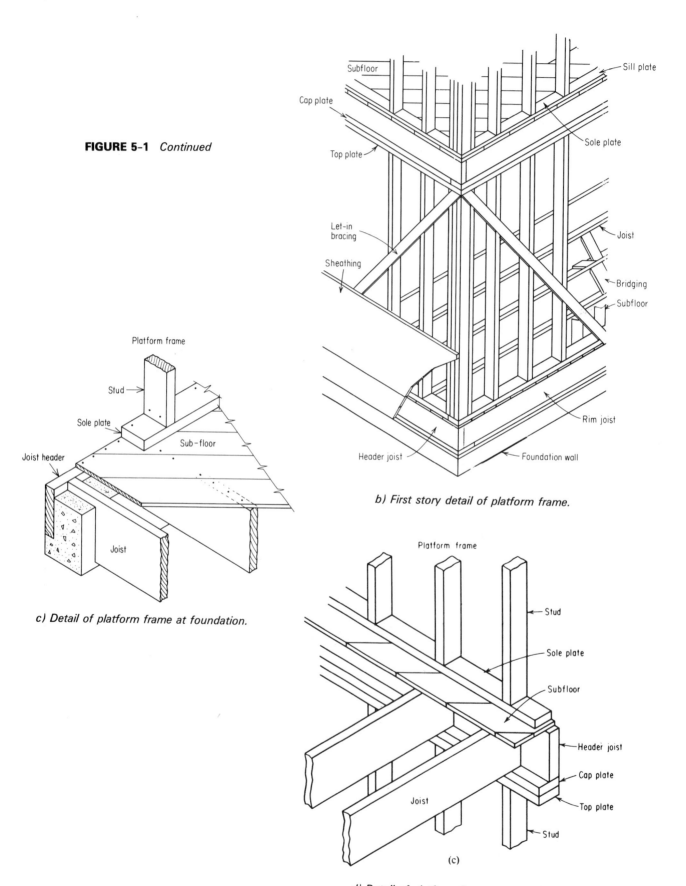

Subfloor

Cap plate

Top plate

Let-in bracing

Sheathing

Header joist

Sill plate

Sole plate

Joist

Bridging

Subfloor

Rim joist

Foundation wall

b) First story detail of platform frame.

Platform frame

Stud

Sole plate

Joist header

Sub-floor

Joist

c) Detail of platform frame at foundation.

Platform frame

Stud

Sole plate

Subfloor

Header joist

Cap plate

Top plate

Stud

Joist

(c)

d) Detail of platform frame at second floor.

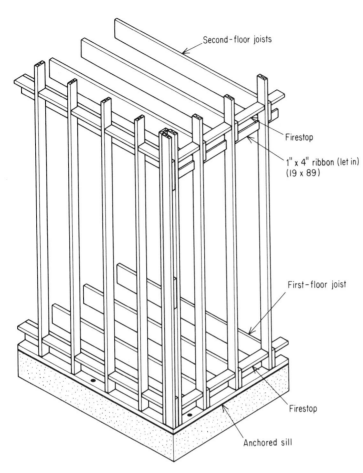

Second-floor joists

Firestop

1" x 4" ribbon (let in)
(19 x 89)

First-floor joist

Firestop

Anchored sill

a) Balloon frame.

FIGURE 5–2

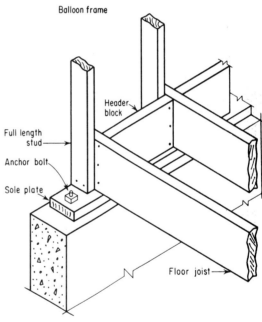

Balloon frame

Header block

Full length stud

Anchor bolt

Sole plate

Floor joist

b) Detail of balloon frame at foundation.

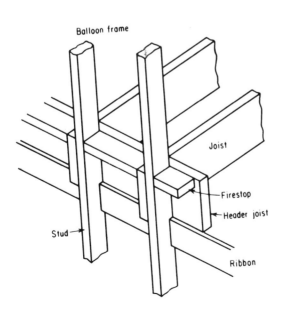

Balloon frame

Joist

Firestop

Header joist

Stud

Ribbon

c) Detail of balloon frame at second floor.

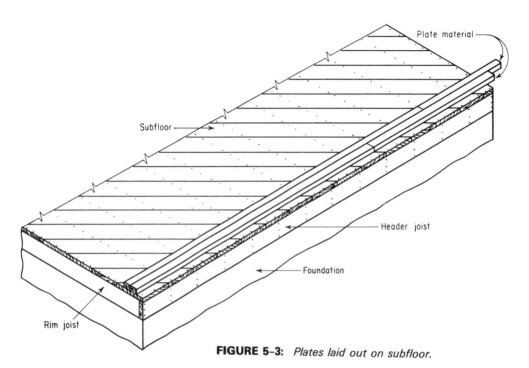

FIGURE 5-3: *Plates laid out on subfloor.*

Figure 5-4: *Center line of opening.*

FIGURE 5-5: *Locating trimmer and regular stud.*

PLATFORM FRAME WALL CONSTRUCTION

Plate Layout

The first step in the construction of a wall is to *lay out* on the sole plate—the base for a platform frame—the exact location of each member required in that wall. They will include *corner posts, regular studs, opening trimmers, partition junctions,* and *cripple studs.* Since the location of the members on the top plate will be identical, the two plates are laid out together. Proceed as follows:

1. Pick out straight stock for the plates; two or more pieces are probably required for each one. Set them out across the subfloor, side by side (see Fig. 5-3).

2. Square off one end of each plate, cut them to length; if more than one piece is needed, the end joints should come in the center of a stud position.

3. Check the plans for the locations of the centers of door and window openings. Mark these on the plates by a centerline, as illustrated in Fig. 5-4.

4. Measure on each side of the centerline one-half the width of the opening, and mark for trimmer studs outside these points (see Fig. 5-5). Use a "T" to indicate trimmer.

5. Outside the trimmer, lay off the position of a full-length stud (see Fig. 5-6). Use an "X" to indicate full-length stud.

6. Check the plans for positions of partitions that intersect the wall; mark the position and framing needed (see Fig. 5-7).

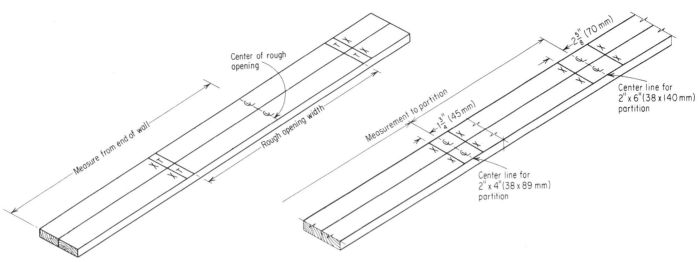

FIGURE 5-6: *Layout for rough opening.*

FIGURE 5-7: *Layout for partition junctions.*

7. Locate the positions of all the regular centers, making sure the first space is ¾ in. (20 mm) smaller so panel sheathing joints will occur on studs [see Fig. 5-8(a)]; stud locations in an opening are marked with a "C" to indicate cripples.

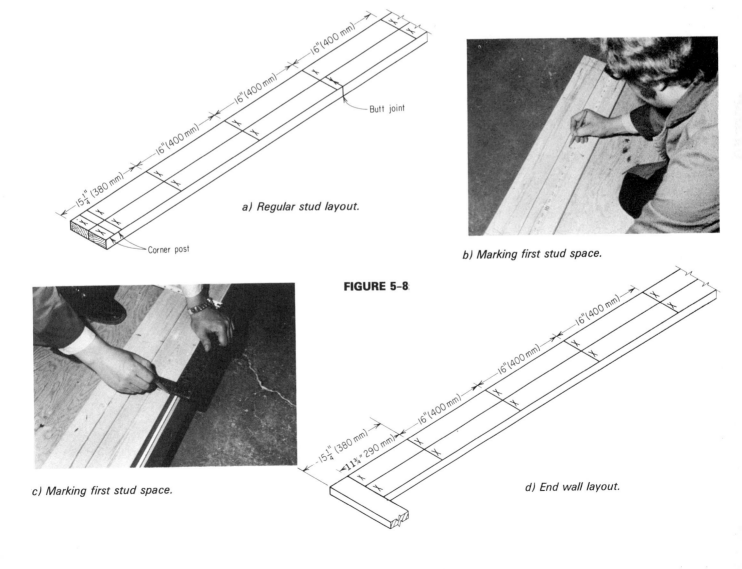

a) Regular stud layout.

b) Marking first stud space.

FIGURE 5-8

c) Marking first stud space.

d) End wall layout.

End wall frames must fit snugly between the sidewalls when they are in place, and therefore the plates must be 7 in. (178 mm) shorter [two plate widths, for a 2 × 4 (38 × 89 mm) frame] than the end wall dimension. Measure 4¼ in. (108 mm) [width of plate + one-half the thickness of the stud less than 16 in. (400 mm)] from the end of the plate to the edge of the first stud [see Fig. 5–8(d)]. The rest of the studs will be on regular centers from that point.

Size and Spacing of Studs

The size and spacing of studs is determined by the *location* of the wall, its *height*, and the *load* which it will have to support. Table 5–1 indicates the size and spacing required for studs in both exterior and interior walls for a variety of loading conditions.

TABLE 5-1: *Size and spacing of studs*

Type of wall	Supported loads (including dead loads)	Min. stud size [in. (mm)]	Max. stud spacing [in. (mm)]	Max. unsupported height [ft-in. (m)]
Interior	No load	2 × 2 (38 × 38)	16 (400)	7–10 (2.4)
		2 × 4 (38 × 89) (flat)	16 (400)	11–10 (3.6)
	Limited attic storage[a]	2 × 3 (38 × 64)	24 (600)	9–10 (3.0)
		2 × 4 (38 × 89)	24 (600)	11–10 (3.6)
	Full attic storage[b] plus 1 floor, or roof load plus 1 floor, or limited attic storage[a] plus 2 floors	2 × 4 (38 × 89)	16 (400)	11–10 (3.6)
	Full attic storage[b] plus 2 floors or roof load plus 2 floors	2 × 4 (38 × 89)	12 (300)	11–10 (3.6)
		2 × 6 (38 × 140)	16 (400)	13–9 (4.2)
	Roof load, full attic storage[b] limited attic storage plus 1 floor	2 × 4 (38 × 89)	24 (600)	11–10 (3.6)
		2 × 3 (38 × 64)	16 (400)	7–10 (2.4)
	Full attic storage[b] plus 3 floors or roof load plus 3 floors	2 × 6 (38 × 140)	12 (300)	13–9 (4.2)
Exterior	Roof with or without attic storage	2 × 3 (38 × 64)	16 (400)	7–10 (2.4)
		2 × 4 (38 × 89)	24 (600)	9–10 (3.0)
	Roof with or without attic storage plus 1 floor	2 × 4 (38 × 89)	16 (400)	9–10 (3.0)
		2 × 6 (38 × 140)	24 (600)	9–10 (3.0)
Exterior	Roof with or without attic storage plus 2 floors	2 × 4 (38 × 89)	12 (300)	9–10 (3.0)
		2 × 6 (38 × 140)	16 (400)	11–10 (3.6)
	Roof with or without attic storage plus 3 floors	2 × 6 (38 × 140)	12 (300)	5–11 (1.8)

[a]Applies to attics not accessible by a stairway.
[b]Applies to attics accessible by a stairway.

CUTTING THE FRAME

The assembly of the frame will be made much simpler if all the pieces have been cut accurately first. Read the drawings carefully and find the length of studs. Check sole and top plate layouts for the number of full-length studs required, taking into account the extras needed at corners, partition junctions, wall openings, etc. Check the plans for the width and height of all exterior openings and for the height of the top of the openings above the subfloor. See Chapter 11 for details of window frame sizes. For standard wood-sash windows, the rough opening should be ¾ in. (20 mm) larger in height and width than the frame. For other types of windows, be sure to determine the proper rough opening sizes required. Rough openings for wooden door frames should also be ¾ in. (20 mm) larger than the frame. See Chapter 11 for details on door-frame sizes.

Count the number of trimmers, lintels, rough sills, and cripples required from the layout. With rough opening sizes and heights above floor, it is not difficult to calculate the length and cut these ready for assembly.

Stud Lengths

Regular. The length of regular studs, corner post members, and partition junction studs will be the finished wall height plus clearance [usually 1 in. (25mm)] less the thickness of *three* plates. For example, if the finished wall height is to be 96 in. (2400 mm), the regular stud length will be 96 in. + 1 in. − (3 × 1½) = 92½ in. [2400 + 25 − (3 × 38) = 2311 mm].

Trimmers. The length of a trimmer stud will be the height of the top of the opening above the floor less the thickness of *one* plate.

Lower cripple studs. The length of the lower cripple stud will be the height of the bottom of the opening above the floor less the thickness of *two* plates.

Upper cripple studs. Although it is possible to calculate the length of the upper cripple studs, it is usually more satisfactory, from a practical standpoint, to measure the distance between lintel and top plate after the remainder of the members in the wall have been assembled (see Fig. 5-9).

Rough Sill Length

The length of the rough sill (see Fig. 5-9) will be equal to the width of the rough opening.

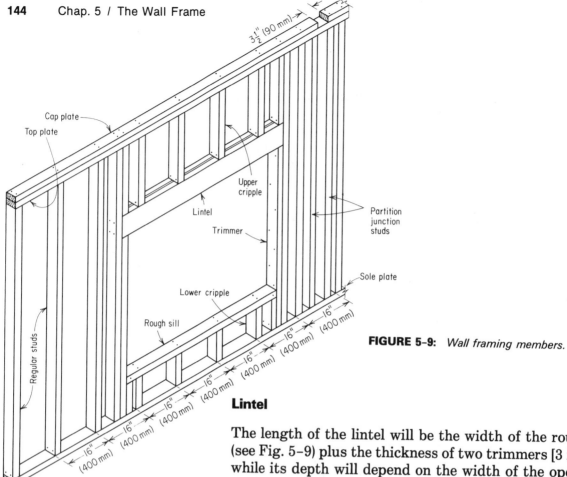

FIGURE 5-9: *Wall framing members.*

FIGURE 5-10a: *Using nailer to assemble lintel. (Courtesy Bostitch Textron)*
b) Double lintel for 2 × 4 (38 × 89 mm) wall frame.
c) Lintel for a 2 × 6 (140 mm) wall frame.

Lintel

The length of the lintel will be the width of the rough opening (see Fig. 5–9) plus the thickness of two trimmers [3 in. (76 mm)], while its depth will depend on the width of the opening. Table 5–2 gives the depth of lintel required and maximum spans allowable for interior and exterior walls under a variety of loading conditions.

The thickness of the lintel will be equal to the thickness of the wall frame and must be made up accordingly. For a 2 × 4 in. (89-mm) frame, it is usually made of two pieces of 1½-in. (38-mm) material nailed together, with the proper thickness of spacer

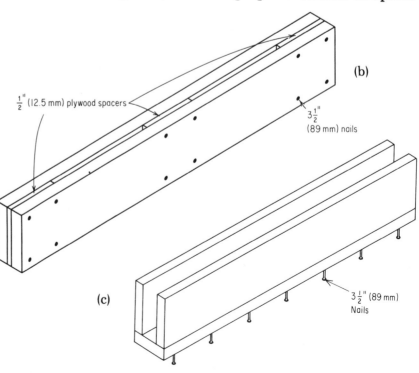

TABLE 5-2: *Maximum spans for wood lintels*

Location of lintels	Supported loads including dead loads and ceiling	Depth of lintel		Maximum allowable span	
		in.	*mm*	*ft–in.*	*m*
Interior walls	Limited attic storage	4	89	4–0	1.22
		6	140	6–0	1.83
		8	184	8–0	2.44
		10	235	10–0	3.05
		12	286	12–6	3.81
	Full attic storage or roof load or limited attic storage plus 1 floor	4	89	2–0	0.61
		6	140	3–0	0.91
		8	184	4–0	1.22
		10	235	5–0	1.52
		12	286	6–0	1.83
	Full attic storage plus 1 floor, or roof load plus 1 floor, or limited attic storage plus 2 or 3 floors	4	89	—	—
		6	140	2–6	0.76
		8	184	3–0	0.91
		10	235	4–0	1.22
		12	286	5–0	1.52
	Full attic storage plus 2 or 3 floors or roof load plus 2 or 3 floors	4	89	—	—
		6	140	2–0	0.61
		8	184	3–0	0.91
		10	235	3–6	1.10
		12	286	4–0	1.22
Exterior walls	Roof with or without attic storage	4	89	3–8	1.12
		6	140	5–6	1.68
		8	184	7–4	2.23
		10	235	9–2	2.75
		12	286	11–0	3.35
	Roof with or without attic storage plus 1 floor	4	89	1–10	0.33
		6	140	4–7	1.23
		8	184	6–5	1.84
		10	235	7–4	2.23
		12	286	8–3	2.52
	Roof with or without attic storage plus 2 or 3 floors	4	89	1–10	0.33
		6	140	3–8	1.12
		8	184	5–6	1.68
		10	235	6–5	1.84
		12	286	7–4	2.23

(Courtesy National Research Council.)

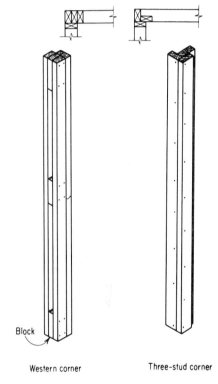

Block

Western corner   Three-stud corner

FIGURE 5–11: *Corner posts.*

between them, as illustrated in Fig. 5–10. For a 2 × 6 (140-mm) frame, two pieces of material could be used, spaced with a 2 × 6 nailed to the bottom edge [see Fig. 5–10(c)].

ASSEMBLING THE FRAME

When the pieces have all been cut, the process of assembly can begin, with the first step being to make up the corner posts and lintels. A number of types of corner posts are used, two of which are illustrated in Fig. 5–11, but for assembly on a flat surface, a partial corner, made by omitting the unit which is part of the end wall frame, is the most practical arrangement. The third member of the corner will be supplied by the end wall frame (see Fig. 5–12). Lintels are normally constructed as illustrated in Fig. 5–10.

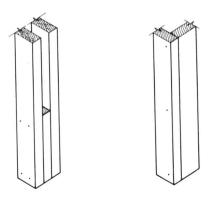

FIGURE 5–12: *Partial corner posts.*

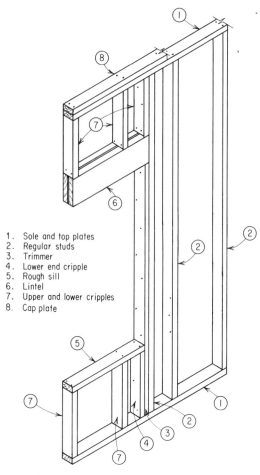

1. Sole and top plates
2. Regular studs
3. Trimmer
4. Lower end cripple
5. Rough sill
6. Lintel
7. Upper and lower cripples
8. Cap plate

FIGURE 5-13: *Order of wall member assembly.*

The assembly of the wall frames can now begin, starting with one of the sidewalls. If the project is a platform frame, the wall will be assembled on the subfloor and raised into position in one or more units, depending on the length. Proceed as follows:

1. Lay out the sole and top plates on edge, spaced stud length apart, with the laid-out faces toward each other. The sole plate should be near the edge of the floor frame on which it will rest.

2. Place a full-length stud at each position indicated on the layout and a corner post at each end.

3. Nail the sole and top plates to the ends of these with two 3½-in. (89-mm) nails in each end.

4. Nail the trimmers to the full-length studs on each side of openings.

5. Nail the two end lower cripple studs to the inside of the trimmers, as shown in Fig. 5-13.

6. Lay the rough sill on the ends of the two lower cripples and nail.

7. Set the lintels on the ends of the trimmers and endnail through the full-length studs.

8. Set the remainder of the cripple studs, upper and lower, in position and nail. The upper cripples must be toenailed to the top edge of the lintel.

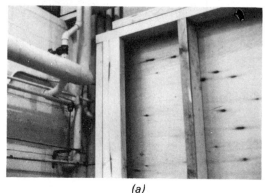

(a)

(b)

FIGURE 5-14: *a) Gap for tying corner together.*
b) Gap for partition connection.

9. Nail on the cap plate. It must be kept back 3½ in. (89 mm) from the ends of the wall and a gap left open at partition junctions (see Fig. 5-14).

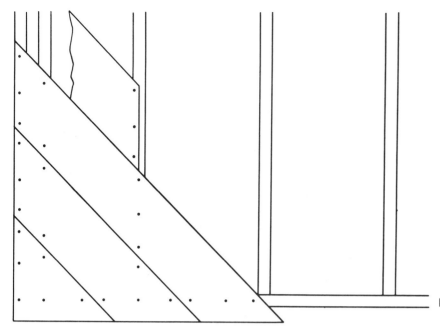

FIGURE 5-15: *Diagonal board sheathing.*

WALL SHEATHING

In the case of a platform frame, it is usually advantageous to sheathe the frame before standing it in place. Begin by making sure that the frame is square. Check the diagonal distances from the bottom corners to the opposite top ones. They should be exactly the same. Tack the wall frame to the subfloor to keep it straight and square.

There are a number of materials available which may be used as sheathing. They include *shiplap, common boards, plywood, exterior fiber board, exterior gypsum board, particle board,* and *insulating sheathing.* Sheathing-grade plywood is probably the most widely used, but all the others do a satisfactory job, and, in an age of energy conservation, styrofoam sheathing is particularly attractive because of its high insulation value.

Common boards must be applied diagonally for best results. Although this method of application uses more material, it gives much greater rigidity than horizontal application. Boards should be applied at an angle of 45° (see Fig. 5-15).

The angle may be obtained by laying the square on a piece of material with the same two figures, e.g., 300, on both tongue and blade touching the near edge (see Fig. 5-16). Mark on either tongue or blade for the angle.

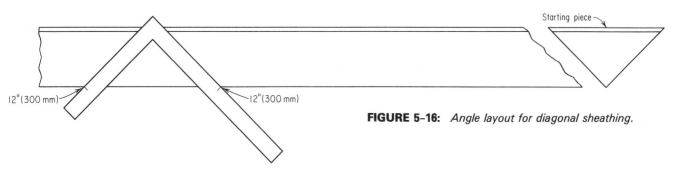

12"(300 mm)   12"(300 mm)   Starting piece

FIGURE 5-16: *Angle layout for diagonal sheathing.*

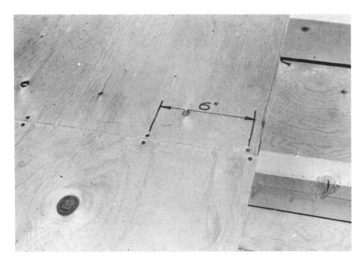
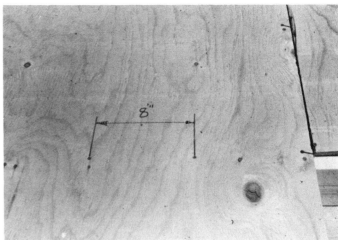

FIGURE 5-17:   *Nail spacing and plywood joint staggering.*

FIGURE 5-18:   *Space between sheets.*

Use two 2¼-in. (60-mm) nails for 6-in. (140-mm) boards and three nails for wider material on each stud (see Fig. 5-15).

Plywood, minimum thickness ⁵⁄₁₆-in. (7 mm), should be applied with the long dimension horizontal for greatest rigidity. Check the accuracy of stud centers when this type of sheathing is used, for the sheets are exactly 96 in. (2400 mm) long. The second course should have the end joint offset from the one below (see Fig. 5-17). Use 2-in. (50-mm) coated or spiral nails, spaced 6 in. along the edge of the sheet and 8 in. on intermediate supports (see Fig. 5-17). A space ¹⁄₁₆ in. wide should be left along panel joints to allow for expansion (see Fig. 5-18). Some framers let the sheathing overhang over the bottom plate to allow for easy alignment (see Fig. 5-19). When the sheathing does not hang over the bottom plate, polyethelene is used to seal the joint (see Fig. 5-20).

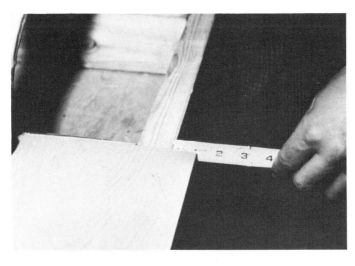

FIGURE 5-19:   *Overhang at bottom of wall.*

FIGURE 5-20:   *Poly for sealing joint at wall and floor.*

Figure 5-21: *Fiber board sheathing.*

Exterior fiber board, a material made from shredded wood fibers impregnated with asphalt, is commonly made in 4 × 8 ft (1200 × 2400 mm) sheets, $\frac{7}{16}$-in. (11 mm) thick. The panels are nailed with $1\frac{3}{4}$-in. (44-mm) broad-headed nails or $1\frac{1}{2}$-in. (38-mm) staples, spaced 6-in. (150 mm) along the edges and 8-in. (200 mm) along intermediate supports. Vertical joints should be offset, similar to plywood (see Fig. 5-21).

Exterior gypsum board is made with a core of gypsum encased in asphalt-impregnated paper. The sheets are 2 × 8 ft × $\frac{1}{2}$-in. thick (600 × 1200 × 12.5 mm), with the long edges having a tongue and groove horizontal joint. The sheets are applied with the long dimension horizontal. Two-inch (50-mm) coated or spiral nails, spaced 6 in. (150 mm) apart on all studs, are used. All vertical joints must be staggered (see Fig. 5-22). Some building codes require diagonal bracing with this type of sheathing.

Particle board sheathing is made of wood particles bonded together with an adhesive, compressed into sheets 4 × 8 ft (1200 × 2400 mm). The panels range from $\frac{1}{4}$ to $1\frac{1}{2}$ in. (6-38 mm) thick, and the minimum thickness will conform to Table 5-3.

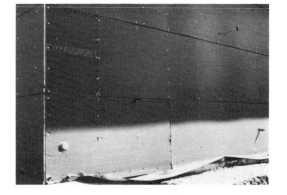

FIGURE 5-22: *Gypsum board exterior sheathing.*

TABLE 5-3: *Thickness of wall sheathing*

	Minimum thickness [in. (mm)]	
Type of sheathing	*With supports 16 in. (400 mm) o.c.*	*With supports 24 in. (600 mm) o.c.*
Lumber	$\frac{11}{16}$ (17.5)	$\frac{11}{16}$ (17.5)
Fiber board	$\frac{3}{8}$ (9.5)	$\frac{7}{16}$ (11)
Gypsum board	$\frac{3}{8}$ (9.5)	$\frac{1}{2}$ (12.5)
Plywood	$\frac{1}{4}$ (6)	$\frac{5}{16}$ (7.5)
Particle board	$\frac{1}{4}$ (6.35)	$\frac{5}{16}$ (8)

FIGURE 5-23: *Particleboard sheathing.*

FIGURE 5-24: *Styrofoam sheathing.*

Particle board is applied by the same method as plywood sheathing. Panels can be placed horizontally or vertically, and they will not be affected by the difference in strength that occurs when plywoods are placed with the grain parallel to studs as opposed to perpendicular (see Fig. 5-23).

Insulating exterior sheathing board is applied to the outside of exterior wall studs much in the same manner as conventional sheathing. Insulating sheathing will increase the "R" values of the wall to meet today's energy-saving requirements. Insulating sheathing includes such materials as styrofoam, fiberglass, and thermax board.

Styrofoam insulating sheathing is a synthetic material made from expanded polystyrene in rigid sheets with a smooth, high-density skin. It has a thermal resistance ("R" value) of 5 per inch (25 mm). The sheets are 2 × 8 ft (600 × 2400 mm), 1 or 1½ in. (25 or 38 mm) thick with butt edges, and are applied horizontally (see Fig. 5-24).

Fasteners may be broad-headed (roofing) nails, staples, or screws, with a spacing of 6 in. (150 mm) for nails with a ⁷⁄₁₆-in. head or staples with a ½-in. crown. The length of the fasteners should be a minimum of ¾ in. (19 mm) longer than the thickness of the sheathing.

The styrofoam should be covered with exterior building paper, over which any type of exterior finish, including wood or metal siding, stucco, or brick veneer may be applied.

Styrofoam will burn, and during shipping, storage, installation, and use it must not be exposed to open flame or other ignition source.

Fiberglass sheathing boards are composed of resin-bonded glass fibers faced with a durable air barrier. They will not rot even after continuous wetting and drying cycles and are not affected by fungi or insects. The glass fiber board is permeable to water vapor and will not trap moisture. The moisture-resistant facing has fair resistance to impact loads.

Glass fiber insulating board is attached to the studs with $\frac{7}{16}$-in. (11-mm) head diameter galvanized roofing nails. The nails should be $\frac{3}{4}$ in. (19 mm) longer than the thickness of the board. Attachment is required at spacings of 6 in. (150 mm) along edges and 12 in. (300 mm) along intermediate supports. If 1-in. (25-mm) head diameter nails or washers are used under the head, the spacing can be increased to 12 in. (300 mm) along the edges and 18 in. (450 mm) along intermediate supports.

Glass fiber sheathing is available in sheets 48 × 96 in. and 48 × 108 in. (1200 × 2400 and 1200 × 2700 mm). One-inch and 1½-in. (25- and 38-mm) thickness sheets are available. Thermal resistance at 75 °F (24 °C) mean temperature is 4.4 (RSI 0.77) for 1-in. (25-mm) thickness and 6.7 (RSI 1.18) for 1½-in. (38-mm) thickness (see Fig. 5–25).

Thermax sheathing consists of a glass-reinforced polyisocyanurate foam plastic core with aluminum foil faces. Since Thermax sheathing has a closed cell structure and aluminum faces that make an efficient barrier to moisture as well as heat flow, there is concern in colder climates for the potential of entrapped water vapor within stud spaces. A vent strip system is used for moisture relief along the top edge of the wall, allowing for the escape of water vapor (see Fig. 5–26). No special fastening is required. The boards are secured in place with galvanized roofing nails having ⅜-in. (9.5-mm) heads long enough to penetrate framing ¾-in. (19 mm). Thermax sheathing is not a structural panel, so wood let-ins or metal strap bracing is used. The panels are available in standard insulation board sizes of 4 × 8 ft (1200 × 2400 mm) and 4 × 9 ft (1200 × 2700 mm) and thicknesses varying from ½ in. (12.7 mm) to 2¼ in. (57.2 mm). Thermal resistance at 75 °F (24 °C) mean temperature ranges from 3.6 (RSI 0.63) for ½-in material to 16.2 (RSI 2.85) for 2¼-in. material (see Fig. 5–27).

Figure 5-25: *Glass fiber sheathing.*

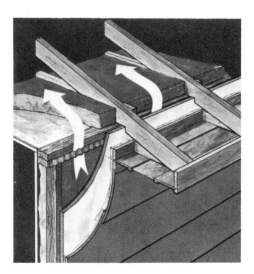

FIGURE 5-26: *Corrugated plastic vent strip along top plate.*

FIGURE 5-27: *Thermax sheathing.*

ERECTING THE FRAME

Usually a one-story wall frame can be raised by hand unless it is particularly long or heavy, in which case a crane of some sort must be used. Be sure that the bottom of the frame is close to the edge of the subfloor so that as the wall is raised the sheathing overhang does not catch on the floor (see Fig. 5-28).

FIGURE 5-28: *Raising a section of wall by hand.*

When the wall is in a vertical position, it must be tied in at the bottom and straightened and braced at the top. Adjust at the bottom until the overhang of the wall sheathing is snug against the face of the floor frame. Nail the sole plate down with two 3½-in. (89-mm) nails between each pair of studs. Be sure the nails reach a joist or the joist header. Plumb the corner studs and tie them in place with diagonal bracing. Walls that do not have the sheathing overhanging will be set flush with the edge of the floor frame.

To straighten the top, run a line very tightly along the top plate, tying it around the ends. Block the line out at the ends with a small piece of ¾-in. (19-mm) material (see Fig. 5-29). Take another piece of ¾-in. (19-mm) material and check at various points to see if the line is the correct distance from the plate. Push the top of the wall in or out, where necessary, to correct any points not in line. Brace by nailing one end of a board to the face of a stud at the top and the other end to a block fastened to the subfloor.

When one side wall has been erected and plumbed, the opposite one may be assembled and erected in the same manner, followed by the end walls, one by one. It is impractical to sheathe the end wall frames completely on the subfloor because the end

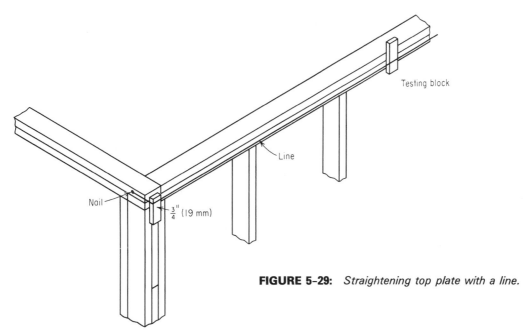

FIGURE 5-29: *Straightening top plate with a line.*

wall sheathing should extend over the corner posts, but the majority of it may be applied at this time. With the end wall erect and the sole plate nailed in position, the end studs can be nailed to the sidewall partial corner posts to make the post complete.

The remainder of the sheathing can now be applied, the wall plumbed and braced, and the end wall cap plate set in place. It will overlap the sidewall top plates to tie sidewalls and end walls securely together (see Fig. 5-30).

When a balloon frame is being erected, special problems are involved. Since the studs rest on the sill plate, there is no sole plate, and they must be held properly spaced by some other means until the frame is erected or until the sheathing is applied. Because of its height, it is usually impractical to try to raise a balloon frame wall, assembled on the floor, by hand. A crane is required to do the job efficiently, and if one is available, assembly on the floor is the best method. Otherwise, the studs have to be placed in position one at a time. If the wall is sheathed on the floor, the ribbon must be put in after the wall is erected. In nearly every case, scaffolding will have to be erected during the building of balloon frame walls.

FIGURE 5-30: *Plywood overlap at exterior corner.*

PARTITIONS

Partitions may be *bearing* or *nonbearing*. Bearing partitions carry part of the roof load, bear one end of the ceiling joists, or both, whereas nonbearing partitions simply enclose space and carry the finishing materials.

Partitions may be laid out, assembled, and erected in the same way as outside walls. The position of the sole plate on the subfloor is first marked by *snapping a chalk line*. This is done by driving a nail at each end of the partition position and stringing a chalked line tightly between them. Pull the line up in the center and let it snap against the floor. This will mark a straight line to which the partition sole plate can be set.

FIGURE 5-31: *Partition connection.*

Plates are laid out and studs and other members cut and assembled. Nail three blocks about 12 in. (300 mm) in length between the sidewall partition junction studs, one at the top and bottom and one in the center (see Fig. 5-31), after which the partition can be raised into place. Plumb the end studs and nail them to the partition blocks. Finally, add a cap plate, with its ends overlapping the sidewall top plate (see Fig. 5-32). Erect the longest partition first, then the cross partitions, and, finally, the partitions forming clothes closets, hallways, etc. Another method of providing backing at the junction of partition and exterior wall is to place horizontal blocking at 16 in. (400 mm) o.c. between regular studding (see Fig. 5-33).

One partition requiring special attention will be the one for the bathroom in which the main plumbing vent stack is located. In some cases, this one will have to be made 6 in. (140 mm) wide to accommodate the size of the stack 2 × 4 (38 × 89 mm) studs staggered on 2 × 6 (38 × 140 mm) plates (see Fig. 5-34) will provide the width, and by weaving an insulation blanket between the studs, some sound control in that wall is also possible. Of course, 2 × 6 (38 × 140 mm) studding may be used if desired.

Broom and clothes closet partitions may sometimes be framed with 2 × 2 (38 × 38) material in order to save space, but only if the partition is short or in a protected location.

FIGURE 5-32: *Cap plate overlapping for tie.*

FIGURE 5-33: *Partition junction*

FIGURE 5-34: *Staggered studs in wide partition.*

Sliding Door Opening

Framing to provide for doors which slide into the wall also requires special consideration. The thickness of the door must first be determined. This portion of the partition frame must be wide enough to accommodate the thickness of the door plus clearance and at least 1½-in. (38-mm) framing on each side (see Fig. 5–35). In some cases, the track may be wider than the thickness of the door.

Openings for Heating and Ventilating

Openings must be made in the frame to accommodate heating system stacks leading to registers, if the registers are of the wall type, as well as openings for air-conditioning and ventilating systems, where they are specified.

If the register stack will fit between a pair of regular studs, all that is necessary is to cut the sole plate out between the studs, cut a hole in the subfloor, and put in a header between the studs at the required height [see Fig. 5–36(a)]. If a wider space is needed, a header is placed in adjacent spaces. The sole plate is cut out and a hole made in the subfloor, as above [see Fig. 5–36(b)]. Placing a piece of screen over the opening during construction will keep debris out of the heating pipes [see Fig. 5–36(c)].

FIGURE 5–35: *Pocket door framing.*

(a)

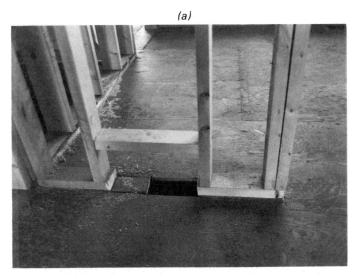

(b)

FIGURE 5–36:
a) Wall openings for registers.
b) Wall openings for ducts and registers.
c) Screen over opening.

(c)

a) Water line.

b) Tissue holder.

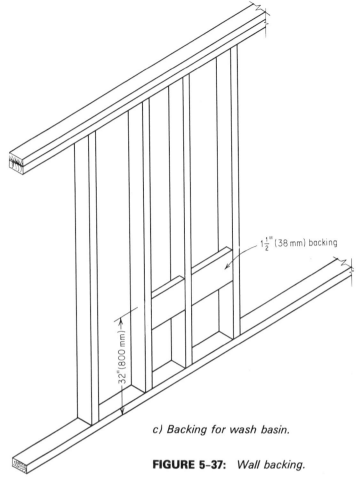

c) Backing for wash basin.

FIGURE 5-37:   *Wall backing.*

Wall Backing

Installations such as wall-type basins, cupboards, etc., which must be fastened to the wall must be provided with some type of solid backing. This may be done by letting a piece of 1½-in. (38-mm) material into the studs, as illustrated in Fig. 5-37).

LIGHT STEEL WALL FRAMING

Light steel wall framing members in several styles, depending on the manufacturer, are made from sheet steel, cold-formed, and consist basically of *studs* and *mounting channels* (see Fig. 5-38).

Studs are made in the form of channels, in two types, *load-bearing* and *non-load-bearing*. Load-bearing studs are made from approximately ⅛-in. (1.5-mm) material, with a 3⅝-in. (92-mm) web and a 1½-in. (38-mm) flange. Non-load bearing studs are made from lighter-gauge steel and in narrower widths. A special load-bearing *thermal stud* is also produced which has five rows of alternately spaced slots in the web to reduce the heat flow.

Mounting channels are made slightly wider to receive the ends of studs (see Fig. 5-38) and are used for top and bottom plates and rough-sill. No cap plate is required (see Fig. 5-38) since all members are spot-welded together.

FIGURE 5-38: *Prefabricated light steel wall frame. (Courtesy U.S. Steel Corp.)*

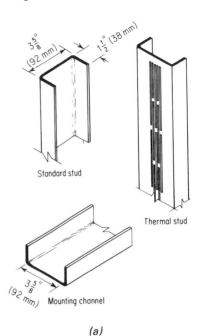

(a)

Lintels are usually made of two wood members, with their top edge enclosed by a length of mounting channel (see Fig. 5-39). Where openings do not reach to the bottom edge of the lintel, another length of mounting channel forms the top of the opening, with cripple studs between it and the lintel (see Fig. 5-38).

Wall frames may be prefabricated and brought to the site ready for erection (see Fig. 5-38), assembled on the subfloor, and raised into position, as in Fig. 5-39. Usually preassembled frames are sheathed before being raised with ⅜-in. (9-mm) plywood sheathing, fastened with steel nails driven by a pneumatic hammer (see Fig. 5-40).

FIGURE 5-40: *Nailing with pneumatic hammer.*

(b)

FIGURE 5-39:
a) Light steel framing members.
b) Steel wall frame raised into position.
(Courtesy Steel Co. of Canada)

FIGURE 5-41: *U-clips for roof members.*
(Courtesy U.S. Steel Corp.)

FIGURE 5-42: *Interior steel partitions.*

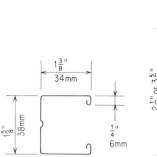

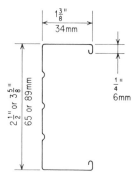

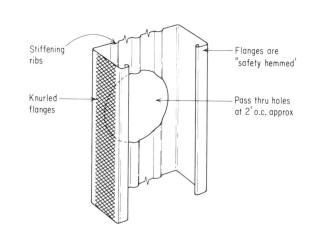

FIGURE 5-43: *Nonload-bearing studs.*

FIGURE 5-44: *Placing electrical wires in metal studs.*

Wall frames are anchored to concrete floors with power-actuated pins, placed 24 in. (600 mm) o.c. Screws are used when fastening to wooden floors. Rafters or trusses are anchored over supports to the top plate with *U-clips*, which are welded to the frame and bolted to the roof members (see Fig. 5-41).

The lighter, non-load-bearing framing material used for interior partitions provides a considerable saving of floor space (see Fig. 5-42). Non-load-bearing studs measure 1⅜ × 3⅝ in. (35 × 92 mm) with a metal thickness of 0.21 in. (53 mm). Studs are also available in 2½- and 1⅝-in. (64- and 40-mm) widths for further space saving (see Fig. 5-43). Openings in the studs are at 24-in. (600-mm) centers for easy placement of electrical wires. Electrical clips are used to protect unarmored wires (see Fig. 5-44).

Mounting channels are tapered for better grip of the studs (see Fig. 5-45) and are fastened to concrete floors with power-actuated pins and with screws to wooden floors (see Fig. 5-46).

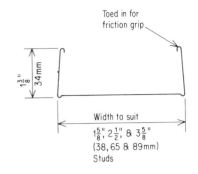

Toed in for
friction grip

$1\frac{3}{8}$" 34 mm

Width to suit
$1\frac{5}{8}$", $2\frac{1}{2}$" & $3\frac{5}{8}$"
(38, 65 & 89 mm)
Studs

FIGURE 5-45: *Mounting channel.*

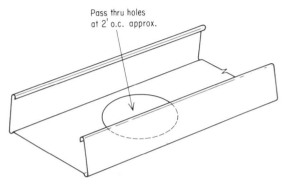

Pass thru holes
at 2' o.c. approx.

When wood trim or base is to be fastened to metal framing, a wood runner is placed under the mounting channel. Studs are fastened to the channel with screws or a crimping tool (see Fig. 5-47). The studs are easily placed in the channel with a twisting action (see Fig. 5-47). Studs should be cut in lengths ¼ in. (6 mm) shorter than the wall height to allow for variations in height (see Fig. 5-48). Mounting channels are fastened in the desired location first, and then the studs are placed individually into the channel (see Fig. 5-47).

(a)

(b)

FIGURE 5-47: *a) Crimping tool.
b) Erecting interior wall frame.
(Courtesy Steel Co. of Canada)*

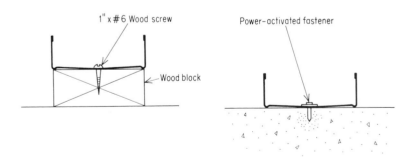

1" x #6 Wood screw

Wood block

Power-activated fastener

FIGURE 5-46: *Fastening mounting channel.*

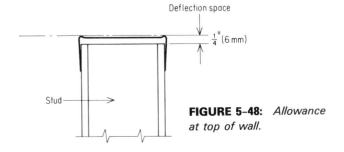

Deflection space

$\frac{1}{4}$" (6 mm)

Stud

FIGURE 5-48: *Allowance
at top of wall.*

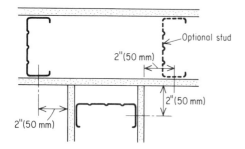
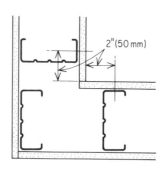
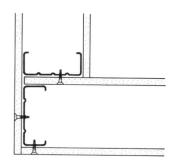

FIGURE 5-49: *Partition connections.*

Wall junctions (intersections and corners) are framed as illustrated in Fig. 5-49. Studs are sometimes placed 2 in. (50 mm) away from the junctions to allow for easy access of screw guns when attaching gypsum board facing. Headers for door openings are cut 12 in. (300 mm) longer than the rough opening to allow for easy attachment (see Fig. 5-50). When wooden frames are to be installed in the opening, wood bucks will facilitate easier installation of the frame (see Fig. 5-51).

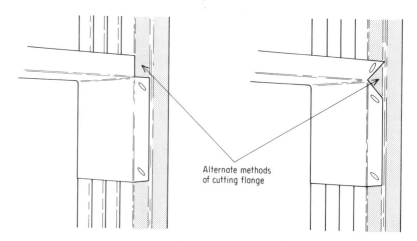

FIGURE 5-50: *Header for door opening.*

FIGURE 5-51: *Wood buck for installation of frame.*

REVIEW QUESTIONS

5-1. There are three basic differences between a balloon frame and a platform frame. Outline these differences briefly.

5-2. What is the purpose of firestops in a wall frame?

5-3. Explain why it is important to have wall backing at some locations in the wall frame.

5-4. Give two reasons for the use of cap plates in wood wall frames.

5-5. The frame for a given window is 62½ in. (1560 mm) wide and 33½ in. (835 mm) high. If 2 × 10s (38 × 235 mm) are used for the lintel and they are placed directly below the top plate, calculate the length of each of the following members, if total wall height is 8'-1" (2425 mm).

    (a) Window header

    (b) Rough sill

    (c) Trimmer

    (d) Bottom cripple stud

5-6. If board sheathing is used on the exterior of a wall frame, why should it be applied diagonally?

5-7. Explain why let-in diagonal bracing may be required when fiber board or gypsum board exterior sheathing is being used.

5-8. Using the following information, lay out a sole plate for an end wall 16 ft (4.85 m) in length:

    (a) Studs 16 in. (400 mm) o.c.

    (b) Exterior door 34 in. (910 mm) × 80 in. (2030 mm) centered 5 ft (1500 mm) from left end

    (c) Window frame 32½ × 45½ in. (800 × 1140 mm), centered 11 ft (3300 mm) from left end

    (d) Partition junction (2 × 4) centered 8 ft, 4 in. (2500 mm) from left end

THE CEILING FRAME

The ceiling frame consists of those members which will carry the materials forming the ceiling immediately below an attic which is not accessible by a stairway. On other levels, the ceiling materials are carried by floor joists.

JOIST SIZES

The size of members used for ceiling joists depends on (1) the span and spacing of the members, (2) the type of material used for the ceiling, and (3) the species and grade of lumber used (see Tables 6-1a and 6-1b).

CUTTING JOISTS

The length of the joists will depend on how they meet over the bearing partition. If two joists butt together over the partition (see Fig. 6-1), they must be cut to exact length from outside of frame to the center of the partition. If they lap one another over the partition, exact lengths are not required (see Fig. 6-2).

At its outer end, the joist must be cut to conform to the rise of the rafters (see Fig. 6-3). To make this cut properly, two things must be known: (1) the *rise* of the rafters (see Fig. 7-3, page 179) and (2) the height of the back of the rafter above the plate (see Fig. 6-4). Having determined these, proceed as follows:

FIGURE 6-1: *Ceiling joists butted over bearing partition.*

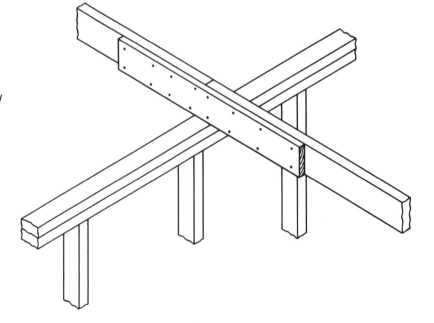

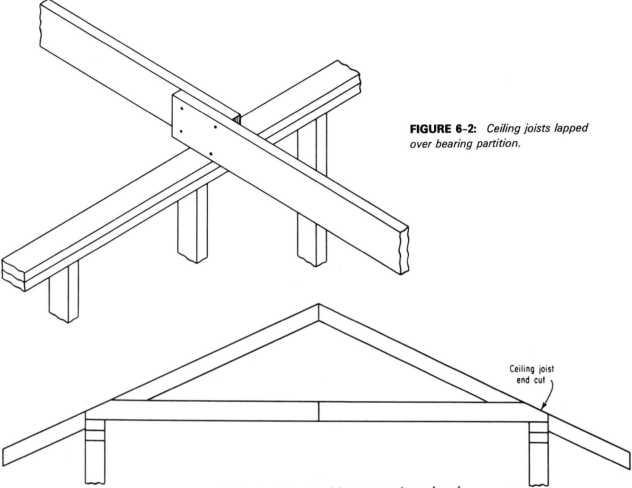

FIGURE 6-2: *Ceiling joists lapped over bearing partition.*

Ceiling joist end cut

FIGURE 6-3: *End of ceiling joists cut to slope of roof.*

1. Check to see which is the crown edge of the stock and make sure that it becomes the *upper* edge.

2. Measure up from the bottom edge of the joist, at the end, a distance equal to the height of the back of the rafter (see Fig. 6-5) and mark point A.

3. Lay the square on the stock with the unit rise on the tongue and 12 in. (250) (unit run) on the blade touching the top edge and the edge of the blade passing through A.

4. Mark on the blade and cut the joist end to that line.

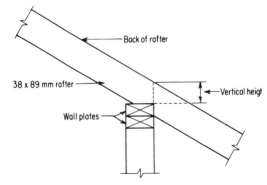

Back of rafter

38 x 89 mm rafter

Vertical height

Wall plates

FIGURE 6-4: *Height of back of rafter above plate.*

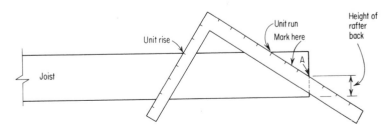

Unit run
Mark here
Unit rise
Height of rafter back
Joist
A

FIGURE 6-5: *Layout for ceiling joist end cut.*

TABLE 6-1a: *Maximum spans for ceiling joists—attic not accessible by a stairway (live load 10 lb/ft²)*

			Live load 10 lb/ft²											
			Gypsum board ceiling, joist spacing						Other ceilings, joist spacing					
			12 in.		16 in.		24 in.		12 in.		16 in.		24 in.	
Species	Grade	Nominal size (in.)	ft	in.	ft	in.	ft	in.	ft	in.	ft	in.	ft	in.
Douglas fir-larch	Select structural	2 × 4	11	6	10	5	9	1	13	2	11	11	10	5
		2 × 6	18	1	16	5	14	4	20	8	18	10	16	5
		2 × 8	23	10	21	8	18	11	27	4	24	10	21	8
		2 × 10	30	5	27	8	24	2	34	10	31	8	27	8
		2 × 12	37	0	33	8	29	4	42	5	38	6	33	8
	No. 1	2 × 4	11	6	10	5	9	1	13	2	11	11	10	5
		2 × 6	18	1	16	5	14	4	20	8	18	10	16	5
		2 × 8	23	10	21	8	18	11	27	4	24	10	21	8
		2 × 10	30	5	27	8	24	2	34	10	31	8	27	8
		2 × 12	37	0	33	8	29	4	42	5	38	6	33	8
	No. 2	2 × 4	11	1	10	1	8	10	12	8	11	7	10	1
		2 × 6	17	6	15	10	13	10	20	0	18	2	15	0
		2 × 8	23	0	20	11	18	3	26	4	23	11	19	9
		2 × 10	29	5	26	8	23	4	33	8	30	7	25	2
		2 × 12	35	9	32	6	28	4	40	11	37	2	30	8
	No. 3	2 × 4	10	8	9	5	7	8	10	11	9	5	7	8
		2 × 6	16	1	13	11	11	4	16	1	13	11	11	4
		2 × 8	21	3	18	4	15	0	21	3	18	4	15	0
		2 × 10	27	1	23	5	19	2	27	1	23	5	19	2
		2 × 12	32	11	28	6	23	3	32	11	28	6	23	3
Hemlock-fir	Select structural	2 × 4	10	10	9	10	8	7	12	5	11	3	9	10
		2 × 6	17	1	15	6	13	6	19	6	17	9	15	6
		2 × 8	22	6	20	5	17	10	25	9	23	5	20	5
		2 × 10	28	8	26	1	22	9	32	10	29	10	26	1
		2 × 12	34	11	31	9	27	8	40	0	36	4	31	9
	No. 1	2 × 4	10	10	9	10	8	7	12	5	11	3	9	10
		2 × 6	17	1	15	6	13	6	19	6	17	7	14	5
		2 × 8	22	6	20	5	17	10	25	9	23	3	19	0
		2 × 10	28	8	26	1	22	9	32	10	29	8	24	3
		2 × 12	34	11	31	9	27	8	40	0	36	1	29	5
	No. 2	2 × 4	10	6	9	6	8	4	12	0	10	11	8	11
		2 × 6	16	6	14	11	12	9	18	1	15	8	12	9
		2 × 8	21	9	19	9	16	10	23	11	20	8	16	10
		2 × 10	27	9	25	2	21	6	30	6	26	5	21	6
		2 × 12	33	9	30	8	26	2	37	1	32	1	26	2
	No. 3	2 × 4	9	2	7	11	6	5	9	2	7	11	6	5
		2 × 6	13	9	11	11	9	9	13	9	11	11	9	9
		2 × 8	18	2	15	9	12	10	18	2	15	9	12	10
		2 × 10	23	2	20	1	16	5	23	2	20	1	16	5
		2 × 12	28	2	24	5	19	11	28	2	24	5	19	11
Tamarack-eastern hemlock	Select structural	2 × 4	10	4	9	4	8	2	11	10	10	9	9	4
		2 × 6	16	3	14	9	12	11	18	7	16	11	14	9
		2 × 8	21	5	19	5	17	0	24	6	22	3	19	5
		2 × 10	27	4	24	10	21	8	31	4	28	5	24	10
		2 × 12	33	3	30	3	26	5	38	1	34	7	30	3
	No. 1	2 × 4	10	4	9	4	8	2	11	10	10	9	9	4
		2 × 6	16	3	14	9	12	11	18	7	16	11	14	9
		2 × 8	21	5	19	5	17	0	24	6	22	3	19	5
		2 × 10	27	4	24	10	21	8	31	4	28	5	24	10
		2 × 12	33	3	30	3	26	5	38	1	34	7	30	3

TABLE 6–1a: *(continued)*

Species	Grade	Nominal size (in.)	Live load 10 lb/ft²											
			Gypsum board ceiling, joist spacing						Other ceilings, joist spacing						
			12 in.		16 in.		24 in.		12 in.		16 in.		24 in.	
			ft	in.	ft	in.	ft	in.	ft	in.	ft	in.	ft	in.
Spruce-pine-fir	No. 2	2 × 4	10	0	9	1	7	11	11	5	10	4	9	1
		2 × 6	15	8	14	3	12	5	17	11	16	4	14	1
		2 × 8	20	8	18	9	16	5	23	8	21	6	18	7
		2 × 10	26	5	24	0	20	11	30	3	27	5	23	8
		2 × 12	32	1	29	2	25	6	36	9	33	5	28	10
	No. 3	2 × 4	9	7	8	8	7	3	10	3	8	10	7	3
		2 × 6	15	0	12	11	10	7	15	0	12	11	10	7
		2 × 8	19	9	17	1	13	11	19	9	17	1	13	11
		2 × 10	25	2	21	10	17	10	25	2	21	10	17	10
		2 × 12	30	8	26	7	21	8	30	8	26	7	21	8
	Select structural	2 × 4	10	2	9	3	8	1	11	8	10	7	9	3
		2 × 6	16	0	14	7	12	9	18	4	16	8	14	7
		2 × 8	21	2	19	3	16	9	24	3	22	0	19	3
		2 × 10	27	0	24	6	21	5	30	11	28	1	24	6
		2 × 12	32	10	29	10	26	1	37	7	34	2	29	10
	No. 1.	2 × 4	10	2	9	3	8	1	11	8	10	7	9	3
		2 × 6	16	0	14	7	12	9	18	4	16	8	13	9
		2 × 8	21	2	19	3	16	9	24	3	22	0	18	2
		2 × 10	27	0	24	6	21	5	30	11	28	1	23	2
		2 × 12	32	10	29	10	26	1	37	7	34	2	28	2
	No. 2	2 × 4	9	10	8	11	7	10	11	3	10	3	8	7
		2 × 6	15	6	14	1	12	4	17	7	15	3	12	5
		2 × 8	20	5	18	7	16	3	23	3	20	1	16	5
		2 × 10	26	1	23	9	20	9	29	8	25	8	21	0
		2 × 12	31	9	28	10	25	2	36	1	31	3	25	6
	No. 3	2 × 4	9	2	7	11	6	5	9	2	7	11	6	5
		2 × 6	13	1	11	4	9	3	13	1	11	4	9	3
		2 × 8	17	4	15	0	12	3	17	4	15	0	12	3
		2 × 10	22	1	19	2	15	7	22	1	19	2	15	7
		2 × 12	26	11	23	3	19	0	26	11	23	3	19	0
Western cedar	Select structural	2 × 4	9	10	8	11	7	9	11	3	10	3	8	11
		2 × 6	15	6	14	1	12	3	17	8	16	1	14	1
		2 × 8	20	5	18	6	16	2	23	4	21	3	18	6
		2 × 10	26	0	23	8	20	8	29	10	27	1	23	8
		2 × 12	31	8	28	9	25	2	36	3	32	11	28	9
	No. 1	2 × 4	9	10	8	11	7	9	11	3	10	3	8	11
		2 × 6	15	6	14	1	12	3	17	8	16	1	14	1
		2 × 8	20	5	18	6	16	2	23	4	21	3	18	6
		2 × 10	26	0	23	8	20	8	29	10	27	1	23	8
		2 × 12	31	8	28	9	25	2	36	3	32	11	28	9
	No. 2	2 × 4	9	6	8	7	7	6	10	10	9	10	8	7
		2 × 6	14	11	13	7	11	10	17	1	15	3	12	5
		2 × 8	19	8	17	11	15	8	22	7	20	1	16	5
		2 × 10	25	2	22	10	19	11	28	9	25	8	21	0
		2 × 12	30	7	27	9	24	3	35	0	31	3	25	6
	No. 3	2 × 4	9	1	7	11	6	5	9	2	7	11	6	5
		2 × 6	13	9	11	11	9	9	13	9	11	11	9	9
		2 × 8	18	2	15	9	12	10	18	2	15	9	12	10
		2 × 10	23	2	20	1	16	5	23	2	20	1	16	5
		2 × 12	28	2	24	5	19	11	28	2	24	5	19	11

TABLE 6-1b: *Maximum spans for ceiling joists—attic not accessible by a stairway (meters) (live load 0.5 kN/m²)*

Lumber species	Grade	Size (mm)	Gypsum board or plastered ceiling, joist spacing			Other ceilings, joist spacing		
			300 mm	400 mm	600 mm	300 mm	400 mm	600 mm
Douglas fir-larch (N)	Select structural	38 × 89	3.40	3.09	2.69	3.89	3.53	3.09
		38 × 140	5.34	4.85	4.24	6.11	5.55	4.85
		38 × 184	7.04	6.40	5.59	8.06	7.32	6.40
		38 × 235	8.98	8.16	7.13	10.28	9.34	8.16
		38 × 286	10.93	9.93	8.67	12.51	11.36	9.93
	No. 1	38 × 89	3.40	3.09	2.69	3.89	3.53	3.09
		38 × 140	5.34	4.85	4.24	6.11	5.55	4.85
		38 × 184	7.04	6.40	5.59	8.06	7.32	6.40
		38 × 235	8.98	8.16	7.13	10.28	9.34	8.16
		38 × 286	10.93	9.93	8.67	12.51	11.36	9.93
	No. 2	38 × 89	3.28	2.98	2.60	3.76	3.41	2.98
		38 × 140	5.16	4.69	4.10	5.91	5.37	4.49
		38 × 184	6.81	6.18	5.40	7.79	7.08	5.92
		38 × 235	8.68	7.89	6.89	9.94	9.03	7.56
		38 × 286	10.56	9.60	8.38	12.09	10.99	9.19
	No. 3	38 × 89	3.15	2.82	2.31	3.26	2.82	2.31
		38 × 140	4.81	4.16	3.40	4.81	4.16	3.40
		38 × 184	6.34	5.49	4.48	6.34	5.49	4.48
		38 × 235	8.09	7.01	5.72	8.09	7.01	5.72
		38 × 286	9.84	8.52	6.96	9.84	8.52	6.96
Hem-fir (N)	Select structural	38 × 89	3.27	2.97	2.60	3.75	3.40	2.97
		38 × 140	5.15	4.67	4.08	5.89	5.35	4.64
		38 × 184	6.78	6.16	5.38	7.77	7.06	6.12
		38 × 235	8.66	7.87	6.87	9.91	9.00	7.81
		38 × 286	10.53	9.57	8.36	12.06	10.95	9.50
	No. 1	38 × 89	3.27	2.97	2.60	3.75	3.40	2.95
		38 × 140	5.15	4.67	4.08	5.89	5.35	4.64
		38 × 184	6.78	6.16	5.38	7.77	6.96	5.68
		38 × 235	8.66	7.87	6.87	9.91	8.89	7.25
		38 × 286	10.53	9.57	8.36	12.06	10.81	8.82
	No. 2	38 × 89	3.16	2.87	2.51	3.62	3.28	2.67
		38 × 140	4.97	4.51	3.87	5.47	4.74	3.87
		38 × 184	6.55	5.95	5.10	7.21	6.25	5.10
		38 × 235	8.36	7.60	6.51	9.21	7.97	6.51
		38 × 286	10.17	9.24	7.92	11.20	9.70	7.92
	No. 3	38 × 89	2.81	2.43	1.98	2.81	2.43	1.98
		38 × 140	4.15	3.59	2.93	4.15	3.59	2.93
		38 × 184	5.47	4.74	3.87	5.47	4.74	3.87
		38 × 235	6.98	6.05	4.94	6.98	6.05	4.94
		38 × 286	8.49	7.36	6.01	8.49	7.36	6.01
Eastern hemlock-tamarack (N)	Select structural	38 × 89	3.12	2.83	2.47	3.57	3.24	2.83
		38 × 140	4.90	4.45	3.89	5.61	5.10	4.45
		38 × 184	6.46	5.87	5.13	7.40	6.72	5.87
		38 × 235	8.25	7.49	6.55	9.44	8.58	7.49
		38 × 286	10.03	9.12	9.96	11.49	10.43	9.12
	No. 1	38 × 89	3.12	2.38	2.47	3.57	3.24	2.83
		38 × 140	4.90	4.45	3.89	5.61	5.10	4.45
		38 × 184	6.46	5.87	5.13	7.40	6.72	5.87
		38 × 235	8.25	7.49	6.55	9.44	8.58	7.49
		38 × 286	10.03	9.12	7.96	11.49	10.43	9.12

TABLE 6–1b: *(continued)*

Lumber species	Grade	Size (mm)	Gypsum board or plastered ceiling, joist spacing			Other ceilings, joist spacing		
			300 mm	*400 mm*	*600 mm*	*300 mm*	*400 mm*	*600 mm*
	No. 2	38 × 89	3.01	2.73	2.38	3.44	3.13	2.73
		38 × 140	4.73	4.29	3.75	5.41	4.92	4.29
		38 × 184	6.23	5.66	4.94	7.13	6.48	5.66
		38 × 235	7.95	7.22	6.31	9.10	8.27	7.22
		38 × 286	9.67	8.79	7.68	11.07	10.06	8.79
	No. 3	38 × 89	2.90	2.63	2.22	3.15	2.72	2.22
		38 × 140	4.55	4.00	3.26	4.62	4.00	3.26
		38 × 184	6.01	5.27	4.30	6.09	5.27	4.30
		38 × 235	7.66	6.73	5.49	7.77	6.73	5.49
		38 × 286	9.32	8.18	6.68	9.45	8.18	6.68
Spruce-pine-fir	Select structural	38 × 89	3.09	2.80	2.45	3.53	3.21	2.80
		38 × 140	4.85	4.41	3.85	5.55	5.05	4.41
		38 × 184	6.40	5.81	5.08	7.32	6.65	5.81
		38 × 235	8.16	7.41	6.48	9.34	8.49	7.41
		38 × 286	9.93	9.02	7.88	11.36	10.33	9.02
	No. 1	38 × 89	3.09	2.80	2.45	3.53	3.21	2.80
		38 × 140	4.85	4.41	3.85	5.55	5.05	4.18
		38 × 184	6.40	5.81	5.08	7.32	6.65	5.51
		38 × 235	8.16	7.41	6.48	9.34	8.49	7.03
		38 × 286	9.93	9.02	7.88	11.36	10.33	8.55
	No. 2	38 × 89	2.98	2.71	2.37	3.41	3.10	2.59
		38 × 140	4.69	4.26	3.72	5.30	4.59	3.75
		38 × 184	6.18	5.62	4.91	6.99	6.05	4.94
		38 × 235	7.89	7.17	6.26	8.92	7.73	6.31
		38 × 286	9.60	8.72	7.62	10.85	9.40	7.67
	No. 3	38 × 89	2.74	2.37	1.94	2.74	2.37	1.94
		38 × 140	4.04	3.50	2.85	4.04	3.50	2.85
		38 × 184	5.33	4.61	3.76	5.33	4.61	3.76
		38 × 235	6.80	5.89	4.80	6.80	5.89	4.80
		38 × 286	8.27	7.16	5.84	8.27	7.16	5.84
Western cedar (N)	Select structural	38 × 89	2.97	2.70	2.36	3.40	3.09	2.70
		38 × 140	4.67	4.24	3.71	5.35	4.86	4.24
		38 × 184	6.16	5.59	4.89	7.05	6.40	5.59
		38 × 235	7.86	7.14	6.24	9.00	8.17	7.14
		38 × 286	9.56	8.68	7.58	10.94	9.94	8.68
	No. 1	38 × 89	2.97	2.70	2.36	3.40	3.09	2.70
		38 × 140	4.67	4.24	3.71	5.35	4.86	4.23
		38 × 184	6.16	5.59	4.89	7.05	6.40	5.58
		38 × 235	7.86	7.14	6.24	9.00	8.17	7.12
		38 × 286	9.56	8.68	7.58	10.94	9.94	8.66
	No. 2	38 × 89	2.87	2.61	2.28	3.29	2.99	2.61
		38 × 140	4.51	4.10	3.58	5.17	4.63	3.78
		38 × 184	5.95	5.41	4.72	6.82	6.10	4.98
		38 × 235	7.60	6.90	6.03	8.70	7.79	6.36
		38 × 286	9.24	8.39	7.33	10.58	9.47	7.73
	No. 3	38 × 89	2.74	2.37	1.94	2.74	2.37	1.94
		38 × 140	4.04	3.50	2.85	4.04	3.50	2.85
		38 × 184	5.33	4.61	3.76	5.33	4.61	3.76
		38 × 235	6.80	5.89	4.80	6.80	5.89	4.80
		38 × 286	8.27	7.16	5.84	8.27	7.16	5.84

When a hip roof is involved, ceiling joists cannot be set close to the end walls, because they would interfere with the end wall rafters. In such a case, the regular ceiling joists must be stopped back far enough for the rafters to clear them (normally one or two joist spacings, depending on the rise of the rafters), and *stub* joists are run at right angles to the end wall plate (see Fig. 6-6), with the end cut exactly the same as for the regular ceiling joists.

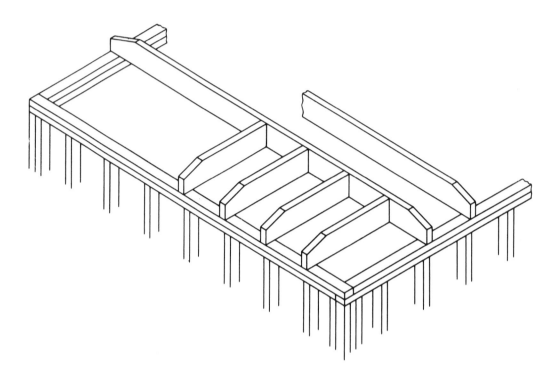

FIGURE 6-6: *Stub ceiling joists.*

ASSEMBLING JOISTS

Normally, ceiling joists run across the narrow dimension of the building, but this need not always be the case. Some joists may run in one direction and others at right angles to them. The main consideration is that there be adequate bearing and support at the ends (see Fig. 6-7).

Sometimes it is desired to have a *clear span ceiling* in some part of the building—a ceiling that runs unbroken from one outside wall to the other. Some means of supporting the inner ends of the joists is required, and this is done by means of a *flush beam*. A flush beam is one which has its bottom edge flush with the bottom of the ceiling joists and is supported at its ends by a bearing wall or bearing partition (see Fig. 6-8).

Ceiling joists are carried by a flush beam on *joist hangers* (see Fig. 6-8) by *endnailing* to the two outer sections of the beam (see Fig. 6-9, page 170.)

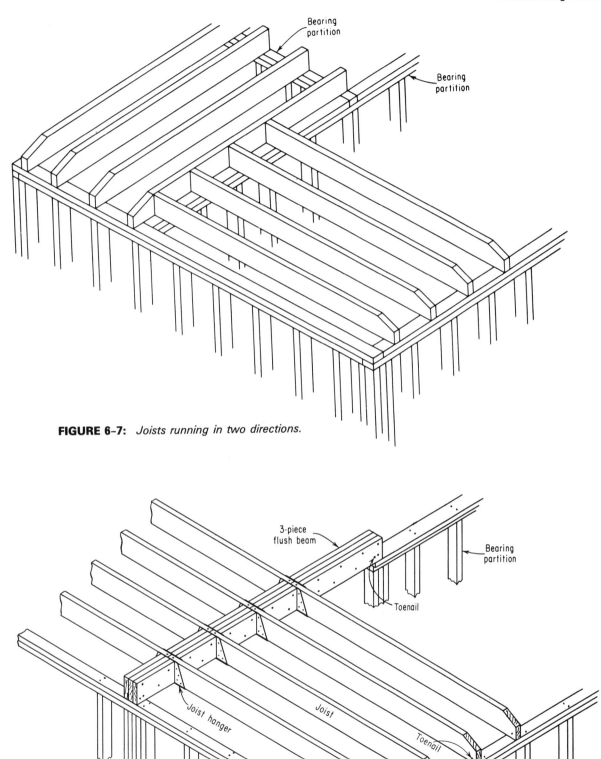

FIGURE 6-7: *Joists running in two directions.*

FIGURE 6-8: *Flush beam with joist hangers.*

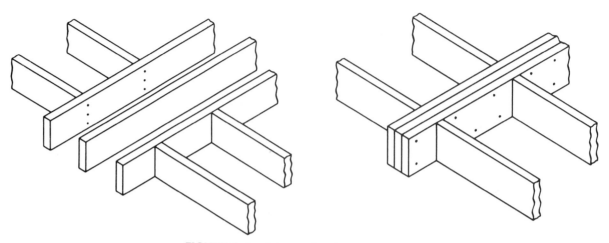

FIGURE 6-9: *Joists end-nailed to flush beam.*

Layout for Ceiling Joists

The first step in the placing of ceiling joists is the layout of their positions on the wall plate. Ceiling joists are normally spaced at 12, 16, or 24 in. (300, 400, or 600 mm) o.c. and, wherever possible, should be located to conform to the spacing of the rafters. This procedure not only facilitates the locating of the rafters but also makes it possible for ceiling joists and rafters to be nailed together.

Layout may begin either from the corners, from the center of the wall, or from some point along it, depending on the type of roof and the dimensions of the building. For a gable roof, the two sidewall plates must be laid out, while in the case of a hip roof, all four plates must be marked. For example, for a gable roof on a building 26 ft, 8 in. (8000 mm) long, in which the ceiling joists and rafters are spaced at 16 in. (400 mm) o.c., proceed as follows:

1. Check the building length to see whether it may be divided evenly by the specified spacing. In this case the length divides evenly into 20 spaces, so layout may begin at a corner.

2. From one, measure along the plate 16 in. (400 mm) to a point A, which will be the *center of the first rafter.*

3. Measure back ¾ in. (19 mm) (one-half the rafter thickness) and square across the plate at that point. That line will represent the *outside edge* of both the *rafter* and the ceiling joist, each sitting on opposite sides of the line. Mark with an "X" the position of the joist on the plate [see Fig. 6–10(a)].

4. Lay the square on the plate with the inner edge of the tongue against the edge of the plate and the *end of the tongue* touching the joist position line just drawn. Mark on the blade and mark an "X" on the same side of the line as above (see Fig.

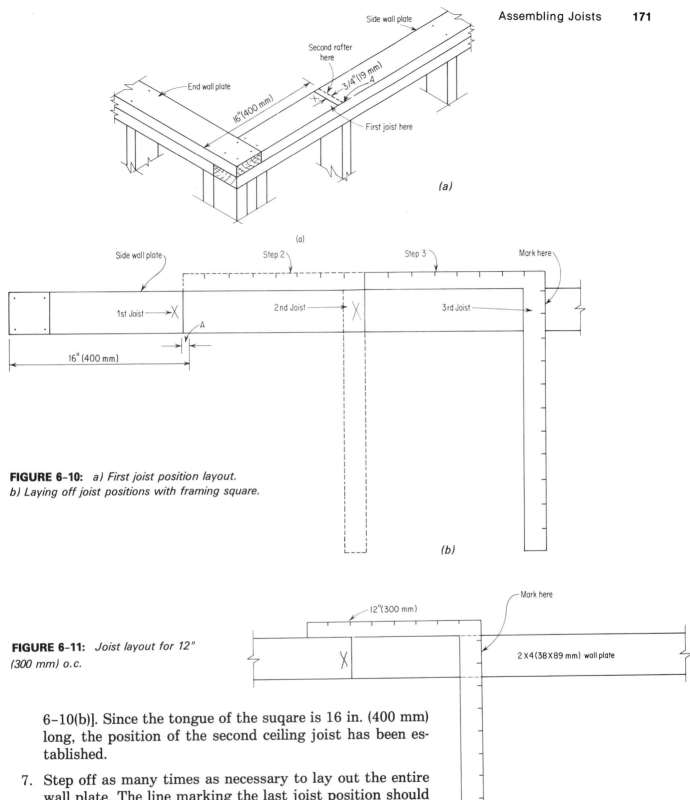

FIGURE 6-10: *a) First joist position layout.*
b) Laying off joist positions with framing square.

FIGURE 6-11: *Joist layout for 12"*
(300 mm) o.c.

6-10(b)]. Since the tongue of the suqare is 16 in. (400 mm) long, the position of the second ceiling joist has been established.

7. Step off as many times as necessary to lay out the entire wall plate. The line marking the last joist position should be 16¾ in. (419 mm) from the end of the plate.

If the spacing is 24 in. (600 mm) o.c., use the blade of the square for measuring, rather than the tongue, while if the spacing is 12 in. (300 mm) o.c., lay the 12-in. (300-mm) mark on the tongue to the position line (see Fig. 6-11).

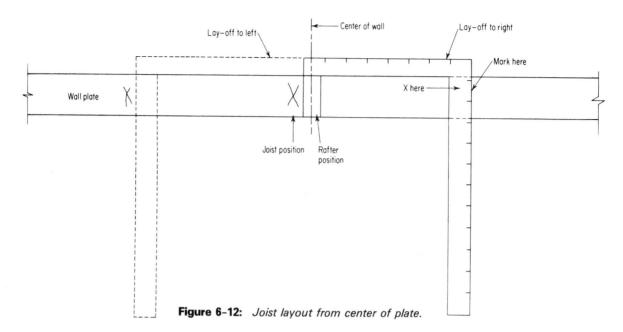

Figure 6-12: *Joist layout from center of plate.*

If the length is not evenly divisible by the joist spacing, it may be preferable to begin layout at the center of the wall and lay out both ways toward the corners (see Fig. 6-12).

In a hip roof, there will normally be an end common rafter (see Fig. 7-2), which will be located in the center of the end wall. There should be a stub joist position next to it, and the remainder of the stub joist positions may be laid out from it with the square as illustrated in Fig. 6-12.

In cases where less than a full joist spacing remains at the corner from the end wall layout, layout for the sidewall joists should begin at a distance from the corner equal to one-half the width of the building and proceed to the corner and to the center of the wall plate. The opposite end is then laid out in the same manner (see Fig. 6-13).

If ceiling joists are to *butt* one another over the center bearing, the layout on both sidewall plates will be identical. However, if the joists are to *lap* at the center (see Fig. 6-2), then the positions on one plate must be *offset* by the thickness of the joist from those on the opposite plate.

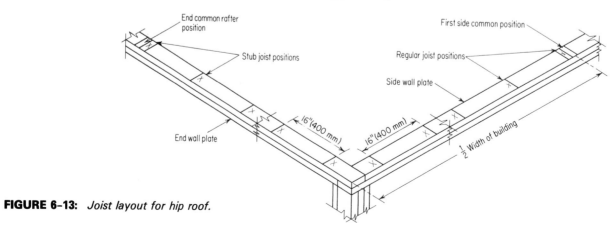

FIGURE 6-13: *Joist layout for hip roof.*

Nailing of Joists

Ceiling joists must be toenailed to the plate with a minimum of two 3¼-in. (82-mm) nails at each end (see Table 7–1, page 193). In addition, if joists butt together at the center, a ¾-in. (19-mm) board, joist width and about 3 ft (1 m) long, is nailed across the joint, as illustrated in Fig. 6–1. If they lap at the center, there must be a minimum of two 3-in. (76-mm) nails at each end of the lap, as shown in Fig. 6–2.

Joist Restraint

Ceiling joists must be restrained from twisting along their bottom edges, and this may be done in several ways. At the ends, the toenails which tie the joists and plates together provide the restraint. Between plates, at intervals not greater than 7 ft (2100 mm), the joists must be provided with *crossbridging, blocking,* or *strapping* in a manner similar to that used for floor joists (see Chapter 4). If the ceiling is *furred* for the application of some ceiling material (see Fig. 6–14), the furring will provide the necessary restraint.

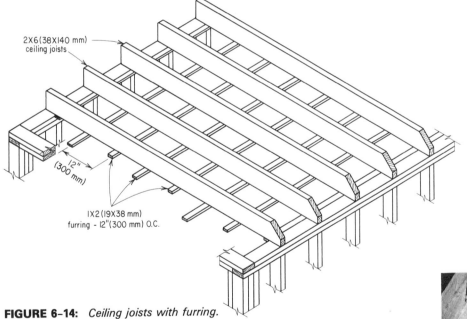

2X6(38X140 mm) ceiling joists

12" (300 mm)

1X2 (19X38 mm) furring - 12"(300 mm) O.C.

FIGURE 6–14: *Ceiling joists with furring.*

Ceiling Backing

Partitions which run parallel to the direction of ceiling joists must be provided with some means of carrying the edges of the ceiling material which meet the partition. This is done by nailing *ceiling backing* to the cap plate (see Fig. 6–15). A piece of 1½-in. (38-mm) material, at least 2 in. (50 mm) wider than the cap plate, is used, allowing it to project the same amount on both sides.

FIGURE 6–15: *Ceiling backing.*

FIGURE 6-16: *Attic access.*

Attic Access

Building codes and fire regulations demand that there be access to the attic, and it is provided by a framed opening or *hatchway* at least 20 × 28 in. (500 × 700 mm) (see Fig. 6-16). For appearance sake, it is usually placed in a relatively inconspicuous location and provided with a trapdoor or other type of cover which opens upward.

Sound Reduction Through Ceilings

In buildings such as apartments of more than one story, a sound transmission problem arises where the floor joists at one level also carry the ceiling material for the rooms below, because sound is transmitted through the floor frame from one level to another.

One method of overcoming the problem is to provide a separate frame to carry the ceiling materials (see Fig. 6-17). The

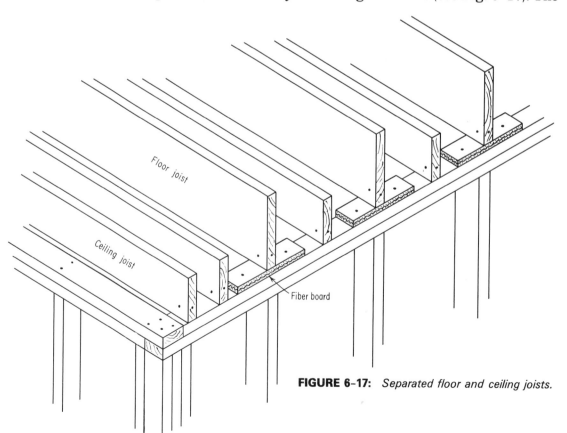

FIGURE 6-17: *Separated floor and ceiling joists.*

ends of the regular floor joists will not rest directly on the plates but will have some type of resilient pad, such as rubber, neoprene, or fiber board under each end. Thus the bottom edges of the joists will be above the level of the top of the plate. Then smaller members, acting as ceiling joists, are set between each pair of floor joists, directly on the plates. They carry the ceiling material, which, in this way, will not be in contact with the floor joists.

REVIEW QUESTIONS

6-1. List three main factors that influence the size of ceiling joists to be used in any particular situation.

6-2. What figures would you use on the framing square to lay out the cut on the outer end of ceiling joists used in conjunction with a 7-in. (175-mm) rise roof. Mark on _____ .

6-3. When is it necesasry to use *stub* ceiling joists?

6-4. What is the minimum ceiling height, according to National Housing Standards, in each of the following areas?

(a) Living room

(b) Bedroom

(c) Laundry area above grade

6-5. (a) What is the purpose of ceiling backing?

(b) What width of ceiling backing should be used on a 2 × 4 (38 × 89 mm) partition?

6-6. (a) What is the main reason for providing access to the attic through the ceiling?

(b) What minimum size of opening is required?

(c) In what area of the ceiling would you suggest that the opening be placed?

6-7. If a building span is 30 ft (8780 mm) and ceiling joists must have at least 24 in. (600 mm) of lap over a center bearing partition, what standard length is required for the joists?

6-8. What depth and standard length of ceiling joist material are required in each of the following situations?

	Joist span	Joist spacing	Ceiling	Material
(a)	15 ft	16 in.	Plastered	Hem-fir, No. 1
(b)	5560 mm	600 mm	Gypsum board	Douglas fir, No. 1
(c)	6020 mm	300 mm	Acoustic tile	Hem-fir, No. 2
(d)	14 ft, 6 in.	16 in.	¼-in. plywood	Spruce-pine-fir, No. 1

6-9. Draw a neat diagram to illustrate the plate layout for joists and rafters if joists are spaced 16 in. (400 mm) o.c. and the rafters are spaced 24 in. (600 mm) o.c.

7

ROOF, GABLE, AND DORMER FRAME

The final step in completing the skeleton of a building is the framing of the roof. The work involved depends on the type of roof to be used, and the student must first become familiar with the types of roof in common use. Each has some special problems involved in its construction.

ROOF SHAPES

The shape of the roof may be one of several common designs, which include *shed* roof, *gable* roof, *hip* roof, *gambrel* roof, *dutch hip* roof, *mansard* roof, *flat* roof and *intersecting* roof (see Fig. 7–1).

The shed roof slopes in only one direction. The gable roof has two slopes, and the continuation of the end walls up to meet the roof is known as a *gable end*. The hip roof has four slopes, which terminate at a point if the plan is square or in a ridge if the plan is rectangular (see Fig. 7–1). A gambrel roof is a modification of a gable, each side having two slopes instead of one. The Dutch hip roof is a combination of the gable and hip roof, with small gable ends and slopes on four sides. The mansard roof is a modification of the hip with two slopes on each side instead of one. The flat roof has no slope at all. An intersecting roof is formed by the meeting of two sloped roofs of one type or another.

ROOF PLAN

By examining a plan view of a roof frame, the different types of rafters involved in forming roofs can be easily distinguished. Figure 7–2 is a plan view of the frame of an intersecting hip roof. Here all types of rafters are involved.

The *common* rafters are those which run at right angles to the wall plate and meet the *ridge* at their top end. In a gable roof, all the rafters are *commons,* while in a hip roof, the two center end rafters and those side rafters which meet the ridge are *commons.* Hip rafters run from the corners of the wall plates to the ridge at an angle of 45° to the plates and form the intersection of two adjacent roof surfaces. *Valley* rafters occur where two roofs intersect. Jack rafters run parallel to the commons but are shorter and are named according to their position. *Hip jacks* run from plate to hip, usually meeting at the hip in pairs. *Valley jacks* run from the valley rafter to the ridge or plate. *Cripple jacks* run from hip to valley or from valley to valley, touching neither plate nor ridge. The part of any rafter that projects beyond the wall plate is known as the *rafter tail.*

(a) Shed roof.

(b) Gable roof.

(c) Hip roof.

(d) Gambrel roof.

(e) Dutch hip roof.

(f) Modified Mansard roof.

(g) Flat roof.

(h) Intersecting roof.

FIGURE 7–1: Roof shapes.

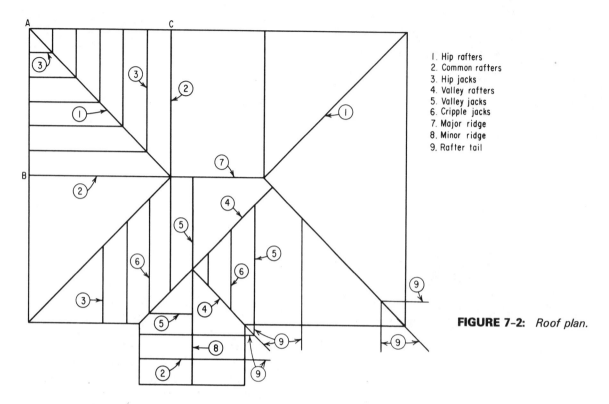

1. Hip rafters
2. Common rafters
3. Hip jacks
4. Valley rafters
5. Valley jacks
6. Cripple jacks
7. Major ridge
8. Minor ridge
9. Rafter tail

FIGURE 7-2: *Roof plan.*

RAFTER TERMS

To be able to understand and solve the problems involved in laying out and cutting a set of rafters, it is necessary to know the terms used in discussing the subject.

Common Rafter

One of the basic terms, which applies to all rafters, is the *span*, the dimension of a building bridged by a pair of common rafters, while that part of the span traversed by one of the pair is called the *run* of the rafter. The length of the rafter, measured on a line from the center of the ridge to the outside edge of the building, is the *line length* of the rafter (see Fig. 7-3). The *total length* of the rafter is the sum of the line length and the *tail length* (see Fig. 7-3). The vertical distance from the level of the wall plates to the line length meeting point of a pair of rafters is the *total rise* of the rafter (see Fig. 7-3).

To make rafter layout simpler, run, rise, and line length are broken down into *units*. The basic one is the *unit run*—a standard length of 12 in. (250 mm) [see Fig. 7-4(a)]. The *unit rise* is the vertical distance which a rafter rises for one unit of run, while the length of rafter resulting from one unit of run is the *unit line length* [see Fig. 7-4(a)].

If a framing square is applied to a common rafter in position, as in Fig. 7-4(b), with the 12-in. (250-mm) mark on the blade coinciding with the outside edge of the wall plate, a figure on the tongue will indicate the *unit rise* of the rafter. In this case, the figure is 6 in. (150 mm) and the diagonal distance between 6 in. (150) on the tongue and 12 in. (250) on the blade is the *unit line length* for that rafter [see Fig. 7–4(b)].

The inclination of a rafter is called the *slope* and can be expressed as a fraction or in degrees. The slope is most commonly expressed as the unit rise per unit run. For example, in Fig. 7–4(b), the slope is ⁶⁄₁₂ (150/250).

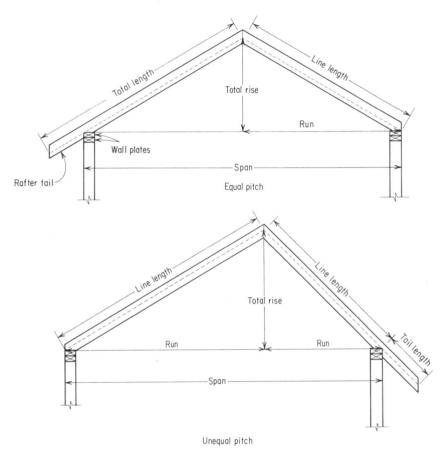

FIGURE 7–3: *Span, run, rise, and line length of a rafter.*

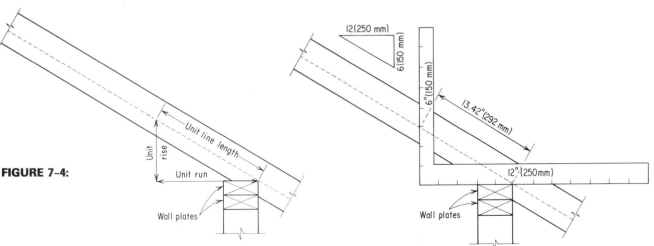

(a) Unit run, unit rise, and unit line length.   *(b) Unit run, unit rise, and unit line length on the framing square.*

FIGURE 7–4:

In order that rafters may seat securely on the wall plate, it is common practice to cut a triangular section from the bottom edge, at the point at which it meets the wall plate, as shown in Fig. 7-5. This notch is called a *birdsmouth*, which is made by making a *seat cut*—cut horizontal when the rafter is in position—and a *plumb cut*—vertical when the rafter is in position (see Fig. 7-5). Common rafters are made to fit together at the top or fit to a ridgeboard (see Fig. 7-6) by making a *plumb cut* at the upper end of each, and the rafter tail may have a single *plumb cut* or a plumb cut and a *seat cut* at the bottom end (see Fig. 7-5). The horizontal distance from that bottom plumb cut to outside of the wall frame is known as the *overhang* (see Fig. 7-4).

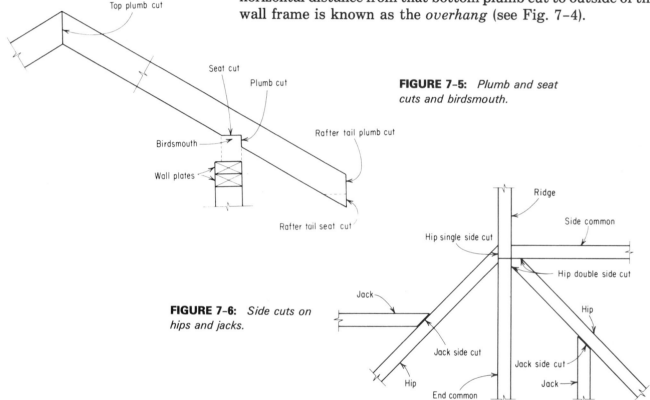

FIGURE 7-5: *Plumb and seat cuts and birdsmouth.*

FIGURE 7-6: *Side cuts on hips and jacks.*

Jack Rafter

Jack rafters run parallel to common rafters and therefore have the same unit run, rise, and line length. They have the same birdsmouth at the plate and must have a top plumb cut. But, since they meet a hip rafter at an angle, each one must also have a *side cut* or *cheek cut* at the top end.

Hip Rafter

Hip rafters also have a birdsmouth at the plate but, since they run at an angle to the plate and the ridge, they must also have a *side cut* at the top end as well as a plumb cut. That side cut may be *single* or *double* (see Fig. 7-6), depending on how the rafter meets the ridge.

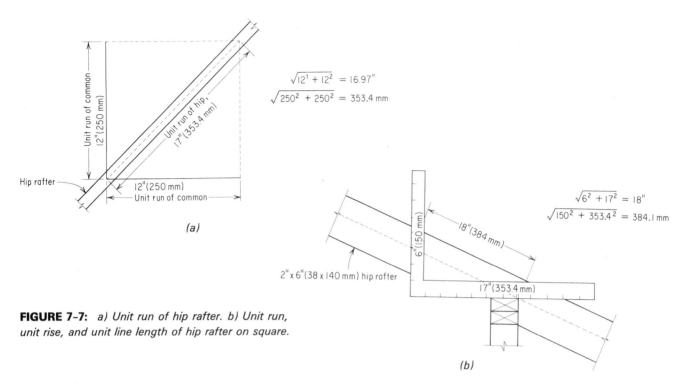

$$\sqrt{12^1 + 12^2} = 16.97''$$
$$\sqrt{250^2 + 250^2} = 353.4 \text{ mm}$$

$$\sqrt{6^2 + 17^2} = 18''$$
$$\sqrt{150^2 + 353.4^2} = 384.1 \text{ mm}$$

FIGURE 7-7: *a) Unit run of hip rafter. b) Unit run, unit rise, and unit line length of hip rafter on square.*

Hip rafters run at an angle of 45° to the wall plate (see Fig. 7-2), and, as a result, their unit run is different from that of a common rafter. As illustrated in Fig. 7-7(a), a hip rafter must traverse the diagonal of a square in order to span one unit of common run. Thus the unit run of a hip rafter will be the length of the diagonal of a square with sides of 12 in. (250 mm)—namely 17 in. (353.5 mm).

If the framing square is applied to a hip rafter in position, as shown in Fig. 7-7(b), the 17-in. (353.5-mm) mark on the blade must coincide with the outside edge of the wall plate to represent the *unit run*, and a figure on the tongue will indicate the *unit rise*. If the rafter has the same rise as the one shown in Fig. 7-4, then that figure will be 6 in. (150). Again, the diagonal distance between 6 in. (150) on the tongue and 17 in. (353.5) on the blade is the *unit line length* for the hip rafter.

Since hip rafters occur at the intersection of two surfaces of a hip roof, the *roof sheathing* on each surface must meet on a hip. To provide a flat surface on which the sheathing ends may rest, the top edges of hips are *backed* (beveled), as shown in Fig. 7-8.

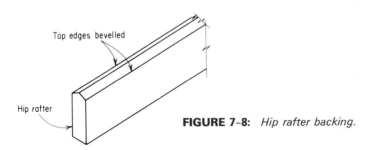

FIGURE 7-8: *Hip rafter backing.*

Shortening

In Fig. 7-9, two common rafters have first been represented by two lines *AC* and *BC*. Then full-width rafters have been superimposed in such a way that the lines *AC* and *BC* pass through the peak of the birdsmouth. In addition, a ridgeboard is inserted between the two rafter ends.

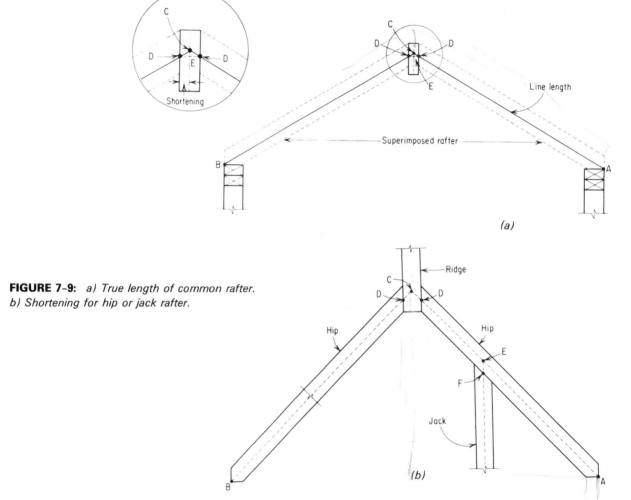

FIGURE 7-9: *a) True length of common rafter. b) Shortening for hip or jack rafter.*

The two lines *AC* and *BC* meet in the center of the ridge at *C*, and the distance from *A* or *B* to *C* is the line length of the rafter (see Fig. 7-3). However, the rafter actually ends at *D*; the distance from *A* or *B* to *D* is the actual length of the rafter; it has been *shortened* by a distance *CD*, which varies, depending on the rise of the rafter. However, if the *shortening* is measured at *right angles to the plumb cut*, its length will remain constant—*one-half the thickness of the ridge*. For a 1½-in. (38-mm) ridge, that distance will be ¾ in. (19 mm).

With regard to a hip rafter, it meets the ridge at an angle of 45° [see Fig. 7-9(b)] and therefore must be *shortened*, in a horizontal direction (at right angles to a plumb cut) by a distance *CD* [see Fig. 7–9(b)], *one-half the diagonal thickness of the ridge*. For a 1½-in. (38-mm) ridge, that will amount to approximately 1 1/16 in. (27 mm).

A jack rafter meets the hip at an angle of 45° in plan [see Fig. 7–9(b)] and must be *shortened* by a distance equal to *EF*, taken at right angles to the plumb cut. For a 1½-in. (38-mm) hip, that distance will also be 1 1/16 (27 mm).

RIDGE

Rather than have a pair of rafters meet together at their top end, it is customary to introduce a ridgeboard between them. This ridgeboard makes it easier to keep the top line of the roof straight and provides support for the roof sheathing between rafters.

The ridge of a gable roof will have the same length as the length of the building plus roof overhang at the gable ends, if any. The length of a hip roof ridge, however, takes a little more consideration.

Turn back to Fig. 7–2 and study it a moment. In plan, the hip rafter is the diagonal of a square, the sides of which are *AB* and *AC, AB* being equal to half the width of the building. The top end of the hip is at a point half the width of the building *along* the length. The hip at the opposite end covers a like distance. The distance between the top ends of the two hips, then, is equal to the length of the building minus the width. In other words, the line length of the ridge is the length of building minus the width.

When a ridgeboard is introduced into the frame, the common rafters are shortened by half the ridge thickness. Consequently, the end commons would not reach the ends of the ridge if they were cut to true length. So the ridge length is increased by half its own thickness at each end. In other words, the *true length of the hip roof ridge is the length of building minus the width plus the thickness of the ridge*.

An intersecting roof contains two ridges. The ridge of the main roof is called the *major* ridge, whereas the ridge in the projection is known as the *minor* ridge.

HOW TO USE THE RAFTER TABLE

Having considered rafter framing terminology, it is now necessary to examine the methods used to determine the lengths of various types of rafters and how to lay out and cut them.

Calculation of lengths may be done in several ways, including the use of the *rafter table and mathematics,* by *scaling on the framing square,* and by *scale drawings.* The layout of lengths may be done by *framing square step-off.* For most situations, the combination of rafter table and math, with tape measurement of lengths, is the simplest and most accurate.

Rafter Table

Figure 7–10(a) illustrates a framing square, with a *rafter table* on the face of the blade. The numbers along the top edge of the blade represent *unit rises,* with calculations for 17 rises ranging from 2 to 18. Figure 7–10(b) illustrates a metric framing square, with calculations for 10 rises ranging from 50 to 500 mm, in increments of 50 mm. The first step in using the rafter table, then, is to locate the number which corresponds to the unit rise of the roof in question.

To find the *unit line length* of a common rafter, look on the first line of the table under the unit rise concerned; the figure found there is the line length for one ft. (250 mm) of common run. Multiply that figure by the number of units of run for that rafter to obtain the line length of the common rafter in in. (mm). The number of units of run for the rafter may be calculated by dividing the *total run* of the rafter in in. by 12 (mm by 250).

The second line gives the *unit line length* of a hip or valley rafter per unit of common run. The line length of the hip or valley rafter will be that figure, multiplied by the number of units of common run in the building.

The third line gives the difference in length of successive jack rafters spaced 16 in. (400 mm) o.c. and is also the line length of the first jack rafter from the corner, spaced 16 in. (400 mm) o.c. Thus, having found the *true* (shortened) length of the first jack rafter from the corner, the length of the next one is found by adding that "difference in length" figure to the shortened length.

The fourth line on the imperial square [Fig. 7–10(a)] gives the difference in length of jack rafters spaced 2 ft o.c.

The fifth line indicates the figure to use on the tongue for the side cut of jack rafters. Again, 12 in. (250) is the blade figure, and the cut is marked on the blade.

The sixth line on the imperial square and the fourth line on the metric square indicate the figure to use on the tongue of the square in laying out the side cuts for hip and valley rafters. The blade figure is always 12 in. (250 mm), and the cut is always marked on the blade.

The sixth line on the metric square indicates the size of the angle formed by a common rafter at the point at which it meets the plate.

Framing Calculations and Layout

Figure 7–11 is a plan view of a hip roof with an intersecting gable roof, with dimensions as indicated. On it are marked a number of typical roof framing members whose length and layout will be demonstrated.

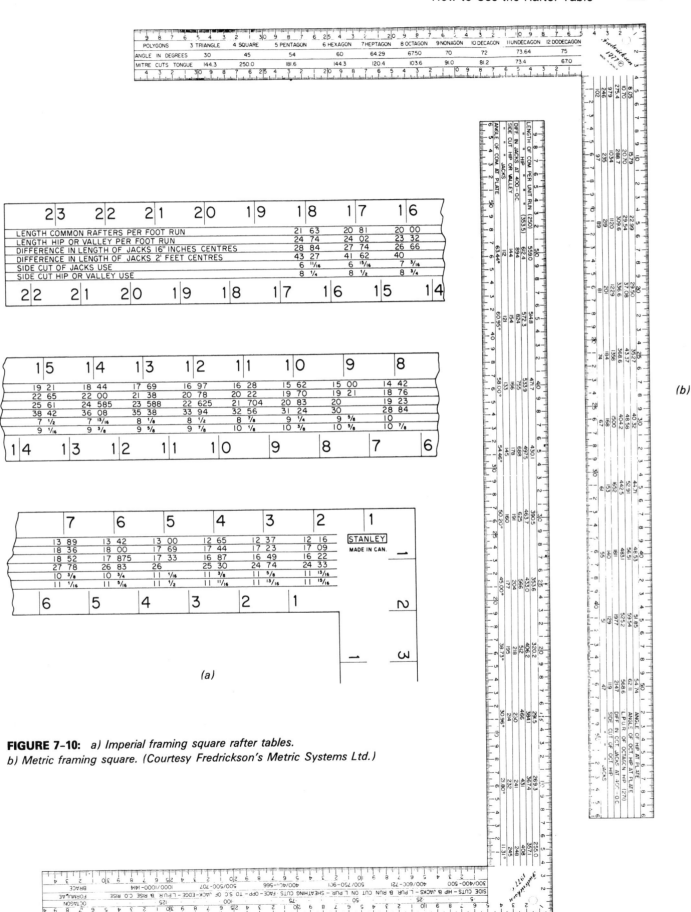

FIGURE 7-10: *a) Imperial framing square rafter tables.*
b) Metric framing square. (Courtesy Fredrickson's Metric Systems Ltd.)

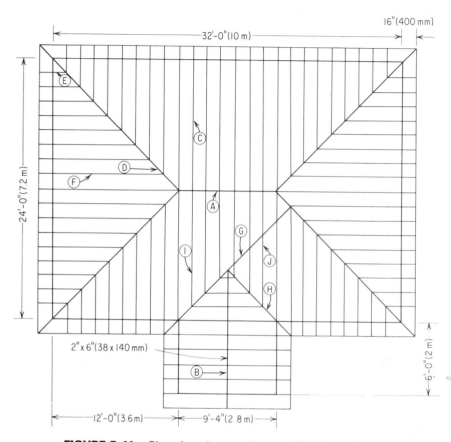

FIGURE 7-11: *Plan view, intersecting roof 6" (150 mm) rise.*

Length of Major Ridge (A)

The length of the ridge for a hip roof is the length of the building minus the width plus the thickness of the rafter or the ridge, whichever is the greatest. In this example the length is 32 ft − 24 ft + 1½ in. = 8 ft, 1½ in. (10,000 mm − 8000 mm + 38 mm = 2038 mm). The size of the material for the ridge will normally be 2 × 6 in. (38 × 140 mm).

Length of Minor Ridge (B)

The length of the minor ridge will be the length of the projection (see Fig. 7–11) plus one-half of its width plus the overhang minus one-half the diagonal thickness of the supporting valley rafter. In this case the length will be 6 ft + 4 ft + 16 in. − 1 1/16 in. = 11 ft, 2 15/16 in. (2000 mm + 1200 mm + 400 mm − 27 mm = 3573 mm).

Length and Layout of Common Rafter (C)

The unit line length for a common rafter with a slope of 6 to 12 as shown under 6 [see Fig. 7–10(a)] is 13.42. The line length of the rafter is equal to the number of units of run (feet) times the unit line length: 12 × 13.42 = 161.04 in. or 13 ft, 5 1/16 in. The corresponding metric slope of 150/250 will result in a unit line length of 291.5 as shown under 15. The number of units of run for the common rafter with a run of 3600 mm will be $3600 \div 250 = 14.4$. The line length of the rafter will be $291.5 \times 14.4 = 4198$ mm.

The line length of the rafter tail will be $^{16}/_{12} \times 13.42 = 17.89$ in. ($^{400}/_{250} \times 291.5 = 466$ mm).

A typical common rafter (C) may now be laid out, cut, and used as a pattern for all the common rafters. Proceed as follows:

1. Pick out a piece of straight 2 × 4 (38 × 89 mm) stock of sufficient length and lay out a plumb cut at the top end. This is done by laying the square on the stock with the rise—6 in. (150 mm)—on the tongue and the unit run—12 in. (250 mm)— on the blade. Mark along the tongue, making sure the square is setting so the crown is on the topside of the rafter [see Fig. 7-12(a)].

2. From the top plumb line, measure 13 ft 5$^1/_{16}$ in. (4198 mm) along the back of the rafter to locate the plumb cut at the birdsmouth [see Fig. 7-12(b)].

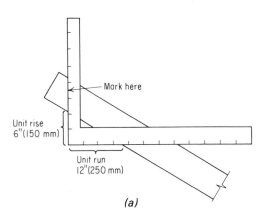

(a)

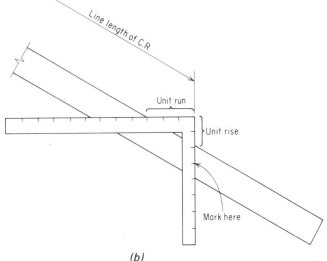

(b)

(c)

FIGURE 7-12: *a) Marking plumb cut at top end of rafter.*
b) Measuring line length of rafter and marking plumb cut at birdsmouth.
c) Locating and marking seat cut at birdsmouth.
d) Measuring tail length and marking tail plumb cut.

3. Measure 2$^1/_2$ in. (63 mm) down from the back of the rafter along the birdsmouth plumb cut to locate the seat of the birdsmouth; draw a line perpendicular to the plumb cut at this point to establish the seat cut [see Fig. 7-12(c)].

4. From the birdsmouth plumb cut, measure 17$^7/_8$ in. (466 mm) along the back of the rafter to the tail plumb cut. Draw a plumb line through this point parallel to the other plumb lines [see Fig. 7-12(d)].

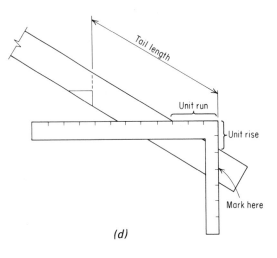

(d)

5. Measure a distance equal to the thickness of the rough fascia perpendicular to the tail plumb cut for the shortening at the tail end. Cut the tail end at this point [see Fig. 7–12(e)].

6. At the top end of the rafter, shorten the rafter a distance equal to half the thickness of the ridge. This measurement is made at right angles to the plumb line at the top end of the rafter. Cut the top end at this point [see Fig. 7–12(f)].

7. Use this cut rafter as the pattern for other rafters. Make sure that the crown (bow) is always placed on the top edge of the rafter [see Fig. 7–12(g)].

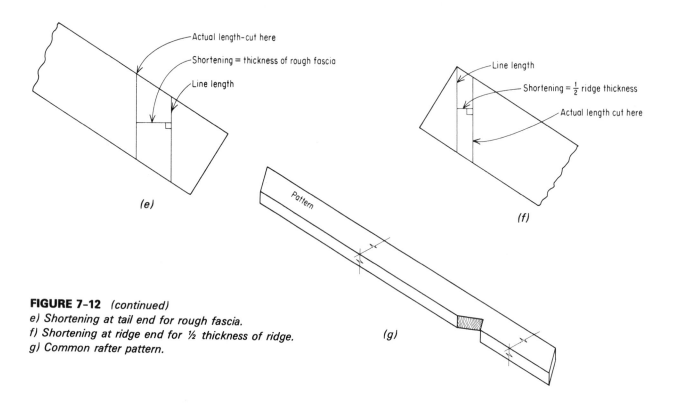

FIGURE 7-12 *(continued)*
e) Shortening at tail end for rough fascia.
f) Shortening at ridge end for ½ thickness of ridge.
g) Common rafter pattern.

Length and Layout of Hip Rafter (D)

The unit line length for a hip rafter with a unit rise of 6 in. is 18 in. [as shown under 6 on the second line of the rafter table in Fig. 7–10(a)]. Since the number of units of run is 12, the line length of the rafter will be $12 \times 18 = 216$ in. $= 18$ ft. The line length of the rafter tail will be $^{16}\!/_{12} \times 18 = 24$ in.

The corresponding metric slope of 150 to 250 will result in a unit line length of 384.1 mm [as shown under 15 on the second line of the rafter table in Fig. 7–10(b)]. The line length is equal to the number of units of run for the common rafter (14.4) times the unit line length ($14.4 \times 384.1 = 5531$ mm). The line length of the rafter tail is $^{400}\!/_{250} \times 384.1 = 615$ mm.

A typical hip rafter (D) may now be laid out using the following procedure:

1. Pick out a straight piece of 2 × 6 (38 × 140 mm) stock of sufficient length; lay out the plumb line as near the top as possible. This is done by using the unit rise (6 in. or 150 mm) on the tongue and the unit run (17 in. or 354 mm) on the blade and marking the plumb along the tongue. Make sure the crown of the 2 × 6 corresponds with the topside of the rafter [see Fig. 7–13(a)].

2. From the top end plumb line, measure 18 ft (5531 mm) along the back of the rafter to the plumb line at the birdsmouth [see Fig. 7–13(b)].

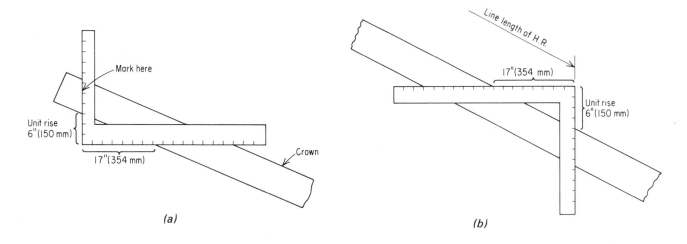

(a)   (b)

3. Measure 2½ in. (63 mm) down from the back of the rafter along a parallel line ¾ in. (19 mm) from the birdsmouth plumb line to locate the seat of the birdsmouth. Draw a line perpendicular to the plumb line at this point to establish the seat cut [see Fig. 7–13(c)].

4. From the birdsmouth plumb line, measure 24 in. (615 mm) along the back of the rafter to establish the tail plumb line [see Fig. 7–13(d)].

FIGURE 7–13: *a) Marking plumb cut at top of rafter.*
b) Marking plumb cut at birdsmouth.
c) Establishing seat cut for birdsmouth of H.R.
d) Establishing tail plumb cut of H.R.

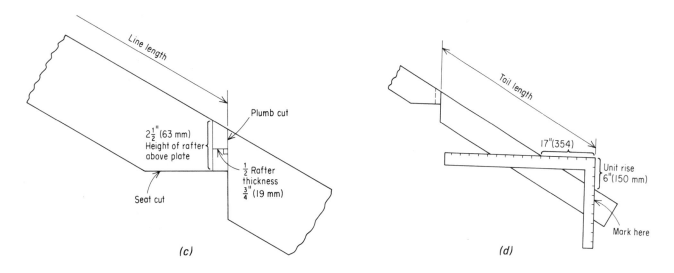

(c)   (d)

5. Measure a distance equal to the diagonal thickness of the rough fascia perpendicular to the tail plumb cut for the shortening at the tail end. This would be $2\frac{1}{8}$ in. for a 2×6 (54 mm for a 38×140) fascia. The end of the rafter tail must fit into the apex of an external angle and therefore must have a *double cut* at the end [see Fig. 7–13(e)]. Lay out the two side cuts by drawing a line parallel to the shortened plumb cut a perpendicular distance of one-half the thickness of the rafter from the plumb cut. Join this line to the midpoint of the top of the plumb cut squared across the edge [see Fig. 7–13(f)].

6. From the top plumb cut line, measure back at right angles a distance equal to one-half the diagonal thickness of the ridge ($1\frac{1}{16}$ in. or 27 mm) and through this point draw a second plumb line. The rafter is now properly *shortened* for the ridge. Square across the top edge of the rafter from this line and mark the center point. Single or double side cut lines will be drawn through this point to a point established by measuring back at right angles a distance equal to one-half the thickness of the rafter [see Fig. 7–13(f)].

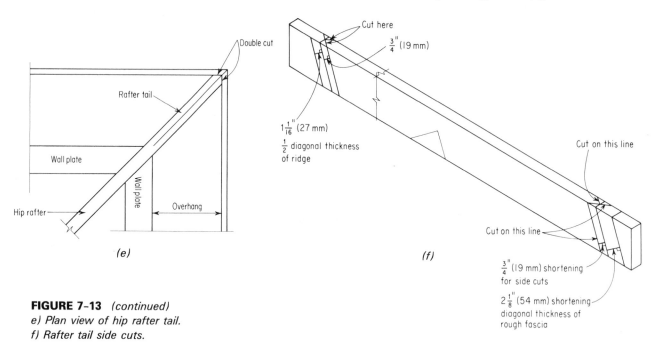

FIGURE 7–13 *(continued)*
e) Plan view of hip rafter tail.
f) Rafter tail side cuts.

Length and Layout of Jack Rafter (F)

The rafter tail and the birdsmouth layout of this rafter are identical to the common rafter. The third line of the rafter table indicates that, for a 6-in. (150-mm) rise, the difference in length of jack rafters 16 in. (400) o.c. is 17.875 in. (466 mm). That is also the amount the longest jack rafter is shorter than the common rafter. To lay out this rafter, proceed as follows:

1. Pick out a piece of stock (2 × 4) and mark a plumb cut as near the top end as possible. This is done by using the unit rise—6 in. (150 mm)—on the tongue and the unit run—12 in. (250 mm)—on the blade. Mark along the tongue, making sure the crown of the rafter is on the top edge [see Fig. 7–12(a)].

2. From the plumb cut, measure the jack rafter line length to locate the birdsmouth. The longest jack rafter line length is equal to the common rafter line length minus the difference in length of adjacent jack rafters: 161.04 in. − 17.875 = 143.165 in. = 11 ft. 11¾₁₆ in. (4198 − 466 mm = 3732 mm) [see Fig. 7–12(b)].

3. The layout of the birdsmouth and the rafter tail are exactly the same as for the common rafter.

4. From the top plumb cut line, measure back at right angles a distance equal to half the diagonal thickness of the hip rafter—1¹⁄₁₆ in. (27 mm)—and through it draw a second plumb line parallel to the first plumb line. With the electric handsaw set at 45° cut along this line. Remember that these rafters are cut in pairs to be placed on opposite sides of the hip rafter (see Fig. 7–14).

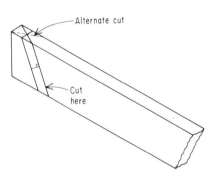

FIGURE 7–14: *Shortening at top of jack rafter.*

Line Length and Layout of Jack Rafter (E)

The jack rafter (E) is the eighth and shortest rafter in the series, and therefore its length will be eight times the difference in length of jack rafters shorter than the common rafter.

Length and Layout of Valley Rafter (G)

The total run of the valley rafter (G) is 16 in. (400) less than that of a common rafter, namely 10 ft, 8 in. (3200 mm). Its line length will be 10.67 × 18 = 192.06 in. or 16 ft, ¹⁄₁₆ in. (³²⁰⁰⁄₂₅₀ × 384.1 = 4916 mm).

The layout for the rafter tail plumb line and top plumb line is the same as for a hip rafter of the same rise. The layout for the birdsmouth differs, as the rafter must fit in an inside corner as opposed to the outside. Figure 7–15 illustrates the layout of the birdsmouth.

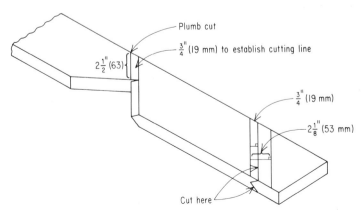

FIGURE 7–15: *Layout of birdsmouth and cut at tail end of rafter.*

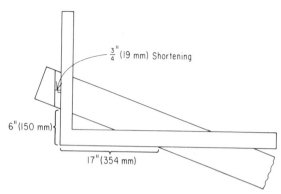

FIGURE 7–16: *Shortening at top of valley rafter (G).*

The valley meets the hip at right angles in the plan view, and therefore the *shortening* will be one-half the thickness of the hip, ¾ in. (19 mm), measured at right angles to the plumb line (see Fig. 7–16).

The end of the valley rafter tail must fit into an internal angle, as illustrated in Fig. 7–17. Figure 7–15 illustrates the layout of the double cut needed.

The layout of valley rafter (H) is exactly the same as valley rafter (G); the only difference is that the run is shorter so it will affect the line length.

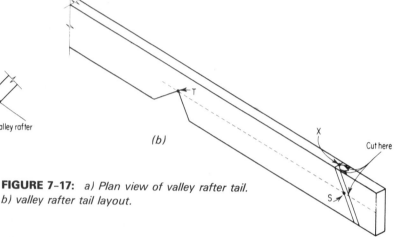

FIGURE 7–17: *a) Plan view of valley rafter tail. b) valley rafter tail layout.*

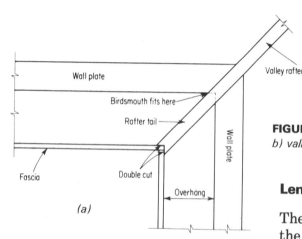

Length and Layout of Valley Jack (I)

The run of valley jack (I) is 16 in. (400 mm) less than the run of the common rafter without a rafter tail. The line length of the rafter will be 161.04 in. − 17.875 = 143.165 in. = 11 ft, 11⁹/₁₆ in. (4198 − 466 = 3732 mm).

The plumb line and shortening at the top end are the same as for a common rafter, while the plumb line, shortening (one-half the diagonal thickness of the valley rafter), and the side cut at the bottom end are identical to the top end of a hip jack.

Line Length and Layout of Cripple Jack (J)

The run of cripple jack (J) is 3 ft, 4 in. (1000 mm). The line length of the rafter will be 3.33 × 13.42 = 44.69 in. or 3 ft, 8¹¹/₁₆ in. ($^{1000}/_{250}$ × 291.5 = 1166 mm). Shortening and side cuts, top and bottom, are the same as for other jack rafters.

ROOF FRAME ASSEMBLY

When all the rafters have been cut, they can then be assembled into a roof frame. In a gable roof, the end rafters are positioned first, at the ends of the wall plates (see Fig. 7–18) and nailed to the plate and to the ridge according to the nailing schedule in Tables 7–1a and 7–1b. The remainder of the rafters are then raised

FIGURE 7–18: *Gable studs.*

and positioned against ceiling joists, properly spaced at the top end and nailed to the plate and the ridge according to the nailing schedule. Rafters are also nailed to the ceiling joists, according to the nailing schedule if the ridge is supported or as specified in Tables 7–1a and 7–1b if the ridge is unsupported. One rafter of each pair may be *endnailed* through the ridge, while the other will be *toenailed* to the ridge.

TABLE 7–1a: *Minimum rafter-to-joist nailing (unsupported ridge) (minimum number of nails at least 3 in. long)*

Roof slope	Rafter spacing (in.)	Rafter tied to every joist — Building width up to 26 ft, 3 in.			Building width up to 32 ft, 2 in.			Rafter tied to joist every 4 ft — Building width up to 26 ft, 3 in.			Building width up to 32 ft, 2 in.		
		21 or less	31	42 or more	21 or less	31	42 or more	21 or less	31	42 or more	21 or less	31	42 or more
1 in 3	16	4	5	6	5	6	8	11	—	—	—	—	—
	24	6	8	9	8	11	—	11	—	—	—	—	—
1 in 2.4	16	4	4	5	5	6	7	7	9	—	9	—	—
	24	5	6	8	7	8	11	7	9	—	—	—	—
1 in 2	16	4	4	4	4	4	5	6	8	9	8	11	—
	24	4	5	6	5	6	8	6	8	9	8	11	—
1 in. 1.71	16	4	4	4	4	4	4	5	6	8	7	8	11
	24	4	4	5	5	6	7	5	6	8	7	8	11
1 in 1.33	16	4	4	4	4	4	4	4	5	6	5	6	7
	24	4	4	4	4	4	5	4	5	6	5	6	7
1 in 1	16	4	4	4	4	4	4	4	4	4	4	4	5
	24	4	4	4	4	4	4	4	4	4	4	4	5

TABLE 7–1b: *Minimum rafter-to-joist nailing (unsupported ridge) (minimum number of nails at least 76 mm long)*

Roof slope	Rafter spacing (mm)	Rafter tied to every joist — Building width up to 8 m			Building width up to 9.8 m			Rafter tied to joist every 1200 mm — Building width up to 8 m			Building width up to 9.8 m		
		1 or less	1.5	2 or more	1 or less	1.5	2 or more	1 or less	1.5	2 or more	1 or less	1.5	2 or more
1 in 3	400	4	5	6	5	7	8	11	—	—	—	—	—
	600	6	8	9	8	—	—	11	—	—	—	—	—
1 in 2.4	400	4	4	5	5	6	7	7	10	—	9	—	—
	600	5	7	8	7	9	11	7	10	—	—	—	—
1 in 2	400	4	4	4	4	4	5	6	8	9	8	—	—
	600	4	5	6	5	7	8	6	8	9	8	—	—
1 in 1.71	400	4	4	4	4	4	4	5	6	8	7	9	11
	600	4	4	5	5	6	7	5	6	8	7	9	11
1 in 1.33	400	4	4	4	4	4	4	4	5	6	5	6	7
	600	4	4	4	4	4	5	4	5	6	5	6	7
1 in 1	400	4	4	4	4	4	4	4	4	4	4	4	5
	600	4	4	4	4	4	4	4	4	4	4	4	5

(Courtesy National Research Council.)

In a hip roof, the common rafters at the ends of the ridge (see Fig. 7–2) will be erected first, with their bottom end against a ceiling joist and nailed, bottom and top, as above. The end commons are raised next, followed by the hip rafters and finally, the jacks, in pairs. End commons and all jacks will be located against a regular or stub joist (see Fig. 6–6) and nailed, top and bottom, according to the schedule or to Tables 7–1a and 7–1b.

GABLE FRAME

In a building with a gable roof, that triangular portion of the end wall which extends from the top of the wall plate to the rafters is known as the *gable*.

The method used to frame the gables will depend on the type of framing system employed. If a *balloon frame* is being used, the end wall studs extend from sill plate to rafters. If *platform framing* is used, the gable is framed separately; those short studs extending from top plate to rafters are *gable studs* (see Fig. 7–18). In either case, the cuts at the top ends of the studs will be the same.

Each gable stud will be laid out in exactly the same way and each will be longer than the preceding one by the line length of the first. Those on the opposite side of center will be marked and cut with the opposite slope. Openings for ventilating louvres are framed as illustrated in Fig. 7–19 and should be as near the ridge as possible.

FIGURE 7-19: *Opening for a louvre.*

RAFTER BRACING

As is the case with other load-bearing members, the free span of rafters is limited, depending on width of rafter, spacing, and species of timber used. Tables 7–2a and 7–2b give maximum free spans for a number of species, with varying widths and spacings. The free span is expressed in terms of the horizontal projection of the rafter.

TABLE 7-2: *Maximum spans for rafters—not supporting ceiling*

Commercial designation	Grade	Nominal size	Live load 50 lb/ft², rafter spacing						Live load 40 lb/ft², rafter spacing						Live load 30 lb/ft², rafter spacing						Live load 20 lb/ft², rafter spacing					
			12 in.		16 in.		24 in.		12 in.		16 in.		24 in.		12 in.		16 in.		24 in.		12 in.		16 in.		24 in.	
			ft	in.	ft	in.	ft	in.	ft	in.	ft	in.	ft	in.	ft	in.	ft	in.	ft	in.	ft	in.	ft	in.	ft	in.
Spruce-pine-fir [includes spruce (all species except coast sitka spruce), jack-pine, lodgepole pine, ponderosa pine, Balsam fir, and alpine fir]	Select structural	2 × 4	7	6	6	10	5	10	8	1	7	4	6	5	8	11	8	1	7	1	10	2	9	3	8	1
		2 × 6	11	10	10	6	8	7	12	9	11	7	9	5	14	0	12	9	10	8	16	0	14	7	12	7
		2 × 8	15	7	13	10	11	4	16	9	15	3	12	6	18	6	16	9	14	1	21	2	19	3	16	7
		2 × 10	19	11	17	8	14	5	21	5	19	6	15	11	23	7	21	5	18	0	27	0	24	6	21	2
		2 × 12	24	2	21	6	17	7	26	1	23	8	19	5	28	8	26	1	21	11	32	10	29	10	25	9
	No. 1	2 × 4	7	6	6	8	5	5	8	1	7	4	6	0	8	11	8	1	6	9	10	2	9	3	8	0
		2 × 6	11	2	9	8	7	10	12	4	10	8	8	8	13	11	12	1	9	9	16	0	14	2	11	7
		2 × 8	14	9	12	9	10	5	16	3	14	11	11	6	18	4	15	11	13	0	21	2	18	9	15	3
		2 × 10	18	6	16	3	13	3	20	9	17	11	14	8	23	5	20	3	16	7	27	0	23	11	19	6
		2 × 12	22	10	19	9	16	2	25	3	21	10	17	10	28	6	24	8	20	2	32	10	29	1	23	9
	No. 2	2 × 4	6	11	6	0	4	11	7	8	6	7	5	5	8	7	7	6	6	1	9	10	8	10	7	2
		2 × 6	10	1	8	9	7	1	11	2	9	8	7	8	12	7	10	11	8	11	14	7	12	10	10	6
		2 × 8	13	4	11	6	9	5	14	8	12	9	10	3	16	2	14	4	11	8	19	11	16	11	13	10
		2 × 10	17	0	14	8	12	0	18	9	16	3	13	3	21	9	18	4	13	11	24	11	21	7	17	8
		2 × 12	20	8	17	11	14	7	22	10	19	9	16	1	25	6	22	4	16	3	30	4	26	3	21	5
	No. 3	2 × 4	5	3	4	6	3	8	5	9	5	0	4	1	6	4	5	8	4	7	7	3	6	8	5	9
		2 × 6	7	6	6	6	5	4	8	3	7	2	5	10	9	4	8	1	6	7	10	7	9	7	7	10
		2 × 8	9	11	8	7	7	0	10	11	9	6	7	9	12	4	10	8	8	9	14	1	12	7	10	3
		2 × 10	12	8	10	11	8	11	14	0	12	1	9	8	15	9	13	8	11	2	18	0	16	1	13	2
		2 × 12	15	5	13	4	10	10	17	0	14	8	12	0	19	3	16	8	13	7	22	0	19	7	16	0
	Construction	2 × 4	5	10	5	1	4	1	6	5	5	7	4	6	7	4	6	4	5	2	8	7	7	5	6	1
	Standard	2 × 4	4	6	3	11	3	2	5	0	4	4	3	6	5	8	4	11	4	0	6	5	5	9	4	8
	Utility	2 × 4	3	0	2	7	2	1	3	4	2	10	2	4	3	9	3	3	2	8	4	5	3	10	3	1
Western cedar (includes western red cedar and pacific coast yellow cedar)	Select structural	2 × 4	7	3	6	7	5	9	7	9	7	1	6	2	8	7	7	10	6	10	9	10	8	11	7	9
		2 × 6	11	5	10	4	8	4	12	3	11	2	9	5	13	6	12	2	10	1	15	6	14	1	12	3
		2 × 8	15	0	13	8	11	2	16	2	14	8	12	6	17	10	16	2	14	1	20	5	18	6	16	2
		2 × 10	19	2	17	5	14	5	20	8	18	9	15	11	22	9	20	8	18	0	26	0	23	8	20	8
		2 × 12	23	4	21	2	17	7	25	2	22	10	19	5	27	8	25	2	21	11	31	8	28	9	25	2
	No. 1	2 × 4	7	3	6	7	5	5	7	9	7	1	6	0	8	7	7	7	6	6	9	6	8	11	7	3
		2 × 6	11	5	9	10	8	0	12	3	10	4	8	9	13	6	12	2	9	9	15	5	13	6	11	7
		2 × 8	15	2	13	0	10	7	16	2	14	4	12	6	17	10	16	0	13	0	20	5	18	6	15	3
		2 × 10	19	0	16	8	13	7	20	8	18	4	15	11	22	9	20	4	16	7	26	0	23	9	19	6
		2 × 12	23	4	20	3	16	6	25	2	22	4	19	5	27	8	25	0	20	2	31	8	28	11	23	9
	No. 2	2 × 4	7	0	6	2	5	3	7	9	6	9	5	6	8	7	7	4	6	6	9	6	8	7	7	0
		2 × 6	10	1	8	9	7	4	11	8	9	8	7	11	13	0	10	11	9	6	14	10	13	1	10	7
		2 × 8	13	4	11	8	9	7	14	9	12	3	10	3	16	7	14	4	11	8	19	7	16	11	13	9
		2 × 10	17	0	14	0	12	0	18	3	16	3	13	3	20	2	18	4	14	4	24	11	21	7	16	9
		2 × 12	20	8	17	11	14	7	22	10	19	9	16	1	24	6	22	4	17	5	30	4	26	3	21	5
	No. 3	2 × 4	5	3	4	6	3	8	5	9	5	0	4	1	6	6	5	8	4	7	7	8	6	8	5	6
		2 × 6	7	10	6	10	5	7	8	8	7	6	6	2	10	0	8	6	6	11	11	6	10	0	8	2
		2 × 8	10	5	9	0	7	4	11	8	9	11	8	1	11	4	11	3	9	2	15	3	13	3	10	9
		2 × 10	13	2	11	5	9	4	14	9	12	9	10	4	17	4	14	4	11	8	19	6	16	11	13	6
		2 × 12	16	2	14	0	11	5	17	9	15	5	12	7	20	2	17	5	14	3	23	9	20	8	16	3
	Construction	2 × 4	6	0	5	3	4	3	6	8	5	9	4	8	7	6	6	6	5	4	8	10	7	8	6	3
	Standard	2 × 4	4	6	3	11	3	2	5	0	4	4	3	6	5	8	4	11	4	0	6	8	5	9	4	8
	Utility	2 × 4	3	0	2	7	2	1	3	4	2	10	2	4	3	9	3	3	2	8	4	5	3	10	3	1

195

TABLE 7-2b: *Maximum spans for rafters—not supporting ceiling*

Lumber species	Grade	Size (mm)	Live load 2.5 kN/m² Rafter spacing			Live load 2.0 kN/m² Rafter spacing			Live load 1.5 kN/m² Rafter spacing			Live load 1.0 kN/m² Rafter spacing		
			300 mm	400 mm	600 mm	300 mm	400 mm	600 mm	300 mm	400 mm	600 mm	300 mm	400 mm	600 mm
Douglas Fir Larch (N)	Sel. Str.	38 × 89	2.50	2.27	1.98	2.69	2.45	2.14	2.97	2.69	2.35	3.40	3.09	2.69
		38 × 140	3.93	3.57	3.09	4.24	3.85	3.36	4.66	4.24	3.70	5.34	4.85	4.24
		38 × 184	5.19	4.71	4.07	5.59	5.08	4.43	6.15	5.59	4.88	7.04	6.40	5.59
		38 × 235	6.62	6.01	5.19	7.13	6.48	5.66	7.85	6.48	5.40	8.16	7.41	6.36
		38 × 286	8.05	7.31	6.32	8.67	7.88	6.88	9.54	7.88	6.57	9.93	9.02	7.73
	No. 1	38 × 89	2.50	2.27	1.98	2.69	2.45	2.14	2.97	2.69	2.35	3.40	3.09	2.69
		38 × 140	3.93	3.50	2.86	4.24	3.85	3.16	4.66	4.24	3.57	5.34	4.85	4.20
		38 × 184	5.19	4.62	3.77	5.59	5.08	4.16	6.15	5.59	4.71	7.04	6.40	5.54
		38 × 235	6.62	5.90	4.81	7.13	6.48	5.31	7.13	7.13	6.01	8.98	8.16	7.07
		38 × 286	8.05	7.17	5.86	8.76	7.88	6.46	8.67	8.67	7.31	10.93	9.93	8.60
	No.2	38 × 89	2.42	2.19	1.79	2.60	2.37	1.97	2.87	2.60	2.23	3.28	2.98	2.60
		38 × 140	3.64	3.15	2.57	4.02	3.48	2.88	4.51	3.93	3.21	5.16	4.63	3.78
		38 × 184	4.80	4.16	3.39	5.30	4.59	3.74	5.94	5.19	4.23	6.81	6.10	4.98
		38 × 235	6.13	5.30	4.33	6.76	5.85	4.78	7.59	6.62	5.40	8.68	7.79	6.36
		38 × 286	7.45	6.45	5.27	8.22	7.12	5.81	9.23	8.05	6.57	10.56	9.47	7.73
	No. 3	38 × 89	1.87	1.62	1.32	2.06	1.78	1.46	2.33	2.02	1.65	2.74	2.38	1.94
		38 × 140	2.75	2.38	1.95	3.04	2.63	2.15	3.44	2.98	2.43	4.04	3.50	2.86
		38 × 184	3.63	3.15	2.57	4.01	3.47	2.83	4.53	3.92	3.20	5.33	4.62	3.77
		38 × 235	4.64	4.01	3.28	5.12	4.43	3.62	5.78	5.01	4.09	6.81	5.89	4.81
		38 × 286	5.64	4.88	3.99	6.22	5.39	4.40	7.03	6.09	4.97	8.28	7.17	5.85
	Const	38 × 89	2.14	1.85	1.51	2.36	2.04	1.67	2.67	2.31	1.89	3.14	2.72	2.22
	Stand	38 × 89	1.61	1.39	1.14	1.77	1.54	1.25	2.01	1.74	1.42	2.36	2.04	1.67
	Util	38 × 89	1.09	0.95	0.77	1.21	1.05	0.85	1.37	1.18	0.96	1.61	1.39	1.14
	Sel. Str.	38 × 89	2.41	2.19	1.84	2.60	2.36	2.03	2.86	2.60	2.27	3.27	2.97	2.60
		38 × 140	3.76	3.26	2.66	4.08	3.59	2.93	4.49	4.06	3.32	5.15	4.67	3.90
		38 × 184	4.96	4.29	3.51	5.38	4.74	3.87	5.93	5.36	4.37	6.78	6.16	5.15
		38 × 235	6.33	5.48	4.47	6.87	6.05	4.94	7.56	6.84	5.58	8.66	7.87	6.57
		38 × 286	7.70	6.67	5.44	8.36	7.36	6.01	9.20	8.32	6.79	10.53	9.57	7.99
	No. 1	38 × 89	2.39	2.07	1.69	2.60	2.29	1.86	2.86	2.58	2.11	3.27	2.97	2.60
		38 × 140	3.49	3.03	2.47	3.85	3.34	2.72	4.36	3.77	3.08	5.13	4.44	3.63
		38 × 184	4.61	3.99	3.26	5.08	4.40	3.59	5.75	4.98	4.06	6.76	5.86	4.78
		38 × 235	5.88	5.09	4.16	6.49	5.62	4.59	7.33	6.35	5.18	8.63	7.47	6.10
		38 × 286	7.15	6.19	5.06	7.89	6.81	5.58	8.92	7.73	6.31	10.50	9.09	7.42
		38 × 89	2.17	1.88	1.53	2.39	2.07	1.69	2.70	2.34	1.91	3.16	2.76	2.25
		38 × 140	3.13	2.71	2.21	3.46	2.99	2.44	3.91	3.39	2.76	4.60	3.98	3.25

Species	Grade	Size												
Hem-Fir (N)	No. 2	38 × 184	4.13	3.58	2.92	4.56	3.95	3.22	5.15	4.46	3.64	6.07	5.25	4.29
		38 × 235	5.27	4.57	3.73	5.82	5.04	4.11	6.58	5.70	4.65	7.74	6.71	5.47
		38 × 286	6.42	5.56	4.54	7.08	6.13	5.00	8.00	6.93	5.66	9.42	8.16	6.66
		38 × 89	1.61	1.39	1.14	1.77	1.54	1.25	2.01	1.74	1.42	2.36	2.04	1.67
		38 × 140	2.38	2.06	1.68	2.62	2.27	1.85	2.97	2.57	2.10	3.49	3.02	2.47
	No. 3	38 × 184	3.13	2.71	2.22	3.46	3.00	2.44	3.91	3.39	2.76	4.60	3.99	3.25
		38 × 235	4.00	3.46	2.83	4.41	3.82	3.12	4.99	4.32	3.53	5.87	5.09	4.15
		38 × 286	4.87	4.21	3.44	5.37	4.65	3.80	6.07	5.26	4.29	7.15	6.19	5.05
		38 × 89	1.85	1.60	1.31	2.04	1.77	1.44	2.31	2.00	1.63	2.72	2.35	1.92
		38 × 140	1.39	1.20	0.98	1.53	1.32	1.08	1.73	1.50	1.22	2.04	1.76	1.44
	Const	38 × 89	0.95	0.82	0.67	1.05	0.90	0.74	1.18	1.02	0.83	1.39	1.21	0.98
	Stand	38 × 89	2.27	2.06	1.78	2.45	2.22	1.94	2.69	2.45	2.14	3.09	2.80	2.45
	Util	38 × 89	3.57	3.15	2.57	3.85	3.48	2.84	4.24	3.85	3.21	4.85	4.41	3.78
	Sel. Str.	38 × 184	4.71	4.16	3.39	5.08	4.59	3.74	5.59	5.08	4.23	6.40	5.81	4.98
		38 × 235	6.01	5.30	4.33	6.48	5.85	4.78	7.13	6.48	5.40	8.16	7.41	6.36
		38 × 286	7.31	6.45	5.27	7.88	7.12	5.81	8.67	7.88	6.57	9.93	9.02	7.73
		38 × 89	2.27	2.00	1.64	2.45	2.21	1.80	2.69	2.45	2.04	3.09	2.80	2.40
		38 × 140	3.39	2.93	2.39	3.74	3.23	2.64	4.22	3.66	2.99	4.85	4.30	3.51
	No. 1	38 × 184	4.46	3.87	3.16	4.93	4.27	3.48	5.57	4.82	3.94	6.40	5.68	4.63
		38 × 235	5.70	4.93	4.03	6.29	5.44	4.44	7.11	6.15	5.02	8.16	7.24	5.91
		38 × 286	6.93	6.00	4.90	7.65	6.62	5.41	8.64	7.49	6.11	9.93	8.81	7.19
		38 × 89	2.10	1.81	1.48	2.31	2.00	1.63	2.60	2.26	1.85	2.98	2.67	2.18
		38 × 140	3.04	2.63	2.15	3.35	2.90	2.37	3.79	3.28	2.68	4.46	3.86	3.15
Spruce-Pine-Fir	No. 2	38 × 184	4.01	3.47	2.83	4.42	3.83	3.12	5.00	4.43	3.53	5.88	5.09	4.16
		38 × 235	5.11	4.43	3.61	5.64	4.88	3.99	6.38	5.52	4.51	7.50	6.50	5.31
		38 × 286	6.22	5.38	4.40	6.86	5.94	4.85	7.76	6.72	5.48	9.13	7.90	6.45
		38 × 89	1.57	1.36	1.11	1.73	1.50	1.22	1.96	1.70	1.38	2.31	2.00	1.63
		38 × 140	2.31	2.00	1.63	2.55	2.21	1.80	2.89	2.50	2.04	3.40	2.94	2.40
	No. 3	38 × 184	3.05	2.64	2.16	3.37	2.92	2.38	3.81	3.30	2.69	4.46	3.88	3.17
		38 × 235	3.89	3.37	2.75	4.30	3.72	3.04	4.86	4.21	3.43	5.72	4.95	4.04
		38 × 286	4.74	4.10	3.35	5.23	4.53	3.69	5.91	5.12	4.18	6.95	6.02	4.92
		38 × 89	1.79	1.55	1.26	1.97	1.71	1.39	2.23	1.93	1.57	2.62	2.27	1.85
		38 × 140	1.34	1.16	0.95	1.48	1.28	1.05	1.67	1.45	1.18	1.97	1.71	1.39
	Const	38 × 89	0.92	0.79	0.65	1.01	0.87	0.71	1.14	0.99	0.81	1.35	1.16	0.95
	Stand	38 × 89	2.19	1.99	1.73	2.36	2.14	1.87	2.59	2.36	2.06	2.97	2.70	2.36
	Util	38 × 89	3.44	3.12	2.60	3.71	3.37	2.87	4.08	3.71	3.24	4.67	4.24	3.71
	Sel. Str.	38 × 184	4.54	4.12	3.43	4.89	4.44	3.79	5.38	4.88	4.27	6.16	5.59	4.89
		38 × 235	5.79	5.26	4.38	6.24	5.66	4.83	6.86	6.23	5.45	7.86	7.14	6.24
		38 × 286	7.04	6.40	5.13	7.58	6.89	5.88	8.35	7.58	6.63	9.56	8.68	7.58
		38 × 89	2.19	1.99	1.65	2.36	2.14	1.83	2.59	2.36	2.06	2.97	2.70	2.36
		38 × 140	3.43	2.97	2.42	3.71	3.28	2.67	4.08	3.70	3.02	4.67	4.24	3.56
	No. 1	38 × 184	4.89	3.92	3.20	4.89	4.32	3.53	5.38	4.88	3.99	6.16	5.59	4.69
		38 × 235	5.77	5.00	4.08	6.24	5.51	4.50	6.86	6.23	5.09	7.86	7.14	5.99
		38 × 286	7.02	6.08	4.96	7.58	6.71	5.48	8.35	7.58	6.19	9.56	8.68	7.29
		38 × 89	2.11	1.84	1.50	2.28	2.03	1.66	2.51	2.28	1.87	2.87	2.61	2.20
		38 × 140	3.06	2.65	2.16	3.38	2.93	2.39	3.82	3.31	2.70	4.50	3.89	3.18

TABLE 7-2b (cont)

Lumber species	Grade	Size (mm)	Live load 2.5 kN/m² Rafter spacing			Live load 2.0 kN/m² Rafter spacing			Live load 1.5 kN/m² Rafter spacing			Live load 1.0 kN/m² Rafter spacing		
			300 mm	400 mm	600 mm	300 mm	400 mm	600 mm	300 mm	400 mm	600 mm	300 mm	400 mm	600 mm
Western Cedars (N)	No. 2	38 × 184	4.04	3.50	2.85	4.46	3.86	3.15	5.04	4.36	3.56	5.93	5.13	4.19
		38 × 235	5.15	4.46	3.64	5.69	4.92	4.02	6.43	5.57	4.54	7.56	6.55	5.35
		38 × 286	6.27	5.43	4.43	6.92	5.99	4.89	7.82	6.77	5.53	9.20	7.97	6.51
		38 × 89	1.57	1.36	1.11	1.73	1.50	1.22	1.96	1.70	1.38	2.31	2.00	1.63
		38 × 140	2.31	2.00	1.63	2.55	2.21	1.80	2.89	2.50	2.04	3.40	2.94	2.40
	No. 3	38 × 184	3.05	2.64	2.16	3.37	2.92	2.38	3.81	3.30	2.69	4.48	3.88	3.17
		38 × 235	3.89	3.37	2.75	4.30	3.72	3.04	4.86	4.21	3.43	5.72	4.95	4.04
		38 × 286	4.74	4.10	3.35	5.23	4.53	3.69	5.91	5.12	4.18	6.95	6.02	4.92
	Const	38 × 89	1.80	1.56	1.27	1.99	1.72	1.40	2.25	1.95	1.59	2.65	2.29	1.87
	Stand	38 × 89	1.34	1.16	0.95	1.48	1.28	1.05	1.67	1.45	1.18	1.97	1.71	1.39
	Util	38 × 89	0.92	0.79	0.65	1.01	0.87	0.71	1.14	0.99	0.81	1.35	1.16	0.95
Eastern Hemlock Tamarack (N)	Sel. Str.	38 × 89	2.30	2.09	1.82	2.47	2.25	1.96	2.72	2.47	2.16	3.12	2.83	2.47
		38 × 140	3.61	3.28	2.86	3.89	3.53	3.09	4.28	3.89	3.40	4.90	4.45	3.89
		38 × 184	4.76	4.33	3.78	5.13	4.66	4.07	5.65	5.13	4.48	6.46	5.87	5.13
		38 × 235	6.08	5.52	4.82	6.55	5.95	5.19	7.21	6.55	5.72	8.25	7.49	6.55
		38 × 286	7.39	6.71	5.87	7.96	7.23	6.32	8.76	7.96	6.95	10.03	9.12	7.96
	No. 1	38 × 89	2.30	2.09	1.82	2.47	2.25	1.96	2.72	2.47	2.16	3.12	2.83	2.47
		38 × 140	3.61	3.28	2.75	3.89	3.53	3.04	4.28	3.89	3.40	4.90	4.45	3.89
		38 × 184	4.76	4.33	3.63	5.13	4.66	4.01	5.65	5.13	4.48	6.46	5.87	5.13
		38 × 235	6.08	5.52	4.64	6.55	5.95	5.12	7.21	6.55	5.72	8.25	7.49	6.55
		38 × 286	7.39	6.71	5.64	7.96	7.23	6.22	8.76	7.96	6.95	10.03	9.12	7.79
	No. 2	38 × 89	2.21	2.01	1.72	2.38	2.17	1.89	2.62	2.38	2.08	3.01	2.73	2.38
		38 × 140	3.48	3.03	2.47	3.75	3.34	2.72	4.13	3.75	3.08	4.73	4.29	3.63
		38 × 184	4.59	3.99	3.26	4.94	4.40	3.59	5.44	4.94	4.06	6.23	5.66	4.78
		38 × 235	5.86	5.09	4.16	6.31	5.62	4.59	6.59	6.31	5.18	7.59	7.22	6.10
		38 × 286	7.12	6.19	5.06	7.68	6.83	5.58	8.45	7.68	6.31	9.67	8.79	7.42
	No. 3	38 × 89	1.80	1.56	1.27	1.99	1.72	1.40	2.25	1.95	1.59	2.65	2.29	1.87
		38 × 140	2.64	2.29	1.87	2.92	2.53	2.06	3.30	2.86	2.33	3.88	3.36	2.74
		38 × 184	3.49	3.02	2.46	3.85	3.33	2.72	4.35	3.77	3.07	5.12	4.43	3.62
		38 × 235	4.45	3.85	3.15	4.91	4.25	3.47	5.55	4.81	3.93	6.53	5.66	4.62
		38 × 286	5.41	4.69	3.83	5.97	5.17	4.22	6.57	5.85	4.77	7.95	6.88	5.62
	Const	38 × 89	2.07	1.79	1.46	2.28	1.97	1.61	2.53	2.23	1.82	2.90	2.63	2.15
	Stand	38 × 89	1.53	1.32	1.08	1.69	1.46	1.19	1.91	1.65	1.35	2.25	1.95	1.59
	Util	38 × 89	1.07	0.92	0.75	1.18	1.02	0.83	1.33	1.15	0.94	1.57	1.36	1.11

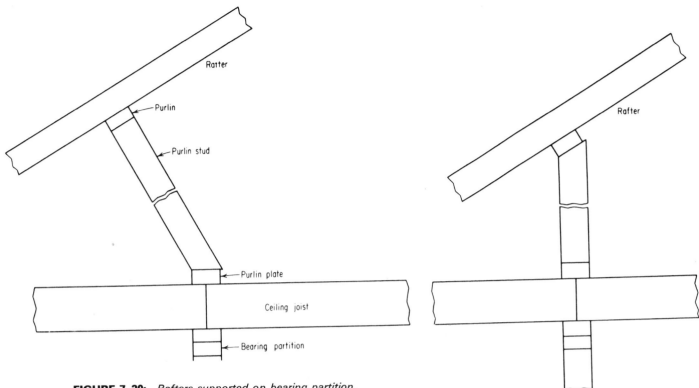

FIGURE 7-20: *Rafters supported on bearing partition.*

FIGURE 7-21: *Vertical purlin stud.*

When the free span of rafters exceeds the maximum allowable, they must be provided with a support. The support is given, where possible, by braces resting on a bearing partition (see Fig. 7-20). First, a *purlin* is nailed to the underside of the rafters, as shown in Fig. 7-20, and a purlin plate to the ceiling joists over the partition. Purlin studs are cut to fit between these, preferably at right angles to the rafters. If they are so placed, the cut at the bottom of the stud will be the same as that on gable studs. If the purlin studs are vertical, this same cut will apply to the top end (see Fig. 7-21).

FIGURE 7-22: *Rafters supported by strongback.*

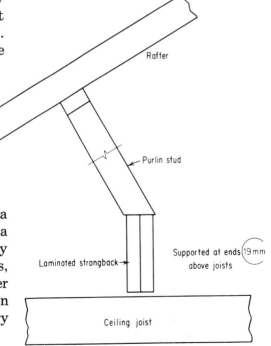

When no bearing partition is available to carry the load, a strongback must be used (see Fig. 7-22). A strongback is a straight timber, either solid or laminated, rigid enough to carry the imposed roof load. This member runs over the ceiling joists, blocked up so that it will clear the joist by ¾ in. (19 mm). Another method is to have the dwarf wall set on the ceiling joists. When this occurs, the size of the ceiling joists must be increased to carry the load.

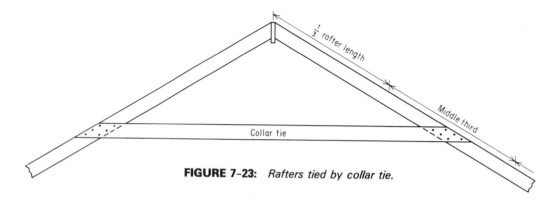

FIGURE 7-23: *Rafters tied by collar tie.*

Collar Ties

Rafters are tied together and stiffened by means of *collar ties*. These are horizontal supports nailed to the rafters, as illustrated in Fig. 7-23, and should be located in the middle third of the rafter length. For rafters on roof slopes of 4 in 12 (1 in 3) or more, intermediate support is generally provided by 2 × 4s (38 × 89 mm). When the length of the ties is over 8 ft (2400 mm), the ties are stiffened by nailing a 1 × 4 (19 × 89 mm) member across the center of their span, from end to end of the roof.

Gable End Projection

Construction of roof projections commonly used at gable ends is shown in Fig. 7-24. Roof overhang projections less than 16 in. (400 mm) over the gable end face will usually terminate with a framing member called a *rake rafter*. When plywood soffit is used, a ladder-type construction is nailed to the rafter located above the gable end wall. Blocking usually spaced 24 in. (600 mm) o.c.

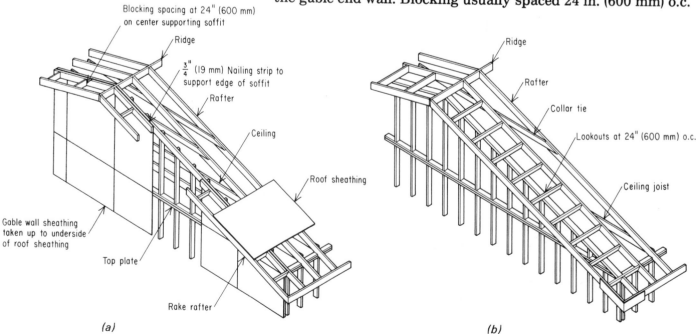

FIGURE 7-24: *a) Narrow overhang construction. b) Typical framing for overhang over 16" (400 mm).*

will act as backing for the plywood soffit. When metal or vinyl is used for soffit material, no special framing is needed for fastening of the soffit. The overhang is finished with a 2 × 6 (38 × 140 mm) rake rafter, supported at the top by the ridge and at the bottom by the rough fascia with the roof sheathing providing the intermediate support.

Roof overhang projections over 16 in. (400 mm) beyond the gable end wall are supported by members called *lookouts*. These will extend back into the roof system a distance equal to the overhang. When the lookouts are placed with their narrow dimension parallel to the roof sheathing, the gable end is built lower to allow room for the lookouts. When the lookouts are placed with the wider dimension parallel to the roof sheathing, the gable end rafter is notched to allow room for the lookouts (see Fig. 7–25).

FIGURE 7–25: *Overhang framing using lookouts on flat.*

DORMER FRAME

Sometimes it is desirable to let light into the building through the roof, and this may be done by means of vertical windows. The house-like structure containing such windows is called *dormer* (see Fig. 7–26). In addition to letting in light, a dormer may increase the usable floor space in the attic.

FIGURE 7–26: *Dormer.*

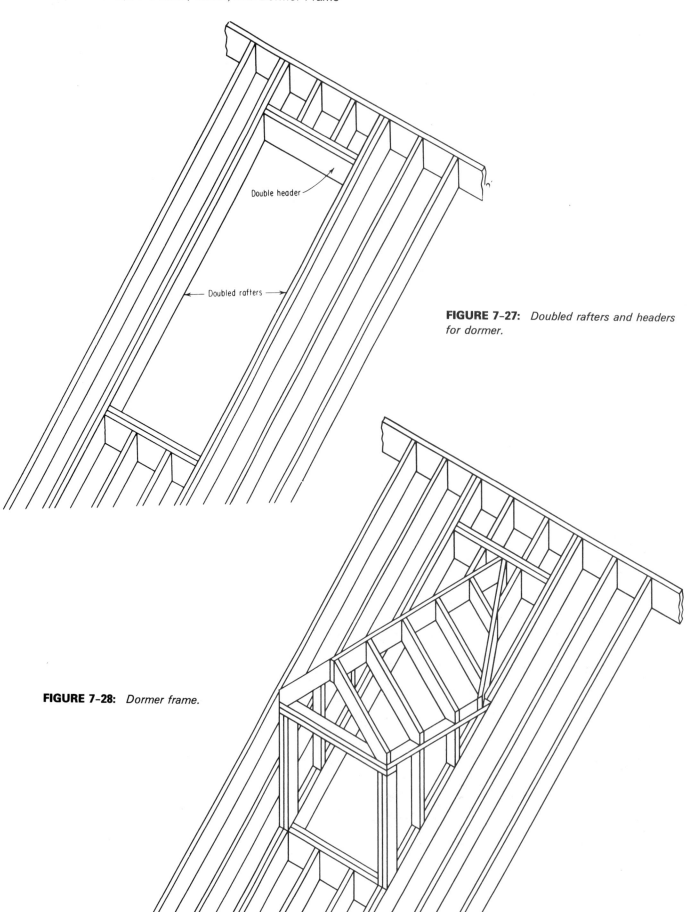

FIGURE 7-27: *Doubled rafters and headers for dormer.*

FIGURE 7-28: *Dormer frame.*

The roof frame must be reinforced to support the dormer. This is done by doubling the rafters on which the sides of the dormer rest and by putting in double headers at the upper and lower ends of the opening (see Fig. 7-27). On this reinforced frame the dormer is built, as illustrated in Fig. 7-28. Framing procedure is the same as for any other wall and roof frame. Cuts on the bottom end of the dormer studs will be the same as for gable studs. If the studs are spaced 16 in. (400 mm) o.c., each will be longer or shorter than the preceding one by the amount of rise for 16 in. 400 mm) of common run.

The rough opening for the dormer window is framed in the same way as the opening for a window in a wall frame. Window headers usually need not be as large, since they carry very little load.

TRUSSED RAFTERS

Trussed rafters are rafters which have been tied together in pairs, along with the bottom chord, to form an individual unit. In addition to the two rafters and the bottom chord, compression and tension webs are incorporated into the unit, making it a truss—a self-supporting structure. Roof trusses are now much more commonly used for roof framing than standard rafters and ceiling joists. They save material and can be put into place much quicker than standard framing. Trusses are usually designed to span from exterior wall to exterior wall, so a great deal of flexibility is available in size and shape of rooms. The entire living area can be designed as a single living area.

A wide variety of roof shapes and types can be framed with trusses. Numerous truss types are available including King post, Fink, Howe, mono, scissor, and flat, as illustrated in Fig. 7-29. Special trusses such as gable end trusses, hip trusses, and girder trusses are used to meet special needs (see Fig. 7-30). Trusses are most adaptable to rectangular houses because of the constant width requiring only one type of truss. This quick erection of the roof frame results in the fast enclosing of the house. Trusses can also be used for L-shaped houses with the use of some special trusses (See Fig. 7-31). For hip roofs, hip trusses are provided for each end, and valley areas require valley trusses (see Figs. 7-30 and 7-31).

Trusses are commonly spaced at 24 in. (600 mm) o.c., allowing efficient use of exterior sheathing or interior finish. Roof trusses must be designed, however, in accordance with accepted engineering practice for the particular details being used. Suppliers of pressed-on metal truss plates have developed an economical design service to fill this need.

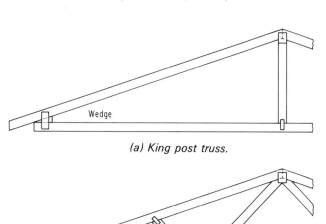

Wedge

(a) King post truss.

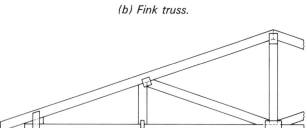

Gusset plate

(b) Fink truss.

(c) Howe truss.

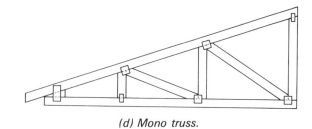

(d) Mono truss.

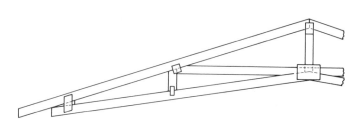

(e) Scissor truss.

FIGURE 7–29: *Truss types.*

(f) Flat truss.

(g) Joist hangers used to support Mono truss.

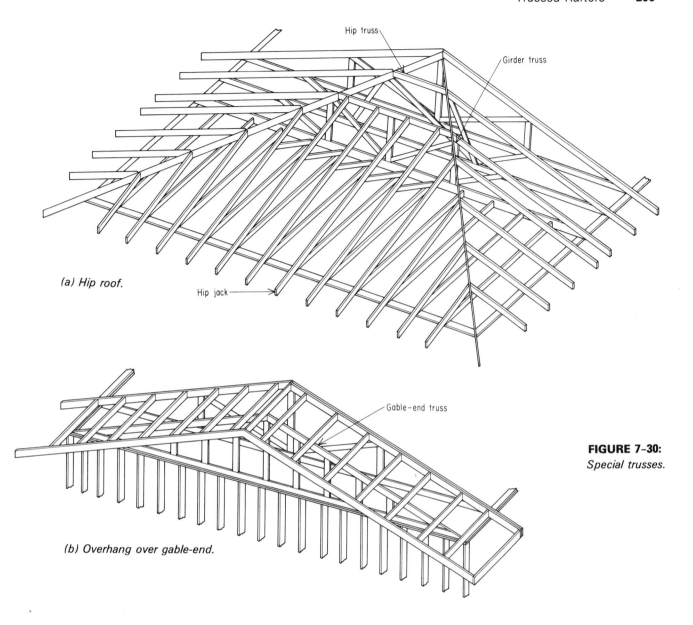

(a) Hip roof.

Hip truss

Girder truss

Hip jack

(b) Overhang over gable-end.

Gable-end truss

FIGURE 7-30:
Special trusses.

(c) Girder truss used in an L-shaped building.

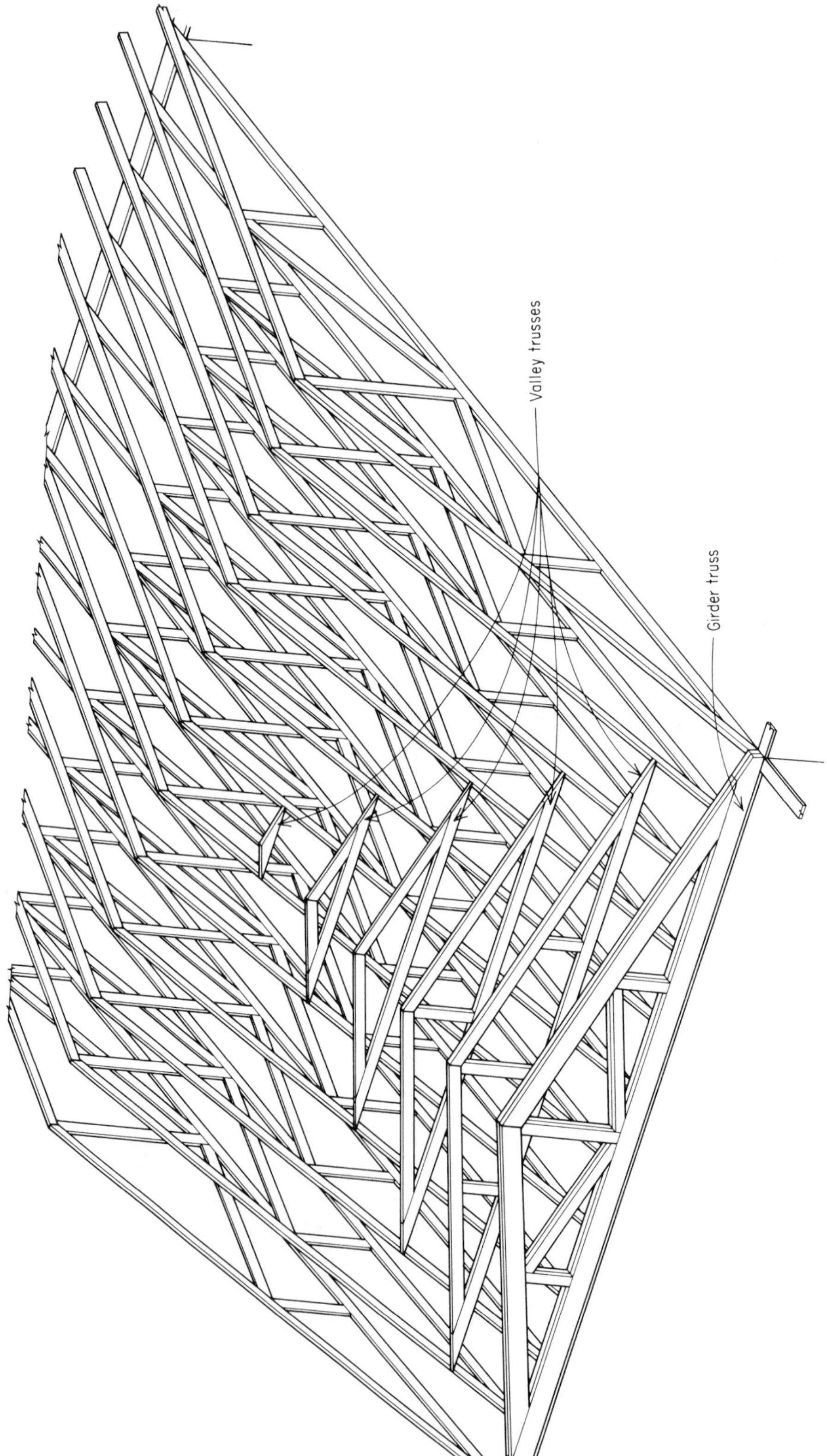

FIGURE 7-31: Framing for an L-shaped roof.

Valley trusses

Girder truss

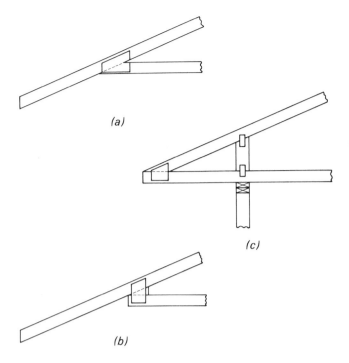

(a)

(c)

(b)

(d)   Truss at exterior wall.

FIGURE 7-32: *Framing eave overhangs.*

Standard Gable Roof

A standard gable roof consists of a number of common trusses set at 24-in. (600-mm) spacings and finished at either end with gable-end trusses (see Fig. 7-30). Eave overhangs may be formed in two ways. One is with the type of truss illustrated in Fig. 7-32(a). Here the overhang will be exactly the same as that formed by an ordinary rafter with a rafter tail. The back of the rafter can be raised further above the wall plate to allow extra room for insulation [see Fig. 7-32(b)]. If the truss is made so that its bottom chord extends to the ends of the rafters, then the truss overhang will be formed by part of the truss end extending beyond the plate line [see Fig. 7-32(c)]. Figure 7-33 illustrates two methods of supporting overhang over the gable end.

FIGURE 7-33: *Overhang support at gable-end.*

FIGURE 7-34: *Bracing gable-end.*

FIGURE 7-35: *Bracing web members.*

FIGURE 7-36: *Trusses delivered to job-site.*

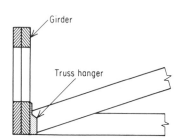

FIGURE 7-37: *Truss hangers supporting trusses.*

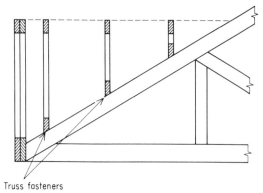

FIGURE 7-38: *Valley trusses nailed in position.*

Erection of trusses will usually begin at one end with the gable end and then continue by adding common trusses and nailing them at prescribed spacing. Bracing is added to keep the gable end plumb (see Fig. 7–34), as well as a strip nailed to chords or web members to keep members straight and at correct spacing (see Fig. 7–35). Trusses are delivered to the job site banded and ready for placement (see Fig. 7–36).

The L-Shaped Roof

The L-shaped roof requires common trusses as well as some speciality trusses (see Fig. 7–31). The girder truss is placed first and braced in a plumb position. Truss hangers are fastened to the lower chord of the girder to support the ends of the common trusses (see Fig. 7–37). If the overhangs of the trusses to be supported by the girder are still in place, cut them off prior to placement in hangers. After all common trusses are nailed into position, valley trusses are nailed in position on the sloped roof (see Fig. 7-38). Care must be taken to keep the peaks of the valley trusses in line with the common trusses. Blocking is nailed between the trusses at the peak to set spacing and provide support for sheathing (see Fig. 7-39).

FIGURE 7-39: *Blocking at peak of roof.*

Hip Roof

Hip roof construction starts with the placement of the hip girder in location as specified on the drawings. Hip jacks are then fastened to the girder and nailed to the end wall plate. Hip trusses are then positioned, followed by the common trusses [see Fig. 7-30(a)]. The other end of the roof is finished in the same way as the first end. Apply all permanent bracing as per the structural truss drawing, and the roof is ready for sheathing.

ROOF SHEATHING

Sheathing for the roof frame may be plywood, particle board, shiplap, or common boards. Minimum thicknesses for roof sheathing materials, based on the rafter spacing, are given in Table 7-3.

TABLE 7-3: *Minimum thicknesses of roof sheathing [inches (mm)]*

Joist or rafter spacing	Minimum plywood thickness		Minimum particle board thickness	Minimum lumber thickness
	Edges supported	Edges unsupported		
12 (300)	$\frac{5}{16}$ (7.5)	$\frac{5}{16}$ (7.5)	$\frac{3}{8}$ (9.5)	$\frac{11}{16}$ (17)
16 (400)	$\frac{5}{16}$ (7.5)	$\frac{3}{8}$ (9.5)	$\frac{3}{8}$ (9.5)	$\frac{11}{16}$ (17)
24 (600)	$\frac{3}{8}$ (9.5)	$\frac{1}{2}$ (12.5)	$\frac{7}{16}$ (11.1)	$\frac{3}{4}$ (19)

(Courtesy National Research Council.)

Shiplap or common boards must be applied solid if composition roofing is to be used, but boards may be spaced if wood shingles or tile roofing are applied in areas not subjected to wind-driven snow (see Fig. 7–40).

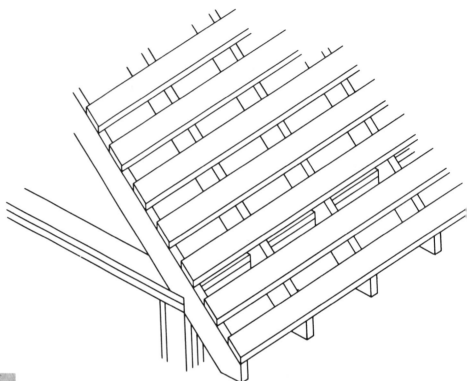

FIGURE 7–40: *Spaced roof sheathing.*

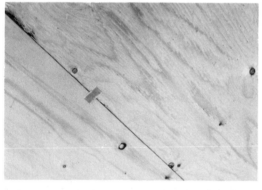

FIGURE 7–41: *Using H-clips.*

When plywood is used for roof sheathing, it should be placed with the face grain perpendicular to the roof framing. The end joints of the panels are staggered to provide a good tie and spaced at least ⅛ in. (2–3 mm) to prevent buckling when expansion occurs. The thickness of plywood or particle board depends on the spacing of the framing members. To prevent damage to the roof covering when thinner material is used, the joints running at right angle to the rafters should be supported by blocking nailed between framing members or by using H-clips (see Fig. 7–41).

Sheathing on roof joists requires special consideration so that condensation doesn't build up in the roof system. Two by fours (38 × 89 mm) are nailed at right angles to the joists to allow a good movement of air between the joist spaces (see Fig. 7–42). The sheathing is then laid with the grain parallel to the roof joists (see Fig. 7–43).

Using roof joists can also require a change in the method used to frame the end walls. As the interior room will have a sloping ceiling, the wall must be framed with a corresponding slope at the top (see Fig. 7–44).

FIGURE 7-42: *Members nailed at right angles to roof joists.*

REVIEW QUESTIONS

7-1. On a simple diagram, illustrate each of the following rafter terms: (a) span, (b) run, (c) unit run, (d) unit rise, (e) total rise, (f) overhang, (g) rafter tail, (h) line length, (i) plumb cut, (j) seat cut.

7-2. What do you understand by "a rafter with a 12-in. (250-mm) rise?"

7-3. By how much do line length and total length of a rafter differ?

7-4. What is the reason for "dropping" a hip rafter?

7-5. Give the length of the ridgeboard in each of the following cases:

(a) Hip roof, plan dimensions 22 ft (6600 mm) × 26 ft (7800 mm) 1½-in. (38 × 89 mm) ridgeboard

(b) Hip roof, plan dimensions 26 ft, 10 in. (7850 mm) × 32 ft, 8 in. (9675 mm), ¾-in. (19 × 140 mm) ridgeboard

7-6. Give the unit length of the common rafter for each of the following rises:

(a) 6 in. (150 mm)

(b) 8 in. (200 mm)

(c) 10 in. (250 mm)

7-7. Calculate the total length of the common rafter in each of the following:

(a) Span—24 ft; overhang—12 in.; rise—8 in.

(b) Span—8000 mmm; overhang—400 mm; rise—150 mm

(c) Span—29 ft; overhang—16 in.; rise—4 in.

(d) Span—9600 mm; overhang—700 mm; rise—100 mm

7-8. Find the unit line length of the hip rafter for each of the rafter rises listed in Question 7-6.

7-9. Calculate the total line length of the hip rafter in each of the cases listed in Question 7-7.

7-10. Explain the difference in the truss framing of a gable roof and a hip roof.

FIGURE 7-43: *Sheathing laid parallel to roof joists.*

FIGURE 7-44: *Wall framing for roof joist system.*

8

ALTERNATIVE ROOF SYSTEMS

When one thinks of a roof on a building in the category of light construction, particularly a house roof, the same picture usually appears. It is the picture of a sloped roof, either gable or hip, made with the conventional roof-framing materials. And, indeed, a great percentage of the roofs, on dwelling houses especially, are made in just that way.

There are alternatives to this roof, and more and more in modern light construction one of these alternatives is being used. Among them are (1) a *flat roof* in which the ceiling joists are also the roof joists, (2) a flat roof carried by laminated beams resting on the top of the walls, (3) a flat roof carried by box beams resting on top of the walls, (4) a *sloped roof with the rafters resting on the ends of ceiling joists* which project beyond the walls, (5) a roof framed by what is known as the *rigid frame* system, (6) the *A-frame*, (7) a *stressed-skin panel roof,* and (8) an *arched roof.*

The roof to be used in a particular situation is a matter of some architectural importance. A number of factors are involved in the choosing of a roof type, and they must all be weighed carefully before making the final choice.

One of the important factors is the overall appearance desired. If the long, low look is being sought, the flat or low-pitched roof will probably do most to enhance it. On the other hand, flat roofs may tend to be troublesome in wet climates or where the snowfall is heavy or wet. If wide overhangs are required with a sloped roof, they are best achieved by setting the rafters on the ends of overhanging joists. In this way, the cornice never drops below the plate level, so it does not reduce the possible window height, but the walls may tend to look higher from the outside.

Special effects are obtained by using roof beams with either a flat or a sloped roof, but the cost of the roof may be increased, because of the quality of roof decking usually required.

From the foregoing it is evident that there are a number of alternatives possible in the selection of a roof type. However, that choice should not be made lightly. Once the choice is made, every effort should be made to see that it is properly executed. The following pages will outline the general appearance and steps in construction of each of the roof types mentioned above.

FLAT ROOF WITH REGULAR JOISTS

We have already seen in Chapter 6 the procedure for framing a ceiling when a regular sloped roof is to be used on the building. If a flat roof is to be used instead, the same general ceiling framing scheme may be used with some important alterations.

TABLE 8-1a: *Maximum spans for roof joists—supporting ceiling*

| | | | Live load 30 lb/ft² | | | | | | Live load 20 lb/ft² | | | | | |
| | | | Gypsum board or plastered ceiling, joist spacing | | | Other ceilings, joist spacing | | | Gypsum board or plastered ceiling, joist spacing | | | Other ceilings, joist spacing | | |
Commercial designation	Grade	Nominal size (in.)	12 in.	16 in.	24 in.	12 in.	16 in.	24 in.	12 in.	16 in.	24 in.	12 in.	16 in.	24 in.
Douglas fir-larch	Select structural	2 × 4	7 9	7 1	6 2	8 11	8 1	7 1	8 11	8 1	7 1	10 2	9 3	8 1
		2 × 6	12 3	11 1	9 8	14 0	12 9	11 1	14 0	12 9	11 1	16 0	14 7	12 9
		2 × 8	16 2	14 8	12 10	18 6	16 9	14 8	18 6	16 9	14 8	21 2	19 3	16 9
		2 × 10	20 7	18 9	16 4	23 7	21 5	18 9	23 7	21 5	18 9	27 0	24 6	21 5
		2 × 12	25 1	22 9	19 11	28 8	26 1	22 9	28 8	26 1	22 9	32 10	29 10	26 1
	No. 1	2 × 4	7 9	7 1	6 2	8 11	8 1	7 1	8 11	8 1	7 1	10 2	9 3	8 1
		2 × 6	12 3	11 1	9 8	14 0	12 9	11 1	14 0	12 9	11 1	16 0	14 7	12 9
		2 × 8	16 2	14 8	12 10	18 6	16 9	14 8	18 6	16 9	14 8	21 2	19 3	16 9
		2 × 10	20 7	18 9	16 4	23 7	21 5	18 9	.23 7	21 5	18 9	27 0	24 6	21 5
		2 × 12	25 1	22 9	19 11	28 8	26 1	22 9	28 8	26 1	22 9	32 10	29 10	26 1
	No. 2	2 × 4	7 6	6 10	5 11	8 7	7 10	6 10	8 7	7 10	6 10	9 10	8 11	7 10
		2 × 6	11 10	10 9	9 4	13 6	12 3	10 2	13 6	12 3	10 9	15 6	14 1	11 9
		2 × 8	15 7	14 2	12 4	17 10	16 2	13 5	17 10	16 2	14 2	20 5	18 7	15 5
		2 × 10	19 11	18 1	15 9	22 9	20 8	17 1	22 9	20 8	18 1	26 1	23 8	19 9
		2 × 12	24 2	22 0	19 2	27 8	25 2	20 9	27 8	25 2	22 0	31 9	28 10	24 0
	No. 3	2 × 4	7 3	6 4	5 2	7 4	6 4	5 2	8 3	7 4	6 0	8 6	7 4	6 0
		2 × 6	10 11	9 5	7 8	10 11	9 5	7 8	12 7	10 11	8 11	12 7	10 11	8 11
		2 × 8	14 4	12 5	10 2	14 4	12 5	10 2	16 7	14 4	11 9	16 7	14 4	11 9
		2 × 10	18 4	15 11	13 0	18 4	15 11	13 0	21 2	18 4	15 0	21 2	18 4	15 0
		2 × 12	22 4	19 4	15 9	22 4	19 4	15 9	25 9	22 4	18 3	25 9	22 4	18 3
Hem-fir	Select structural	2 × 4	7 6	6 10	5 11	8 7	7 10	6 10	8 7	7 10	6 10	9 10	8 11	7 10
		2 × 6	11 10	10 9	9 4	13 6	12 3	10 6	13 6	12 3	10 9	15 6	14 1	12 2
		2 × 8	15 7	14 2	12 4	17 10	16 2	13 11	17 10	16 2	14 2	20 5	18 7	16 0
		2 × 10	19 11	18 1	15 9	22 9	20 8	17 9	22 9	20 8	18 1	26 1	23 8	20 6
		2 × 12	24 2	22 0	19 2	27 8	25 2	21 7	27 8	25 2	22 0	31 9	28 10	24 11
	No. 1	2 × 4	7 6	6 10	5 11	8 7	7 10	6 7	8 7	7 10	6 10	9 10	8 11	7 7
		2 × 6	11 10	10 9	9 4	13 6	11 11	9 9	13 6	12 3	10 9	15 6	13 9	11 3
		2 × 8	15 7	14 2	12 4	17 10	15 9	12 10	17 10	16 2	14 2	20 5	18 2	14 10
		2 × 10	19 11	18 1	15 9	22 9	20 1	16 5	22 9	20 8	18 1	26 1	23 3	18 11
		2 × 12	24 2	22 0	19 2	27 8	24 6	20 0	27 8	25 2	22 0	31 9	28 3	23 1
	No. 2	2 × 4	7 3	6 7	5 9	8 3	7 5	6 1	8 3	7 6	6 7	9 6	8 7	7 0
		2 × 6	11 5	10 4	8 8	12 3	10 7	8 8	13 0	11 10	10 0	14 2	12 3	10 0
		2 × 8	15 0	13 8	11 5	16 2	14 0	11 5	17 2	15 7	13 2	18 8	16 2	13 2
		2 × 10	19 2	17 5	14 7	20 8	17 11	14 7	22 11	19 11	16 10	23 10	20 8	16 10
		2 × 12	23 4	21 2	17 9	25 2	21 9	17 9	26 8	24 3	20 6	29 0	25 2	20 6
	No. 3	2 × 4	6 2	5 4	4 4	6 2	5 4	4 4	7 2	6 2	5 0	7 2	6 2	5 0
		2 × 6	9 4	8 1	6 7	9 4	8 1	6 7	10 9	9 4	7 7	10 9	9 4	7 7
		2 × 8	12 4	10 8	8 8	12 4	10 8	8 8	14 3	12 4	10 0	14 3	12 4	10 0
		2 × 10	15 8	13 7	11 1	15 8	13 7	11 1	18 2	15 8	12 10	18 2	15 8	12 10
		2 × 12	19 1	16 7	13 6	19 1	16 7	13 6	22 1	19 1	15 7	22 1	19 1	15 7
Spruce-pine-fir	Select structural	2 × 4	7 1	6 5	5 7	8 1	7 4	6 5	8 1	7 4	6 5	9 3	8 5	7 4
		2 × 6	11 1	10 1	8 10	12 9	11 7	10 1	12 9	11 7	10 1	14 7	13 3	11 7
		2 × 8	14 8	13 4	11 7	16 9	15 3	13 4	16 9	15 3	13 4	19 3	17 5	15 3
		2 × 10	18 9	17 0	14 10	21 5	19 6	17 0	21 5	19 6	17 0	24 6	22 3	19 6
		2 × 12	22 9	20 8	18 1	26 1	23 8	20 8	26 1	23 8	20 8	29 10	27 1	23 8
	No. 1	2 × 4	7 1	6 5	5 7	8 1	7 4	6 5	8 1	7 4	6 5	9 3	8 5	7 4
		2 × 6	11 1	10 1	8 10	12 9	11 5	9 4	12 9	11 7	10 1	14 7	13 2	10 9
		2 × 8	14 8	13 4	11 7	16 9	15 1	12 4	16 9	15 3	13 4	19 3	17 5	14 3
		2 × 10	18 9	17 0	14 10	21 5	19 3	15 8	21 5	19 6	17 0	24 6	22 3	18 2
		2 × 12	22 9	20 8	18 1	26 1	23 5	19 1	26 1	23 8	20 8	29 10	27 1	22 1
	No. 2	2 × 4	6 10	6 2	5 5	7 10	7 1	5 9	7 10	7 1	6 2	8 11	8 2	6 8
		2 × 6	10 9	9 9	8 5	11 11	10 4	8 7	12 4	11 2	9 9	13 9	11 11	9 9
		2 × 8	14 2	12 10	11 1	15 9	13 8	11 1	16 3	14 9	12 10	18 2	15 9	12 10
		2 × 10	18 1	16 5	14 2	20 1	17 5	14 2	20 9	18 10	16 5	23 3	20 1	16 5
		2 × 12	22 0	20 0	17 3	24 6	21 2	17 3	25 2	22 11	20 0	28 3	24 6	20 0
	No. 3	2 × 4	6 2	5 4	4 4	6 2	5 4	4 4	7 2	6 2	5 0	7 2	6 2	5 0
		2 × 6	8 11	7 8	6 3	8 11	7 8	6 3	10 3	8 11	7 3	10 3	8 11	7 3
		2 × 8	11 9	10 2	8 3	11 9	10 2	8 4	13 7	11 9	9 7	13 7	11 9	9 7
		2 × 10	15 0	13 0	10 7	15 0	13 0	10 7	17 4	15 0	12 3	17 4	15 0	12 3
		2 × 12	18 3	15 9	12 10	18 3	15 9	12 10	21 1	18 3	14 10	21 1	18 3	14 10
Western cedars	Select structural	2 × 4	6 10	6 2	5 5	7 9	7 1	6 2	7 9	7 1	6 2	8 11	8 1	7 1
		2 × 6	10 9	9 9	8 6	12 3	11 2	9 9	12 3	11 2	9 9	14 1	12 9	11 2
		2 × 8	14 2	12 10	11 2	16 2	14 8	12 10	16 2	14 8	12 10	18 6	16 10	14 8
		2 × 10	18 0	16 5	14 4	20 8	18 9	16 5	20 8	18 9	16 5	23 8	21 6	18 9
		2 × 12	21 11	19 11	17 5	25 2	22 10	19 11	25 2	22 10	19 11	28 9	26 2	22 10
	No. 1	2 × 4	6 10	6 2	5 5	7 9	7 1	6 2	7 9	7 1	6 2	8 11	8 1	7 1
		2 × 6	10 9	9 9	8 6	12 3	11 2	9 9	12 3	11 2	9 9	14 1	12 9	11 0
		2 × 8	14 2	12 10	11 2	16 2	14 8	12 10	16 2	14 8	12 10	18 6	16 10	14 6
		2 × 10	18 0	16 5	14 4	20 8	18 9	16 5	20 8	18 9	16 5	23 8	21 6	18 7
		2 × 12	21 11	19 11	17 5	25 2	22 10	19 11	25 2	22 10	19 11	28 9	26 2	22 7
	No. 2	2 × 4	6 7	6 0	5 2	7 6	6 9	5 11	7 6	6 10	6 0	8 7	7 10	6 10
		2 × 6	10 4	9 5	8 2	11 10	10 4	8 5	11 10	10 9	9 5	13 7	11 11	9 9
		2 × 8	13 8	12 5	10 10	15 8	13 8	11 1	15 8	14 2	12 5	17 11	15 9	12 10
		2 × 10	17 5	15 10	13 10	19 11	17 5	14 2	19 11	18 1	15 10	22 10	20 1	16 5
		2 × 12	21 2	19 3	16 10	24 3	21 2	17 3	24 3	22 1	19 3	27 9	24 6	20 0
	No. 3	2 × 4	6 2	5 4	4 4	6 2	5 4	4 4	7 2	6 2	5 0	7 2	6 2	5 0
		2 × 6	9 4	8 1	6 7	9 4	8 1	6 7	10 9	9 4	7 7	10 9	9 4	7 7
		2 × 8	12 4	10 8	8 8	12 4	10 8	8 8	14 3	12 4	10 0	14 3	12 4	10 0
		2 × 10	15 8	13 7	11 1	15 8	13 7	11 1	18 2	15 8	12 10	18 2	15 8	12 10
		2 × 12	19 1	16 7	13 6	19 1	16 7	13 6	22 1	19 1	15 7	22 1	19 1	15 7

TABLE 8-1a: *(Cont'd.)*

Commercial designation	Grade	Nominal size (in.)	Live load 50 lb/ft² — Gypsum board or plastered ceiling, joist spacing 12 in.	16 in.	24 in.	Live load 50 lb/ft² — Other ceilings, joist spacing 12 in.	16 in.	24 in.	Live load 40 lb/ft² — Gypsum board or plastered ceiling, joist spacing 12 in.	16 in.	24 in.	Live load 40 lb/ft² — Other ceilings, joist spacing 12 in.	16 in.	24 in.
Douglas fir-larch	Select structural	2 × 4	6 7	5 11	5 2	7 6	6 10	5 11	7 1	6 5	5 7	8 1	7 4	6 5
		2 × 6	10 4	9 4	8 2	11 10	10 9	9 4	11 1	10 1	8 10	12 9	11 7	10 1
		2 × 8	13 7	12 4	10 9	15 7	14 2	12 4	14 8	13 4	11 7	16 9	15 3	13 4
		2 × 10	17 4	15 9	13 9	19 11	18 1	15 9	18 9	17 0	14 10	21 5	19 6	17 0
		2 × 12	21 2	19 2	16 9	24 2	22 0	19 2	22 9	20 8	18 1	26 1	23 8	20 8
	No. 1	2 × 4	6 7	5 11	5 2	7 6	6 10	5 11	7 1	6 5	5 7	8 1	7 4	6 5
		2 × 6	10 4	9 4	8 2	11 10	10 9	9 2	11 1	10 1	8 10	12 9	11 7	10 1
		2 × 8	13 7	12 4	10 9	15 7	14 2	12 1	14 8	13 4	11 7	16 9	15 3	13 3
		2 × 10	17 4	15 9	13 9	19 11	18 1	15 6	18 9	17 0	14 10	21 5	19 6	16 11
		2 × 12	21 2	19 2	16 9	24 2	22 0	18 10	22 9	20 8	18 1	26 1	23 8	20 7
	No. 2	2 × 4	6 4	5 9	5 0	7 3	6 7	5 9	6 10	6 2	5 5	7 10	7 1	6 2
		2 × 6	9 11	9 0	7 11	11 5	10 2	8 3	10 9	9 9	8 6	12 3	11 1	9 1
		2 × 8	13 2	11 11	10 5	15 0	13 5	10 11	14 2	12 10	11 3	16 2	14 8	12 0
		2 × 10	16 9	15 3	13 4	19 2	17 1	13 11	18 1	16 5	14 4	20 8	18 9	15 3
		2 × 12	20 5	18 6	16 2	23 4	20 9	17 0	22 0	20 0	17 5	25 2	22 9	18 7
	No. 3	2 × 4	6 0	5 2	4 3	6 0	5 2	4 3	6 7	5 8	4 8	6 7	5 8	4 8
		2 × 6	8 11	7 8	6 3	8 11	7 8	6 3	9 9	8 5	6 10	9 9	8 5	6 10
		2 × 8	11 9	10 2	8 3	11 9	10 2	8 3	12 10	11 1	9 1	12 10	11 1	9 1
		2 × 10	15 0	13 0	10 7	15 0	13 0	10 7	16 5	14 2	11 7	16 5	14 2	11 7
		2 × 12	18 3	15 9	12 10	18 3	15 9	12 10	20 0	17 3	14 1	20 0	17 3	14 1
Hem-fir	Select structural	2 × 4	6 4	5 9	5 0	7 3	6 7	5 9	6 10	6 2	5 5	7 10	7 1	6 2
		2 × 6	9 11	9 0	7 11	11 5	10 4	8 7	10 9	9 9	8 6	12 3	11 2	9 5
		2 × 8	13 2	11 11	10 5	15 0	13 8	11 4	14 2	12 10	11 3	16 2	14 9	12 5
		2 × 10	16 9	15 3	13 4	19 2	17 5	14 6	18 1	16 5	14 4	20 8	18 9	15 10
		2 × 12	20 5	18 6	16 2	23 4	21 3	17 7	22 0	20 0	17 5	25 2	22 10	19 3
	No. 1	2 × 4	6 4	5 9	5 0	7 3	6 7	5 4	6 10	6 2	5 5	7 10	7 1	5 10
		2 × 6	9 11	9 0	7 11	11 3	9 9	7 11	10 9	9 9	8 6	12 3	10 8	8 8
		2 × 8	13 2	11 11	10 5	14 10	12 10	10 6	14 2	12 10	11 3	16 2	14 1	11 6
		2 × 10	16 9	15 3	13 4	18 11	16 5	13 5	18 1	16 5	14 4	20 8	18 0	14 8
		2 × 12	20 5	18 6	16 2	23 1	20 0	16 4	22 0	20 0	17 5	25 2	21 10	17 10
	No. 2	2 × 4	6 1	5 6	4 10	7 0	6 1	4 11	6 7	5 11	5 2	7 6	6 8	5 5
		2 × 6	9 7	8 8	7 1	10 0	8 8	7 1	10 4	9 5	7 9	11 0	9 6	7 9
		2 × 8	12 8	11 5	9 4	13 2	11 5	9 4	13 8	12 5	10 3	14 6	12 6	10 3
		2 × 10	16 2	14 7	11 11	16 10	14 7	11 11	17 5	15 10	13 1	18 6	16 0	13 1
		2 × 12	19 8	17 9	14 6	20 6	17 9	14 6	21 2	19 3	15 11	22 6	19 5	15 11
	No. 3	2 × 4	5 0	4 4	3 7	5 0	4 4	3 7	5 6	4 9	3 11	5 6	4 9	3 11
		2 × 6	7 7	6 7	5 4	7 7	6 7	5 4	8 4	7 3	5 11	8 4	7 3	5 11
		2 × 8	10 0	8 8	7 1	10 0	8 8	7 1	11 0	9 6	7 9	11 0	9 6	7 9
		2 × 10	12 10	11 1	9 1	12 10	11 1	9 1	14 1	12 2	9 11	14 1	12 2	9 11
		2 × 12	15 7	13 6	11 0	15 7	13 6	11 0	17 1	14 10	12 1	17 1	14 10	12 1
Spruce-pine-fir	Select structural	2 × 4	5 11	5 5	4 9	6 10	6 2	5 5	6 5	5 10	5 1	7 4	6 8	5 10
		2 × 6	9 4	8 6	7 5	10 9	9 9	8 3	10 1	9 2	8 0	11 7	10 6	9 1
		2 × 8	12 4	11 3	9 10	14 2	12 10	10 11	13 4	12 1	10 7	15 3	13 10	12 0
		2 × 10	15 9	14 4	12 6	18 1	16 5	13 11	17 0	15 5	13 6	19 6	17 8	15 3
		2 × 12	19 2	17 5	15 3	22 0	20 0	17 0	20 8	18 9	16 5	23 8	21 6	18 7
	No. 1	2 × 4	5 11	5 5	4 9	6 10	6 2	5 3	6 5	5 10	5 1	7 4	6 8	5 9
		2 × 6	9 4	8 6	7 5	10 9	9 4	7 7	10 1	9 2	8 0	11 7	10 3	8 4
		2 × 8	12 4	11 3	9 10	14 2	12 4	10 0	13 4	12 1	10 7	15 3	13 6	11 0
		2 × 10	15 9	14 4	12 6	18 1	15 8	12 10	17 0	15 5	13 6	19 6	17 3	14 1
		2 × 12	19 2	17 5	15 3	22 0	19 1	15 7	20 8	18 9	16 5	23 8	20 11	17 1
	No. 2	2 × 4	5 9	5 3	4 7	6 7	5 9	4 9	6 2	5 7	4 11	7 1	6 4	5 2
		2 × 6	9 1	8 3	6 10	9 9	8 5	6 10	9 9	8 10	7 6	10 8	9 3	7 6
		2 × 8	11 11	10 10	9 1	12 10	11 1	9 1	12 10	11 8	9 11	14 1	12 2	9 11
		2 × 10	15 3	13 10	11 7	16 5	14 2	11 7	16 5	14 11	12 8	18 0	15 7	12 8
		2 × 12	18 7	16 10	14 1	20 0	17 3	14 1	20 0	18 2	15 5	21 10	18 11	15 5
	No. 3	2 × 4	5 0	4 4	3 7	5 0	4 4	3 7	5 6	4 9	3 11	5 6	4 9	3 11
		2 × 6	7 3	6 3	5 1	7 3	6 3	5 1	7 11	6 10	5 7	7 11	6 10	5 7
		2 × 8	9 7	8 3	6 9	9 7	8 3	6 9	10 6	9 1	7 5	10 6	9 1	7 5
		2 × 10	12 3	10 7	8 8	12 3	10 7	8 8	13 5	11 7	9 5	13 5	11 7	9 5
		2 × 12	14 10	12 10	10 6	14 10	12 10	10 6	16 4	14 1	11 6	16 4	14 1	11 6
Western cedars	Select structural	2 × 4	5 9	5 2	4 6	6 7	6 0	5 2	6 2	5 7	4 11	7 1	6 5	5 7
		2 × 6	9 0	8 2	7 2	10 4	9 5	8 2	9 9	8 10	7 9	11 2	10 1	8 10
		2 × 8	11 11	10 10	9 5	13 8	12 5	10 10	12 10	11 8	10 2	14 8	13 4	11 3
		2 × 10	15 3	13 10	12 1	17 5	15 10	13 10	16 5	14 11	13 0	18 9	17 1	14 11
		2 × 12	18 6	16 10	14 8	21 2	19 3	16 10	19 11	18 1	15 10	22 10	20 9	18 1
	No. 1	2 × 4	5 9	5 2	4 6	6 7	6 0	5 2	6 2	5 7	4 11	7 1	6 5	5 7
		2 × 6	9 0	8 2	7 2	10 4	9 5	7 9	9 9	8 10	7 9	11 2	10 1	8 6
		2 × 8	11 11	10 10	9 5	13 8	12 5	10 3	12 10	11 8	10 2	14 8	13 4	11 3
		2 × 10	15 3	13 10	12 1	17 5	15 10	13 1	16 5	14 11	13 0	18 9	17 1	14 4
		2 × 12	18 6	16 10	14 8	21 2	19 3	16 10	19 11	18 1	15 10	22 10	20 9	17 6
	No. 2	2 × 4	5 6	5 0	4 5	6 4	5 9	4 10	6 0	5 5	4 9	6 10	6 2	5 3
		2 × 6	8 9	7 11	6 10	9 9	8 5	6 10	9 5	8 6	7 5	10 8	9 3	7 6
		2 × 8	11 6	10 5	9 1	12 10	11 1	9 1	12 5	11 3	9 10	14 1	12 2	9 11
		2 × 10	14 8	13 4	11 7	16 5	14 2	11 7	15 10	14 4	12 7	18 0	15 7	12 8
		2 × 12	17 10	16 3	14 1	20 0	17 3	14 1	19 3	17 6	15 3	21 10	18 11	15 5
	No. 3	2 × 4	5 0	4 4	3 7	5 0	4 4	3 7	5 6	4 9	3 11	5 6	4 9	3 11
		2 × 6	7 7	6 7	5 4	7 7	6 7	5 4	8 4	7 3	5 11	8 4	7 3	5 11
		2 × 8	10 0	8 8	7 1	10 0	8 8	7 1	11 0	9 6	7 9	11 0	9 6	7 9
		2 × 10	12 10	11 1	9 1	12 10	11 1	9 1	14 1	12 2	9 11	14 1	12 2	9 11
		2 × 12	15 7	13 6	11 0	15 7	13 6	11 0	17 1	14 10	12 1	17 1	14 10	12 1

TABLE 8-1b: *Maximum spans for roof joists—supporting ceiling*

Lumber species	Grade	Size (mm)	Live load 2.5 kN/m² Gypsum board or plastered ceiling 300 mm	400 mm	600 mm	Other ceilings 300 mm	400 mm	600 mm	Live load 2.0 kN/m² Gypsum board or plastered ceiling 300 mm	400 mm	600 mm	Other ceilings 300 mm	400 mm	600 mm
Douglas fir-larch (N)	Select structural	38 × 89	1.98	1.80	1.57	2.27	2.06	1.80	2.14	1.94	1.70	2.45	2.22	1.94
		38 × 140	3.12	2.83	2.48	3.57	3.24	2.83	3.36	3.05	2.57	3.85	3.50	3.05
		38 × 184	4.11	3.74	3.27	4.71	4.28	3.74	4.43	4.03	3.52	5.08	4.61	4.03
		38 × 235	5.25	4.77	4.17	6.01	5.46	4.77	5.66	5.14	4.49	6.48	5.88	5.14
		38 × 286	6.39	5.80	5.07	7.31	6.64	5.80	6.88	6.25	5.46	7.88	7.16	6.25
	No. 1	38 × 89	1.98	1.80	1.57	2.27	2.06	1.80	2.14	1.94	1.70	2.45	2.22	1.94
		38 × 140	3.12	2.83	2.48	3.57	3.25	2.76	3.36	3.05	2.67	3.85	3.50	3.03
		38 × 184	4.11	3.74	3.27	4.71	4.28	3.64	4.43	4.03	3.52	5.08	4.61	3.99
		38 × 235	5.25	4.77	4.17	6.01	5.46	4.65	5.66	5.14	4.49	6.48	5.88	5.10
		38 × 286	6.39	5.80	5.07	7.31	6.64	5.66	6.88	6.25	5.46	7.88	7.16	6.20
	No. 2	38 × 89	1.92	1.74	1.52	2.20	1.99	1.72	2.07	1.88	1.64	2.37	2.15	1.88
		38 × 140	3.02	2.74	2.39	3.45	3.04	2.49	3.75	2.95	2.58	3.72	3.34	2.72
		38 × 184	3.98	3.61	3.16	4.55	4.02	3.28	4.29	3.89	3.40	4.91	4.40	3.59
		38 × 235	5.08	4.61	4.03	5.81	5.12	4.18	5.47	4.97	4.14	6.26	5.61	4.58
		38 × 286	6.17	5.61	4.90	7.07	6.23	5.09	6.65	6.04	5.28	7.62	6.81	5.57
	No. 3	38 × 89	1.80	1.56	1.27	1.80	1.56	1.27	1.98	1.71	1.40	1.98	1.71	1.40
		38 × 140	2.66	2.30	1.88	2.66	2.30	1.88	2.92	2.52	2.06	2.92	2.52	2.06
		38 × 184	3.51	3.04	2.48	3.51	3.04	2.48	3.84	3.33	2.72	3.84	3.33	2.72
		38 × 235	4.48	3.88	3.17	4.48	3.88	3.17	4.91	4.25	3.47	4.91	4.25	3.47
		38 × 286	5.45	4.72	3.85	5.45	4.72	3.85	5.97	5.17	4.22	5.97	5.17	4.22
Hem-fir (N)	Select structural	38 × 89	1.91	1.74	1.52	2.19	1.99	1.74	2.06	1.87	1.63	2.36	2.14	1.87
		38 × 140	3.01	2.73	2.19	3.44	3.13	3.57	3.57	2.94	2.57	3.71	3.37	2.81
		38 × 184	3.97	3.60	3.15	4.54	4.12	3.39	4.27	3.88	3.39	4.89	4.44	3.71
		38 × 235	5.06	4.60	4.02	5.79	5.26	4.32	5.45	4.95	4.33	6.24	5.67	4.73
		38 × 286	6.16	5.59	4.89	7.05	6.40	5.26	6.63	6.03	5.26	7.59	6.90	5.76
	No. 1	38 × 89	1.91	1.74	1.52	2.19	1.99	1.63	2.06	1.87	1.63	2.36	2.14	1.79
		38 × 140	3.01	2.73	2.39	3.38	2.92	2.39	3.24	2.94	2.57	3.70	3.20	2.61
		38 × 184	3.97	3.60	3.15	4.45	3.85	3.15	4.27	3.88	3.39	4.88	4.22	3.45
		38 × 235	5.06	4.60	4.01	5.68	4.92	4.01	5.45	4.95	4.33	6.22	5.39	4.40
		38 × 286	6.16	5.59	4.88	6.91	5.98	4.88	6.63	6.03	5.26	7.57	6.55	5.35
	No. 2	38 × 89	1.85	1.68	1.46	2.09	1.81	1.48	1.99	1.81	1.58	2.28	1.99	1.62
		38 × 140	2.90	2.62	2.14	3.03	2.62	2.14	3.13	2.84	2.34	3.32	2.87	2.34
		38 × 184	3.83	3.46	2.82	3.99	3.46	2.82	4.13	3.75	3.09	4.37	3.79	3.09
		38 × 235	4.89	4.41	3.60	5.10	4.41	3.60	5.27	4.78	3.95	5.58	4.83	3.95
		38 × 286	5.95	5.37	4.38	6.20	5.37	4.38	6.41	5.82	4.80	6.79	5.88	4.80
	No. 3	38 × 89	1.55	1.34	1.10	1.55	1.34	1.10	1.70	1.47	1.20	1.70	1.47	1.30
		38 × 140	2.30	1.99	1.62	2.30	1.99	1.62	2.52	2.18	1.78	2.52	2.18	1.78
		38 × 184	3.03	2.62	2.14	3.03	2.62	2.14	3.32	2.87	2.34	3.32	2.87	2.34
		38 × 235	3.87	3.35	2.73	3.87	3.35	2.72	4.23	3.67	2.99	4.23	3.67	2.99
		38 × 286	4.70	4.07	3.32	4.70	4.07	3.32	5.15	4.46	3.64	5.15	4.46	3.64
Eastern hemlock-tamarack (N)	Select structural	38 × 89	1.82	1.65	1.44	2.09	1.89	1.65	1.98	1.78	1.56	2.25	2.04	1.78
		38 × 140	2.86	2.60	2.27	3.28	2.98	2.60	3.09	2.80	2.45	3.53	3.21	2.80
		38 × 184	3.78	3.43	3.00	4.33	3.93	3.43	4.07	3.70	3.23	4.66	4.23	3.70
		38 × 235	4.82	4.38	3.83	5.52	5.01	4.38	5.19	4.72	4.12	5.95	5.80	4.72
		38 × 286	5.87	5.33	4.65	6.71	6.10	5.33	6.32	5.74	5.01	7.23	6.57	5.74
	No. 1	38 × 89	1.82	1.65	1.44	2.09	1.98	1.65	1.96	1.78	1.56	2.25	2.04	1.78
		38 × 140	2.86	2.60	2.27	3.28	2.98	2.43	3.09	2.80	2.45	3.53	3.21	2.80
		38 × 184	3.78	3.43	3.00	4.33	3.93	3.43	4.07	3.70	3.23	4.66	4.23	3.70
		38 × 235	4.82	4.38	3.83	5.52	5.01	4.38	5.19	4.72	4.12	5.95	5.40	4.72
		38 × 286	5.87	5.33	4.65	6.71	6.10	5.33	6.32	5.74	5.01	7.23	6.57	5.74
	No. 2	38 × 89	1.76	1.59	1.39	2.01	1.83	1.59	1.89	1.72	1.50	2.17	1.97	1.72
		38 × 140	2.76	2.51	2.19	3.16	2.87	2.39	2.98	2.70	2.36	3.41	3.09	2.61
		38 × 184	3.64	3.31	2.98	4.17	3.79	3.15	3.92	3.56	3.11	4.49	4.08	3.45
		38 × 235	4.65	4.22	3.69	5.32	4.83	4.01	5.01	4.55	3.97	5.73	5.21	4.40
		38 × 286	5.65	5.14	4.49	6.47	5.88	4.88	6.09	5.53	4.83	6.97	6.33	5.35
	No. 3	38 × 89	1.69	1.51	1.23	1.74	1.51	1.23	1.82	1.65	1.35	1.91	1.65	1.35
		38 × 140	2.55	2.21	1.80	2.55	2.21	1.80	2.80	2.42	1.98	2.80	2.42	1.98
		38 × 184	3.37	2.92	2.38	3.37	2.92	2.38	3.69	3.20	2.61	3.69	3.20	2.61
		38 × 235	4.30	3.72	3.04	4.30	3.72	3.04	4.71	4.08	3.33	4.71	4.08	3.33
		38 × 286	5.32	4.53	3.70	5.23	4.53	3.70	5.75	4.96	4.05	5.73	4.96	4.05
Spruce pine-fir	Select structural	38 × 89	1.80	1.64	1.43	2.06	1.87	1.68	1.94	1.76	1.54	2.22	2.02	1.76
		38 × 140	2.83	2.58	2.25	3.25	2.95	2.49	3.05	2.77	2.42	3.50	3.18	2.72
		38 × 184	3.74	3.40	2.97	4.28	3.89	3.28	4.03	3.66	3.20	4.61	4.19	3.59
		38 × 235	4.77	4.33	3.79	5.46	4.96	4.18	5.14	4.67	4.08	5.88	5.35	4.58
		38 × 286	5.80	5.27	4.61	6.64	6.04	5.09	6.25	5.68	4.96	7.16	6.50	5.57
	No. 1	38 × 89	1.80	1.64	1.43	2.06	1.87	1.58	1.94	1.76	1.54	2.22	2.02	1.73
		38 × 140	2.83	2.58	2.25	3.25	2.83	2.31	3.05	2.77	2.42	3.50	3.10	2.53
		38 × 184	3.74	3.40	2.97	4.28	3.71	3.05	4.03	3.66	3.20	4.61	4.09	3.14
		38 × 235	4.77	4.33	3.79	5.46	4.77	3.91	5.14	4.67	4.08	5.88	5.22	4.26
		38 × 286	5.80	5.27	4.61	6.14	5.80	4.73	6.25	5.68	4.96	7.16	6.35	5.18
	No. 2	38 × 89	1.74	1.58	1.38	1.99	1.75	1.43	1.88	1.70	1.49	2.15	1.92	1.57
		38 × 140	2.74	2.49	2.07	2.94	2.54	2.07	2.95	2.68	2.27	3.28	2.28	2.27
		38 × 184	3.61	3.28	2.73	3.87	3.35	2.73	3.89	3.54	3.00	4.24	3.67	3.00
		38 × 235	4.61	4.19	3.49	4.94	4.28	3.49	4.97	4.51	3.82	5.41	4.68	3.82
		38 × 286	5.61	5.10	4.25	6.01	5.20	4.25	6.04	5.49	4.65	6.58	5.70	4.65
	No. 3	38 × 89	1.52	1.31	1.07	1.52	1.31	1.07	1.66	1.44	1.17	1.66	1.44	1.17
		38 × 140	2.23	1.93	1.58	2.23	1.93	1.58	2.45	2.12	1.73	2.45	2.12	1.73
		38 × 184	2.95	2.55	2.08	2.95	2.55	2.08	3.23	2.80	2.28	3.23	2.80	2.28
		38 × 235	3.76	3.26	2.66	3.76	3.26	2.66	4.12	3.57	2.91	4.12	3.57	2.91
		38 × 286	4.58	3.96	3.23	4.58	3.96	3.23	5.01	4.34	3.54	5.01	4.34	3.54

TABLE 8-1b: *(Cont'd.)*

Lumber species	Grade	Size (mm)	Live load 2.5 kN/m² Gypsum board or plastered ceiling 300 mm	400 mm	600 mm	Other ceilings 300 mm	400 mm	600 mm	Live load 2.0 kN/m² Gypsum board or plastered ceiling 300 mm	400 mm	600 mm	Other ceilings 300 mm	400 mm	600 mm
Western cedars (N)	Select structural	38 × 89	1.73	1.58	1.38	1.99	1.80	1.58	1.87	1.70	1.48	2.14	1.94	1.70
		38 × 140	2.73	2.48	2.16	3.12	2.84	2.48	2.94	2.67	2.33	3.37	3.06	2.67
		38 × 184	3.60	3.27	2.86	4.12	3.74	3.27	3.88	3.52	3.08	4.44	4.03	3.52
		38 × 235	4.59	4.17	3.64	5.26	4.78	4.17	4.95	4.50	3.93	5.66	5.15	4.50
		38 × 286	5.59	5.08	4.43	6.40	5.81	5.08	6.02	5.47	4.78	6.89	6.26	5.47
	No. 1	38 × 89	1.73	1.58	1.38	1.99	1.80	1.58	1.87	1.70	1.48	2.14	1.94	1.70
		38 × 140	2.73	2.48	2.16	3.12	2.84	2.34	2.94	2.67	2.33	3.37	3.06	2.57
		38 × 184	3.60	3.27	2.86	4.12	3.74	3.09	3.88	3.52	3.08	4.44	4.03	3.38
		38 × 235	4.59	4.17	3.64	5.26	4.78	3.94	4.95	4.50	3.93	5.66	5.15	4.32
		38 × 286	5.59	5.08	4.43	6.40	5.81	4.79	6.02	5.47	4.78	6.89	6.26	5.25
	No. 2	38 × 89	1.68	1.52	1.33	1.92	1.74	1.45	1.81	1.64	1.43	2.07	1.88	1.59
		38 × 140	2.64	2.40	2.09	2.96	2.56	2.09	2.84	2.58	2.25	3.24	2.81	2.29
		38 × 184	3.48	3.16	2.76	3.90	3.38	2.76	3.75	3.41	2.97	4.27	3.70	3.02
		38 × 235	4.44	4.03	3.52	4.98	4.31	3.52	4.78	4.35	3.80	5.45	4.72	3.85
		38 × 286	5.40	4.91	4.28	6.06	5.24	4.28	5.82	5.29	4.62	6.63	5.74	4.69
	No. 3	38 × 89	1.52	1.31	1.07	1.52	1.31	1.07	1.66	1.44	1.17	1.66	1.44	1.17
		38 × 140	2.23	1.93	1.58	2.23	1.93	1.58	2.45	2.12	1.73	2.45	2.12	1.73
		38 × 184	2.95	2.55	2.08	2.95	2.55	2.08	3.23	2.80	2.28	3.23	2.80	2.28
		38 × 235	3.76	3.26	2.66	3.76	3.26	2.66	4.12	3.57	2.91	4.12	3.57	2.91
		38 × 286	4.58	3.96	3.23	4.58	3.96	3.23	5.01	4.34	3.54	5.01	4.34	3.54

TABLE 8-1b: *(Cont'd.)*

Lumber species	Grade	Size (mm)	Live Load 1.5 kN/m² Gypsum board or plastered ceiling 300 mm	400 mm	600 mm	Other ceilings 300 mm	400 mm	600 mm	Live load 1.0 kN/m² Gypsum board or plastered ceiling 300 mm	400 mm	600 mm	Other ceilings 300 mm	400 mm	600 mm
Douglas fir-larch (N)	Select structural	38 × 89	2.35	2.14	1.87	2.69	2.45	2.14	2.69	2.45	2.14	3.09	2.80	2.45
		38 × 140	3.70	3.36	2.94	4.24	3.85	3.36	4.24	3.85	3.36	4.85	4.41	3.85
		38 × 184	4.88	4.43	3.87	5.59	5.08	4.43	5.59	5.08	5.43	6.40	5.81	5.08
		38 × 235	6.23	5.66	4.94	7.13	6.48	5.66	7.13	6.48	5.66	8.16	7.41	6.48
		38 × 286	5.57	6.88	6.01	8.67	7.88	6.88	8.67	7.88	6.88	9.93	9.02	7.88
	No. 1	38 × 89	2.35	2.14	1.87	2.69	2.45	2.14	2.69	2.45	2.14	3.09	2.80	2.45
		38 × 140	3.70	3.36	2.94	4.24	3.85	3.36	4.24	3.85	3.36	4.85	4.41	3.85
		38 × 184	4.88	4.43	3.87	5.59	5.08	4.43	5.59	5.0	4.43	6.40	5.81	5.08
		38 × 235	6.23	5.66	4.94	7.13	6.48	5.66	7.13	6.48	5.66	8.16	7.41	6.48
		38 × 286	7.57	6.88	6.01	8.67	7.88	6.88	8.67	7.88	6.88	9.93	9.02	7.88
	No. 2	38 × 89	2.27	2.07	1.80	2.60	2.17	2.07	2.60	2.37	2.07	2.98	2.71	2.37
		38 × 140	3.58	3.25	2.84	4.10	3.72	3.04	4.10	3.72	3.25	4.69	4.26	3.52
		38 × 184	4.72	4.39	3.74	5.40	4.91	4.02	5.40	4.91	4.29	6.18	5.67	4.64
		38 × 235	6.02	5.47	4.78	6.89	6.26	5.12	6.89	6.26	5.47	7.89	7.17	5.92
		38 × 286	7.32	6.65	5.81	8.38	7.62	6.23	8.38	7.62	6.65	9.60	8.72	7.20
	No. 3	38 × 89	2.18	1.91	1.56	2.21	1.91	1.56	2.50	2.21	1.80	2.55	2.21	1.80
		38 × 140	3.26	2.82	2.30	3.26	2.82	2.30	3.77	3.26	2.66	3.77	3.26	2.66
		38 × 184	4.30	3.72	3.04	4.30	3.72	3.04	4.96	4.30	3.51	4.96	4.30	3.51
		38 × 235	5.49	4.75	3.88	5.49	4.75	3.88	6.34	5.49	4.48	6.34	5.49	4.48
		38 × 286	6.67	5.78	4.72	6.67	5.78	4.72	7.71	6.67	5.45	7.71	6.67	5.45
Hem-fir (N)	Select structural	38 × 89	2.27	2.06	1.80	2.60	2.36	2.06	2.60	2.36	2.06	2.97	2.70	2.36
		38 × 140	3.57	3.24	2.83	4.08	3.71	3.15	4.08	3.71	3.24	4.67	4.25	3.63
		38 × 184	4.70	4.27	3.73	5.38	4.89	4.15	5.18	4.89	4.27	6.16	5.60	4.79
		38 × 235	6.00	5.45	4.76	6.87	6.24	5.29	6.87	6.24	5.45	7.87	7.15	6.11
		38 × 286	7.30	6.63	5.79	8.36	7.59	6.44	8.36	7.59	6.63	9.57	8.69	7.44
	No. 1	38 × 89	2.27	2.06	1.80	2.60	2.36	2.00	2.60	2.16	2.06	2.97	2.70	2.31
		38 × 140	3.51	3.24	2.83	4.08	3.58	2.92	4.08	3.71	3.24	4.67	4.14	3.18
		38 × 184	4.70	4.27	3.73	5.38	4.72	3.85	5.18	4.89	4.27	6.15	5.45	4.45
		38 × 235	6.00	5.45	4.76	6.87	6.02	4.92	6.87	6.24	5.45	7.87	6.96	5.68
		38 × 286	7.30	6.63	5.79	8.36	7.33	5.98	8.36	7.59	6.63	9.57	8.46	6.91
	No. 2	38 × 89	2.19	1.99	1.74	2.51	2.22	1.81	2.51	2.28	1.99	2.87	2.56	2.09
		38 × 140	3.44	3.13	2.62	3.71	3.21	2.62	3.94	3.58	3.03	4.28	3.71	3.03
		38 × 184	4.54	4.13	3.46	4.89	4.24	3.46	5.20	4.72	3.99	5.65	4.89	3.99
		38 × 235	5.80	5.27	4.42	6.42	5.40	4.41	6.64	6.03	5.10	7.21	6.24	5.10
		38 × 286	7.05	6.41	5.17	7.59	6.57	5.37	8.07	7.33	6.20	8.77	7.59	6.20
	No. 3	38 × 89	1.90	1.65	1.34	1.90	1.65	1.34	2.20	1.90	1.55	2.20	1.90	1.55
		38 × 140	2.81	2.44	1.99	2.81	2.44	1.99	3.25	2.81	2.10	3.25	2.81	2.30
		38 × 184	3.71	3.21	2.62	3.71	3.21	2.62	4.28	3.71	3.03	4.28	3.71	3.03
		38 × 235	4.73	4.10	3.35	4.73	4.10	3.35	5.47	4.73	3.87	5.47	4.73	3.87
		38 × 286	5.76	4.99	4.07	5.76	4.99	4.07	6.65	5.76	4.70	6.65	5.76	4.70

TABLE 8-1b: *(Cont'd.)*

Lumber species	Grade	Size (mm)	Live Load 1.5 kN/m²						Live load 1.0 kN/m²					
			Gypsum board or plastered ceiling			Other ceilings			Gypsum board or plastered ceiling			Other ceilings		
			300 mm	400 mm	600 mm	300 mm	400 mm	600 mm	300 mm	400 mm	600 mm	300 mm	400 mm	600 mm
Eastern hemlock-tamarack (N)	Select structural	38 × 89	2.16	1.96	1.71	2.47	2.25	1.96	2.47	2.25	1.96	2.83	2.57	2.25
		38 × 140	3.40	3.09	2.70	3.89	3.53	3.09	3.89	3.53	3.09	4.45	4.05	3.53
		38 × 184	4.48	4.07	3.55	5.13	4.66	4.07	5.13	4.66	4.07	5.87	5.33	4.66
		38 × 235	5.72	5.19	4.54	6.55	5.95	5.19	6.55	5.95	5.19	7.49	6.81	5.95
		38 × 286	6.95	6.32	5.52	7.96	7.23	6.32	7.96	7.23	6.32	9.12	8.28	7.23
	No. 1	38 × 89	2.16	1.96	1.71	2.47	2.25	1.96	2.47	2.25	1.96	2.81	2.57	2.25
		38 × 140	3.40	3.09	2.70	3.89	3.53	3.09	3.89	3.53	3.09	4.45	4.05	3.53
		38 × 184	4.48	4.07	3.55	5.13	4.66	4.07	5.13	4.66	4.07	5.87	5.33	4.66
		38 × 235	5.72	5.19	4.54	6.55	5.95	5.19	6.55	5.95	5.19	7.49	6.81	5.95
		38 × 286	6.95	6.32	5.52	7.96	7.23	6.32	7.96	7.23	6.12	9.12	8.28	7.23
	No. 2	38 × 89	2.08	1.89	1.65	2.48	2.17	1.89	2.38	2.17	1.89	2.73	2.48	2.17
		38 × 140	3.28	2.98	2.60	3.75	3.41	2.92	3.75	3.41	2.98	4.29	3.90	3.38
		38 × 184	4.32	3.92	3.43	4.94	4.49	3.85	4.94	4.49	3.92	5.66	5.14	4.45
		38 × 235	5.51	5.01	4.37	6.31	5.73	4.92	6.31	5.73	5.01	7.22	6.56	5.68
		38 × 286	6.70	6.09	5.32	7.68	6.97	5.98	7.68	6.97	6.09	8.79	7.98	6.91
	No. 3	38 × 89	2.01	1.82	1.51	2.13	1.85	1.51	2.30	2.09	1.74	2.46	2.13	1.74
		38 × 140	3.13	2.71	2.21	3.13	2.71	2.21	3.61	3.13	2.55	3.61	3.13	2.55
		38 × 184	4.13	3.57	2.92	4.13	3.57	2.92	4.77	4.13	3.17	4.77	4.11	3.37
		38 × 235	5.27	4.56	3.72	5.27	4.56	3.72	6.08	5.27	4.30	6.08	5.27	4.30
		38 × 286	6.42	5.55	4.53	6.41	5.55	4.53	7.40	6.41	5.23	7.40	6.41	5.23
Spruce-pine-fir	Select structural	38 × 89	2.14	1.94	1.70	2.45	2.22	1.94	2.45	2.22	1.94	2.80	2.55	2.22
		38 × 140	3.36	3.05	2.67	3.85	3.50	3.04	3.85	3.50	3.05	4.41	4.00	3.50
		38 × 184	4.43	4.03	3.52	5.08	4.61	4.02	5.08	4.61	4.03	5.81	5.28	4.61
		38 × 235	5.66	5.14	4.49	6.48	5.88	5.12	6.48	5.88	5.14	7.41	6.74	5.88
		38 × 286	6.88	6.25	5.46	7.88	7.16	6.23	7.88	7.16	6.25	9.02	8.19	7.16
	No. 1	38 × 89	2.14	1.94	1.70	2.45	2.22	1.94	2.45	2.22	1.94	2.80	2.55	2.22
		38 × 140	3.36	3.05	2.67	3.85	3.47	2.83	3.85	3.50	3.05	4.41	4.00	3.27
		38 × 184	4.43	4.03	3.52	5.08	4.57	3.73	5.08	4.61	4.03	5.81	5.28	4.31
		38 × 235	5.66	5.14	4.49	6.48	5.84	4.77	6.48	5.88	5.14	7.41	6.74	5.50
		38 × 286	6.88	6.25	5.46	7.88	7.10	5.80	7.88	7.16	6.25	9.02	8.19	6.70
	No. 2	38 × 89	2.07	1.88	1.64	2.37	2.15	1.75	2.37	2.15	1.88	2.71	2.46	2.02
		38 × 140	3.25	2.95	2.54	3.59	3.11	2.54	3.72	3.38	2.93	4.15	3.59	2.93
		38 × 184	4.29	3.89	3.35	4.74	4.10	3.35	4.91	4.46	3.87	5.47	4.74	3.87
		38 × 235	5.47	4.97	4.28	6.05	5.24	4.28	6.26	5.69	4.94	6.99	6.05	4.94
		38 × 286	6.65	6.04	5.20	7.36	6.35	5.20	7.62	6.92	6.01	8.50	7.36	6.01
	No. 3	38 × 89	1.86	1.61	1.31	1.86	1.61	1.31	2.15	1.86	1.52	2.15	1.86	1.52
		38 × 140	2.74	2.37	1.93	2.74	2.37	1.93	3.16	2.74	2.23	3.16	2.74	2.23
		38 × 184	3.61	3.13	2.55	3.61	3.13	2.55	4.17	3.61	2.95	4.17	3.61	2.95
		38 × 235	4.61	3.99	3.26	4.61	3.99	3.26	5.32	4.61	3.76	5.32	4.61	3.76
		38 × 286	5.61	4.85	3.96	5.61	4.85	3.96	6.47	5.61	4.58	6.47	5.61	4.58
Western cedars (N)	Select structural	38 × 89	2.06	1.87	1.63	2.36	2.14	1.87	2.36	2.14	1.87	2.70	2.45	2.14
		38 × 140	3.24	2.94	2.57	3.71	3.37	2.94	3.71	3.37	2.94	4.24	3.85	3.37
		38 × 184	4.27	3.88	3.39	4.89	4.44	3.88	4.89	4.44	3.88	5.59	5.08	4.44
		38 × 235	5.45	4.95	4.32	6.24	5.66	4.95	6.24	5.66	4.95	7.14	6.49	5.66
		38 × 286	6.63	6.02	5.26	7.58	6.89	6.02	7.58	6.89	6.02	8.68	7.89	6.89
	No. 1	38 × 89	2.06	1.87	1.63	2.36	2.14	1.87	2.36	2.14	1.87	2.70	2.45	2.14
		38 × 140	3.24	2.94	2.57	3.71	3.37	2.87	3.71	3.17	2.94	4.24	3.85	3.31
		38 × 184	4.27	3.88	3.39	4.89	4.44	3.78	4.89	4.44	3.88	5.59	5.08	4.37
		38 × 235	5.45	4.95	4.32	6.24	5.66	4.83	6.24	5.66	4.95	7.14	6.49	5.58
		38 × 286	6.63	6.02	5.26	7.58	6.89	5.87	7.58	6.89	6.02	8.68	7.89	6.78
	No. 2	38 × 89	1.99	1.81	1.58	2.28	2.07	1.78	2.28	2.07	1.81	2.61	2.37	2.05
		38 × 140	3.13	2.84	2.48	3.58	3.14	2.56	3.58	3.25	2.84	4.10	3.62	2.96
		38 × 184	4.13	3.75	3.27	4.72	4.14	3.38	4.72	4.29	3.75	5.41	4.78	3.90
		38 × 235	5.27	4.78	4.18	6.03	5.28	4.31	6.03	5.48	4.78	6.90	6.10	4.98
		38 × 286	6.41	5.82	5.08	7.33	6.42	5.24	7.33	6.66	5.82	8.39	7.42	6.06
	No. 3	38 × 89	1.86	1.61	1.31	1.86	1.61	1.31	2.15	1.86	1.52	2.15	1.86	1.52
		38 × 140	2.74	2.37	1.93	2.74	2.37	1.93	3.16	2.74	2.23	3.16	2.74	2.23
		38 × 184	3.61	3.13	2.55	3.61	3.13	2.55	4.17	3.61	2.95	4.17	3.61	2.95
		38 × 235	4.61	3.99	3.26	4.61	3.99	3.26	5.32	4.61	3.76	5.32	4.61	3.76
		38 × 286	5.61	4.85	3.96	5.61	4.85	3.96	6.47	5.61	4.58	6.47	5.61	4.58

In the first place, the size of the ceiling joists will have to be increased. As we pointed out in Chapter 6, the size depends on the free span. In general, 2 ×6 (38 × 140 mm) members are commonly used for ceiling joists. If the ceiling joists are also to be the roof joists, the size must be increased, because of the additional loads of roofing materials, snow, and live loads. Size 2 × 8, 2 × 10, and 2 × 12 (38 × 184 mm, 38 × 325, and 38 × 286 mm) members are commonly used, the choice in a particular case depending on the above-mentioned factors. See Tables 8–1a and 8–1b for sizes and spans.

Roof Cornice

It was also noted in Chapter 6 that the outer ends of ceiling joists are cut off flush with the outside of the plate and that the top edge at the end is cut in the plane of the roof. If they are acting as roof joists, the ends will be cut off square, and the joist ends will extend beyond the wall plate if the building is to have a cornice (see Fig. 8–1).

The cornice on the end walls will be formed by stub joists extending over the end wall plates and anchored at their inner ends to a regular joist. Remember that these stub joists should extend at least as far inside the wall plate as they project beyond it. For example, if the joists are on 16-in. (400-mm) centers, the last regular joist should be kept back two spaces or 32 in. (800 mm) wide (see Fig. 8–2)

To frame the cornice at the corner of the building, proceed as follows:

1. Keep the first stub joist as far from the corner as the last regular joist.

2. Run a joist diagonally from the junction of stub and regular joists across the wall plate at the corner (see Fig. 8–2). Its length may be determined by calculating the length of the diagonal of the rectangular square *ABCD*, Fig. 8–2.

3. Cut short joists to run from the diagonal at regular 16-in (400-mm) centers (see Fig. 8–2).

Note. Each end of the diagonal joist requires two cheek cuts (see Fig. 8–2), and the length of the joist must be measured down the center of the top edge. The short joists mentioned in step 3 above will have a single cheek cut, all of which may be laid out with the framing square.

Each stub joist and the diagonal joist must be nailed to the regular joist with three 3-in. (75-mm) nails and toenailed to the plate as well.

Sloped Roof Joists

It is sometimes desirable to have enough slope to this flat roof that water will drain in one direction. It is generally considered

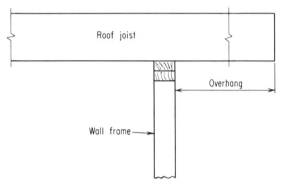

FIGURE 8-1: *Roof joist overhang.*

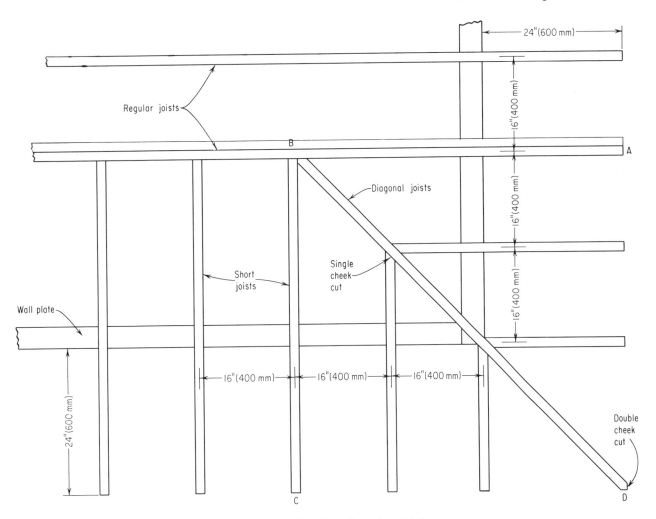

FIGURE 8-2: *Plan view of roof joists.*

that a slope of 1:25 is sufficient. This will allow the water to run but will not be noticeable as a slope on the roof. One method to provide for roof drainage is to slope each joist. This can be achieved by building the supporting walls of different heights or by tapering each joist or by adding a tapered strip to the top of each joist. Where insulation for the roof is placed between the roof joists, a ventilated space of at least 3½ in. (89 mm) must be provided between the top of the joists and the underside of the roof sheathing. This can be achieved by placing 2 × 4 stringers on edge at right angles over the top of the joists (see Fig. 8-3).

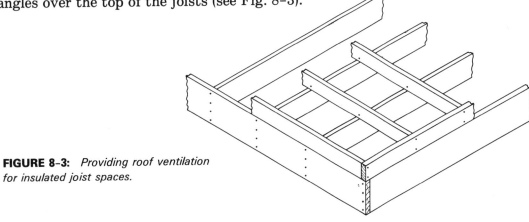

FIGURE 8-3: *Providing roof ventilation for insulated joist spaces.*

Joist Blocking

After all the joists are in place, the next step in the construction of this flat roof is the placing of blocking between every pair of joists. There are at least two important reasons for using this blocking. One is that it provides a nailing surface for the inner edge of the plancier between joists. Another is that when loose fill insulation is used, the blocking retains the loose fill in its place.

To provide a nailing surface, the blocking must project beyond the outer edge of the top plate at least half its own thickness (see Fig. 8–4). Be sure that the blocking pieces are not as wide as the joists. There should be ample room for air circulation between each pair of joists for their entire length. For example, if the joists are 2 × 8 (38 × 184 mm), the blocks should not be more than 2 × 6 (38 × 140 mm).

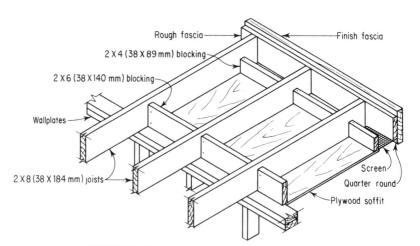

FIGURE 8–4: *Roof joist blocking and ventilation.*

Roof Ventilation

Since the spaces between the pairs of the roof joists normally will have no connection with one another, ventilation must be provided around the entire perimeter of the roof. The best place to provide this ventilation is at the outer ends of the joists, and it can be done by providing a strip of fine screen, about 5 in. (125 mm) wide, along the outer edge of the soffit. The 2 × 4 (38 × 89 mm) blocking is nailed between the joists, 3 in. (75 mm) from their ends and a rough fascia nailed to the ends of the joists. The screen is stapled across the space between the bottom edges of the rough fascia and the blocking and the soffit material cut to the proper width and nailed in place (see Fig. 8–4).

Finish Fascia

The finish fascia is the final trim on the roof overhang, made from either ¾ or 1½ in. (19- or 38-mm) material (see Fig. 8–4). It should project below the bottom edge of the rough fascia about 1 in. (25 mm) and far enough above the top edge to cover the thickness of the roof decking and insulation. This projection is to provide a gravel stop base—that is, to prevent the gravel from the built-up roof from rolling off. If the roof is sloped, the gravel stop is provided on three sides only. The water drains off along the fourth side. Finally, quarter-round is nailed behind the dropped edge of the fascia to cover the outer edge of the screen and to finish off the trim.

Cant Strips

Cant strips are triangular-shaped pieces of wood installed at the junction of a flat roof deck and a wall to prevent cracking of the roofing which is applied over it (see Fig. 8–5). The *cant strip* is occasionally installed at the edge of the roof to serve as a gravel stop, though a metal gravel stop is more commonly used for this purpose (see Fig. 8–6).

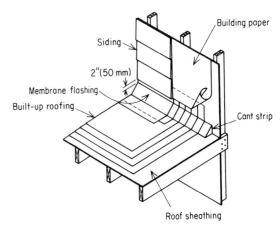

FIGURE 8-5: *Junction of flat roof and wall.*

Roof Decking

The roof decking will usually be plywood, or tongue-and-grooved decking, ¾ or 1½ in. (19 or 38 mm) in thickness. In any case it must be well nailed to the joists and adequately supported to provide a solid base for the built-up roofing.

Flat Roof Insulation

The insulation to be used with a flat roof is a very important consideration. There are three main alternatives. One is to apply rigid insulation over the roof deck and under the built-up roof. A second is to apply insulation batts between the joists from inside the building. The third is to use loose fill insulation between the joists.

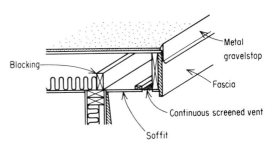

FIGURE 8-6: *Metal gravel stop.*

Several types of rigid insulation are available, among them wood fiber board insulation (usually laminated) and expanded foam roof insulation. Whichever type is used, it must be secured firmly to the roof decking either by nailing or by cementing. Manufacturer's instructions usually explain exactly how the product should be applied.

If batt insulation is used, it is applied between the joists from the inside in the same way that batt insulation is applied to conventional ceiling joists. In this case, the built-up roof is applied directly to the roof deck.

The most effective method of introducing loose fill insulation into a flat roof system is by blowing it into place. Some type of fibrous material is used, and a blower places the insulation in spaces otherwise very difficult to reach.

Roof Flashing

When the insulation is applied on top of the roof deck, the next operation is the installation of flashing. If the insulation is below, the flashing is applied next to the deck. Flashing is metal covering designed to prevent water from entering the roof structure along the edge of the roof (see Fig. 8–6). Part of the metal sheet lies flat on the deck or insulation, and the remainder is bent up to form a gravel stop and down the outside face of the fascia far enough to cover all joints between deck and joists.

Roof Drainage

If the roof is to have cant strip all around, it is necessary to provide some means of drainage. This may be done by means of vents in the roof connected to pipes leading to the sewage disposal system. It may also be done by providing *scupper boxes* at convenient positions around the edge of the roof. These are connected to downspouts which carry away the water.

A scupper box is simply a flat metal box, open at one end, with a drainage hole at the opposite end. A section of the cant strip is cut away, and the scupper box is set into the opening, open end in. The other end projects beyond the edge of the roof far enough that a downspout may be connected to the drainage hole. The flat projecting bottom is covered by the roofing material; when water collects on the roof, it simply drains into the scupper boxes and out through the downspouts.

Built-up Roofing

The final step in the construction of a flat roof is the application of the roofing material. A built-up roof—so called because it is built up from several layers of material—is usually used on flat decks. The materials consist of heavy tarred paper, asphalt in liquid form, and pea gravel. The amount of material used determines the approximate life of the roof, usually 10, 15, or 20 years.

Figure 8–7 illustrates the procedure for laying a 20-year roof over a wood deck. Here are the steps in this procedure:

1. Lay down one layer of sheathing paper, weighing at least 5 lb/100 ft² (0.25 kg/m²). Nail as required to hold with 1-in. (25-mm) roofing nails.

2. Lay down two layers of asphalt-impregnated 15 lb/100 ft² (0.75 kg/m²) felt paper, lapping each sheet 19 in. (475 mm) over the preceding one and nailing sufficiently to hold in place. Start with a half roll, obtained simply by cutting a roll in two (see Fig. 8–7, which illustrates this half-roll starter).

3. Lay down three additional layers of the same paper as above, securing each by a layer of hot tar or asphalt, mopped full width under it. This step is begun by cutting a roll of paper into a 12-in. and a 24-in. (300-mm and a 600-mm) length. Lay down the 12-in. (300-mm) strip first, as illustrated in Fig. 8–7, then the 24-in. (600-mm) strip over it, then a full-length roll. Notice that these three layers all start at the same edge, or,

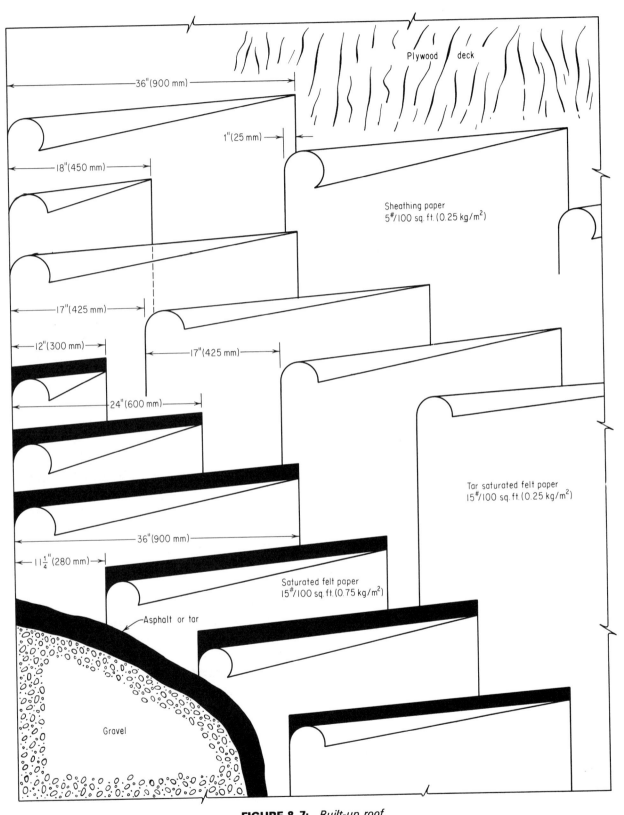

Plywood deck

36"(900 mm)

1"(25 mm)

18"(450 mm)

Sheathing paper
5#/100 sq. ft. (0.25 kg/m²)

17"(425 mm)

12"(300 mm)

17"(425 mm)

24"(600 mm)

Tar saturated felt paper
15#/100 sq. ft. (0.25 kg/m²)

36"(900 mm)

11¼"(280 mm)

Saturated felt paper
15#/100 sq. ft. (0.75 kg/m²)

Asphalt or tar

Gravel

FIGURE 8-7: *Built-up roof.*

in other words, there are now three mopped layers of paper over the first 12-in. (300 mm) of roof. Start the next full roll 11¼ in. (280 mm) up from the starting edge and continue similarly with each succeeding roll. The tar or asphalt, as hot as possible, is mopped on with a string or rag mop, and the paper is laid in the mopping immediately. It should be brushed into place with a stiff-bristled push broom to make sure that no air spaces remain under the paper.

4. Cover the entire surface with a uniform coating of hot tar or asphalt at the rate of 150 lb/100 ft² (7kg/m²) and embed in it 40 lb (19 kg) of pea gravel per 10 ft² (m²). The gravel must be applied while the coating is hot and should be rolled to ensure that it is well embedded. Of course, you must perform this operation one small section at a time, so that the tar will remain hot until the gravel is applied.

5. Check to make sure that the roofing is well up over the flashing and that there is a good seal where flashing and roofing meet, so that water cannot run under the paper.

If the roof has rigid insulation on top of the deck, the procedure must, of necessity, be somewhat difficult. Since paper cannot be nailed to the insulation, it must be mopped into place. The operation consists entirely of laying four layers of 15 lb/100 ft² (0.75 kg/m²) tarred felt paper directly on top of the insulation if a 20-year roof is required.

Start at the edge with a quarter-length roll, apply a half-length roll on top of it, then a three-quarter-length roll, and finally a full roll. Start the next full roll 8½ in. (210 mm) up from the edge, and leave 9 in. (225 mm) exposed with each succeeding roll.

The mopping procedure and the application of an asphalt coat and gravel on top are done exactly the same as above.

Built-up Roofing Types

Built-up roofing may be of several different types. The one described above is a *graveled, asphalt base* roofing. An *asphalt base* may be used *without gravel* or surfaced with *wide selvage asphalt roofing*. A *coal tar* base may be used *with a gravel surface*, or *cold process roofing* may be employed. With all of these there are minimum and maximum slope limits, and these are outlined in Table 8–2.

TABLE 8–2: *Roofing types and slope limits [based on 12-in. (250-mm) run]*

Type of roofing	Minimum rise	Maximum rise
Asphalt base (graveled)	0 in. (0 mm)	3 in. (60 mm)
Asphalt base (without gravel)	⅜ in. (10 mm)	6 in. (120 mm)
Asphalt base (surfaced with wide selvage asphalt roofing	1½ in. (40 mm)	no limit
Coal tar base (graveled)	0 in. (0 mm)	⅜ in. (10 mm)
Cold process roofing	⅜ in. (10 mm)	9 in. (185 mm)

Great care should be taken in melting and handling the tar or asphalt. It should be melted in a fairly flat vat or tub by any convenient heating arrangement. When melted and hot, it is dipped from the vat with a long-handled dipper into 5-gal pails, ready for hoisting to the roof. The hoist may be simple, but make sure that it is strong enough to do the job safely.

FLAT ROOF WITH LAMINATED BEAMS

An open *style* ceiling may be produced in a flat roof by using heavy roof beams to carry the roof decking and leaving the underside of the decking and the beams exposed to view. It is necessary, therefore, that both decking and beams be of such a grade of material as to make that exposure practical.

The roof beams are normally laminated members—that is, they are made up of several pieces glued together to form a solid unit. The lamination may be of two or more pieces on edge or of several pieces laminated on the flat [see Fig. 8-8(a) and (b)]. Material to be used will usually have to be clear or nearly clear stock, since the finished beam will form part of the interior finish. In some cases, regular stock may be used and the two sides and the bottom of the beam covered with a thin veneer of the required type or texture.

The sizes of the beams to be used will depend on their span and their spacing. The size required for a particular job should never be decided arbitrarily. Consult an engineer or other building authority to determine the width and depth best suited to the case.

Laminated beams are made to order by manufacturers specializing in this type of construction. Facilities for applying great pressure are required, since the parts of the beam are glued together, no nails or other fasteners being used. In addition, suitable means of planing and sanding the beams must be available, since they represent a finished product.

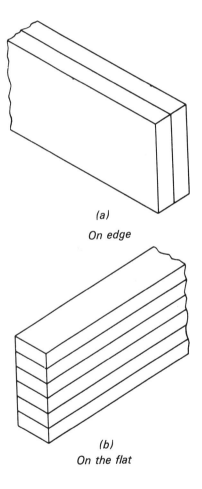

(a)
On edge

(b)
On the flat

FIGURE 8-8: *Laminated roof beams.*

(c)

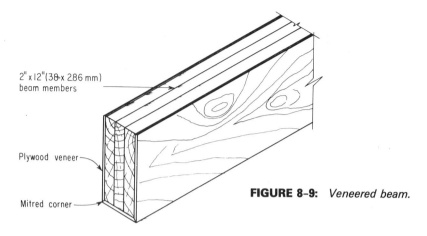

2" x 12"(38 x 286 mm)
beam members

Plywood veneer

Mitred corner

FIGURE 8-9: *Veneered beam.*

If the beams are to be covered, it is possible that they may be made on the job, in most cases of two or three pieces of dry material, dressed flat and laminated to stand on edge (see Fig. 8-9). A good-quality glue must be used and the manufacturer's instructions followed explicitly as to coating procedure, curing time, and temperature.

The covering is usually a thin veneer of plywood on three sides, with the bottom edges mitered for best appearance (see Fig. 8-9). The end joints in the plywood must be well fitted in order that the vertical lines will be as inconspicuous as possible.

The sizes of the beams required must be calculated for each job. Size will depend, as stated previously, on span and spacing as well as on the type of roofing used. The longer the span, of course, the greater the load imposed on the beam for any given type of roofing. And the farther apart the beams are spaced, the more of the roof must be carried by each beam. Nothing can be said, therefore, about the size of beam that should be used. Generally, however, they will be relatively narrow in width in comparison to their depth. Remember that the strength of a beam varies directly as its width. In other words, if you double the width of a beam, you double the strength. On the other hand, the strength of a beam varies as the square of its depth. That means that if you double the depth, you then increase the strength four times.

The number of beams used for the roof, or, in other words, the spacing of the beams, will depend on the span, the weight of the roofing material, and the type of roof decking used. The maximum spacing allowed by most codes is about 84 in. (2100 mm) if 2 in. (38-mm) decking is to be used. Check the local building code for greater detail.

Beam Anchors

When laminated beams are used in a flat roof design, they will normally bear on the wall plates and must be anchored to them. This may be done with *steel angles*, metal *saddles,* or *wooden dowels* if visible anchors cannot be used (see Fig. 8-10).

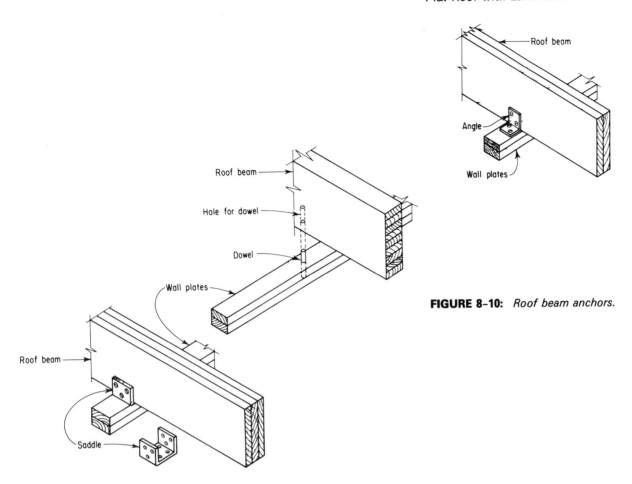

FIGURE 8-10: *Roof beam anchors.*

Roof Overhang

Roof overhang is obtained at the *sides* of the building by allowing the roof beams to project beyond the plane of the wall as far as required (see Fig. 8-11). At the *ends,* the overhang is obtained by allowing the roof decking to project beyond the end beams (see Fig. 8-11). It is important to make sure that this end overhang is not so great that the weight of the roofing, snow, etc., will cause the projecting decking to sag.

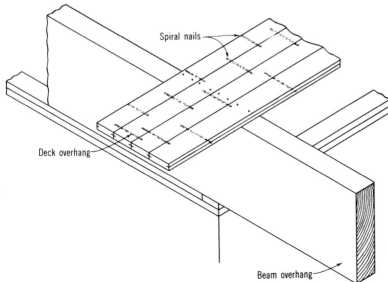

FIGURE 8-11: *Roof overhang with laminated beams.*

Roof Decking

Roof decking over roof beams usually consists of 2- or 3-in. (38- or 64-mm) tongue-and-grooved or double tongue-and-grooved (depending on load and span) cedar, redwood, or pine decking, 6 or 8 in. (140 or 184 mm) wide. The underside of this decking will probably be exposed, so that certain precautions must be taken in selecting and laying it.

First, the material must be dry so that shrinkage and cracking are eliminated or kept at a minimum. Second, the appearance of the decking must be such that it harmonizes the rest of the interior finish. Third, the pieces must be drawn tightly together by edge nailing through predrilled holes with 8-in. (200-mm) spiral nails, spaced about 30 in. (750 mm) o.c. and then face-nailed to the beams (see Fig. 8–11). In some cases, clear material may be required, while in others a knotty appearance may be specified.

The insulation for such a roof is commonly applied on top of the deck, and the procedure for applying insulation, flashing, cant strips, and roofing is the same as previously described for flat roof. The fascia, which may be relatively narrow, is secured to the edge of the roof deck and may have a beveled edge extending above the roof level to serve as a cant strip.

Closures

The spaces between the roof beams, the decking, and the wall plates must be closed, and a number of methods may be used to accomplish this. One method is to frame the space with regular 2 × 4 (38 × 89 mm) framing and then use the same material to cover the frame, inside and out, as is used for the rest of the wall.

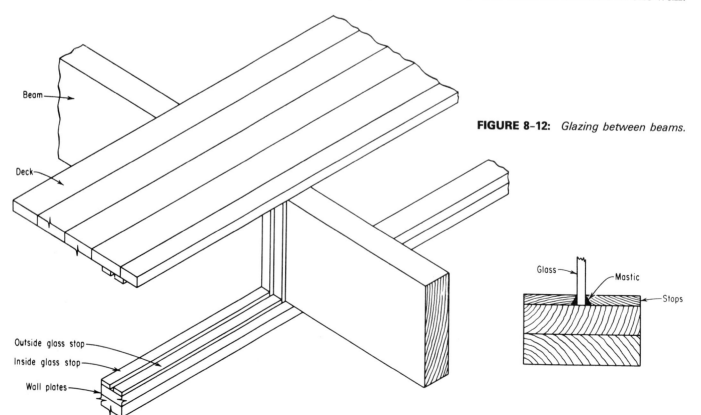

FIGURE 8-12: *Glazing between beams.*

Another method is to use glass, and one method of glazing the spaces is illustrated in Fig. 8–12. Insulated *sandwich panels*, consisiting of a 1-in (25-mm) thickness of rigid insulation between two layers of plywood, may be substituted for the glass panels.

Beam finishes

On the inside, the beams and underside of the roof deck are often treated with some transparent finish such as varnish or lacquer to preserve the natural appearance. Many other alternatives are possible, however. The underside of the deck may be covered with ceiling tile and the beams left in their natural color. Another plan is to cover deck and beams with ceiling tile. The whole surface may be painted to match the rest of the interior decoration. Finishing is largely a matter of individual preference, and each case must be decided on its own merits.

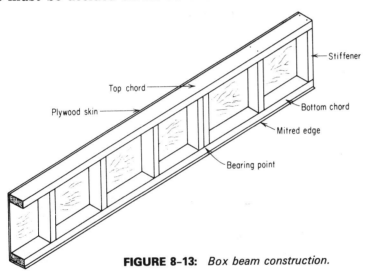

FIGURE 8–13: *Box beam construction.*

FLAT ROOF WITH BOX BEAMS

In place of the laminated beams described in the foregoing section, *box beams* may be used for the same purpose. Such a member is called a box beam because in form it is essentially a hollow box (see Fig. 8–13). It consists of *top and bottom chords* and vertical *stiffeners* and, if it is a deep beam, *diagonal stiffeners* as well. This type of beam has certain advantages over solid beams; it is relatively light in weight and can be made on the job with no special tools or equipment.

The size of beam to use will, of course, depend on individual circumstances. The span, spacing, and roof load all have to be taken into consideration in designing the size. In addition, the ability of the beam to carry a load safely will depend on how the plywood skin is applied to the frame. There are a number of alternative methods, the results of each varying from the others. For example, the plywood may be attached by screws, nails, or glue. Combinations of two of these may also be used, namely, screws and glue or nails and glue.

The type of glue used will also affect the strength of the beam. Some glues require more pressure during their setting period than others. Some require rigid temperature controls for proper results. Some set up rigidly, whereas others are more resilient, allowing slight movement under load.

The thickness of the plywood skin is, of course, of primary importance. Other things being equal, it is the thickness of the skin that determines the ability of the beam to carry the load.

It is evident that it is difficult to compile a simple table, from which, knowing the load, one could pick the width and depth of beam to use under given conditions. The safe way is to consult an engineer or other authority about the size required for a particular job, using specified material. Some quite extensive research has been carried on with this type of beam.

FLAT ROOF WITH OPEN WEB JOISTS

Open web joists are often used for a flat roof system in light construction, particularly in buildings with masonry walls. They may be entirely of steel (see Fig. 8–14), or they may be made with wood top and bottom chords and tubular steel web [see Fig. 8–8(a)].

FIGURE 8-14: *Open web steel joists. (Courtesy Bethlehem Steel Corp.)*

Open web steel joists are lightweight trusses which provide a light but strong structural framework for roofs and floors in a variety of modern buildings. This type of joist is made in two general types—*shortspan*, in lengths up to 48 ft (14.4 m), and *longspan*, in lengths up to 96 ft (28.8 m). End bearing pads support the joists on a wall. They may be welded to a plate cast into a concrete or masonry wall or held in place with lag screws to a wooden plate.

Immediately after joists are placed, bridging should be installed and welded at the intersections. (see Fig. 8–15). Bridging anchors are required to secure the ends of bridging lines to walls, and the type of anchor to be used will depend on the wall. Figure 8–16 illustrates the anchoring of a bridging line to a masonry wall. Where a ceiling is to be applied to the underside of joists, bottom chord extensions are available to provide ceiling support up to the walls (see Fig. 8–17).

FIGURE 8-15: *Joist bridging.*
(Courtesy Bethlehem Steel Corp.)

FIGURE 8-16: *Steel joist bridging anchors.*
(Courtesy Bethlehem Steel Corp.)

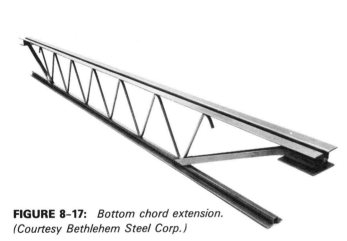

FIGURE 8-17: *Bottom chord extension.*
(Courtesy Bethlehem Steel Corp.)

FIGURE 8-18: *Header and tail joist.*

Openings in the roof frame are framed with headers and tail joists (see Fig. 8-18), and roof overhangs may be provided for by top chord extensions. For short overhangs and moderate loads, the top chord itself may be extended (see Fig. 8-19). Where the overhang is long and loads heavier, joists may have a one-channel or two-channel *outrigger* welded to the top chord (see Fig. 8-20).

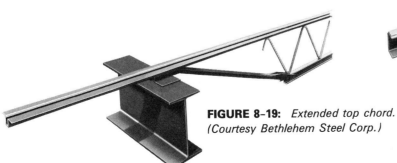

FIGURE 8-19: *Extended top chord.*
(Courtesy Bethlehem Steel Corp.)

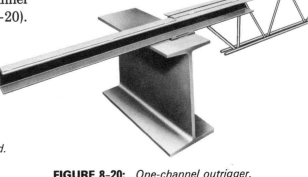

FIGURE 8-20: *One-channel outrigger.*
(Courtesy Bethlehem Steel Corp.)

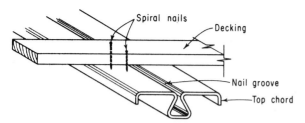

FIGURE 8-21: *Decking nailed to top chord.*

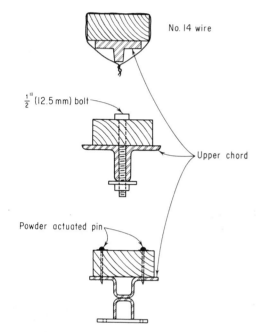

FIGURE 8-22: *Wood pads fastened to top chord of open web joists.*

FIGURE 8-23: *Pipes and ducts installed in floor systems. (Courtesy Bethlehem Steel Corp.)*

Wooden roof decking may be fastened to steel joists in one of two ways. Some joists have the top chord made in such a way that nails may be driven directly into a nail groove provided (see Fig. 8-21). Others require a wooden pad to be fastened to the top chord; this may be done in several ways, as illustrated in Figure 8-22.

Open web joists with wood top and bottom chords are used in the same way as all-steel joists except that decking can be fastened directly to the top chord and a ceiling attached directly to the bottom chord.

One of the main advantages of using open web joists is the ease with which pipes, ducts, and conduit may be installed within the system (see Fig. 8-23). Other advantages include ease of erection, permanence, rigidity, and adaptability. Either wood, concrete, or steel decks may be applied over steel joists with relative ease.

SLOPED ROOF ON PROJECTING JOISTS

This system of constructing a roof is a combination of two ideas, a cornice produced by overhanging roof joists and a sloping roof. The first step is the construction of the frame for a flat roof, as described in the section on flat roofs; regular joists are used. If a hip roof is planned or if the roof is to have an overhanging gable, the joists will project over the wall plates the required amount on all four sides. However, if a regular gable roof is to be used, the joists will all run in one direction, the two end ones being flush with the outside of the end wall frames.

Hip Roof

First, consider a hip roof with this type of construction. The joists are cut and assembled exactly the same as for the flat, projecting roof (see Fig. 8-2). A rough fascia, of the same material, is nailed to the ends of the joists all around. The dimensions of this flat area are the span and the length of the hip roof which will be constructed on it (see Fig. 8-24).

The lengths of all the rafters—commons, hips, and jacks— and the ridge are calculated in the same way as for a conventional roof, except that there are no rafter tails to consider. The lengths may be measured along the top edge of the rafter, rather than along a measuring line. Plumb cuts at the top end of the rafters, cheek cuts, and deductions are made as described in Chapter 7.

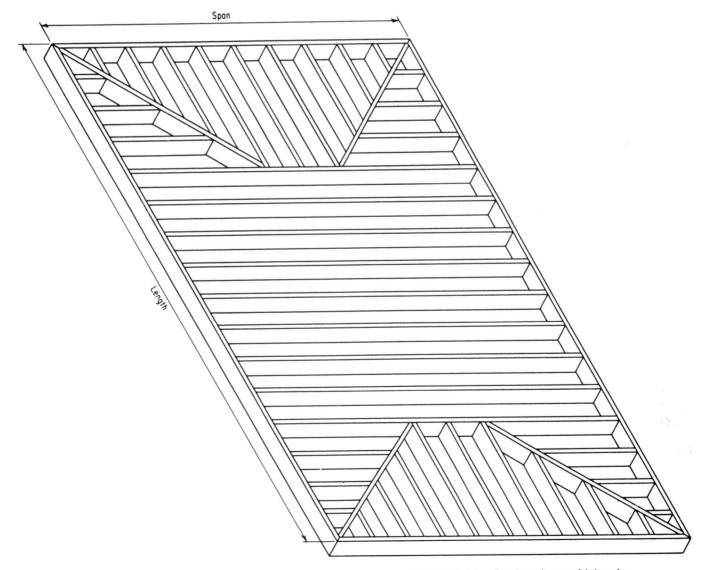

FIGURE 8-24: *Overhanging roof joist plan.*

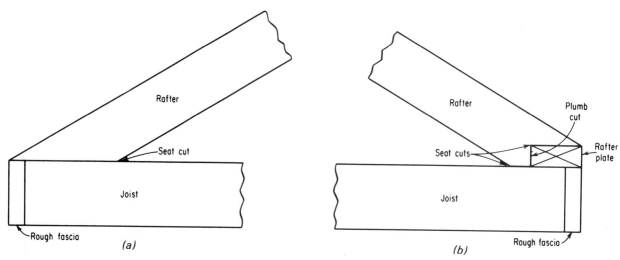

FIGURE 8-25: *Rafter seats.*

The cut of the bottom end of the rafters will depend on the method used to attach the rafters to the joists. One method is to set a rafter on the top edge of a joist, with the end out flush with the outer face of the rough fascia. In this case the bottom cut will be a seat cut right across the rafter [see Fig. 8-25(a)]. Another method is to nail a rafter plate all around the edge of the deck, with its outer edge flush with the face of the rough fascia [see Fig. 8-25(b)].

Gable Roof

When a gable roof is planned, all the rafters, of course, are commons, and the width of the deck becomes the span for the rafters. For an illustration of an overhanging gable, see Fig. 8-26. Either of the methods described above may be used to attach rafters to the joists for both types of gable roof.

FIGURE 8-26: *Overhanging gable.*

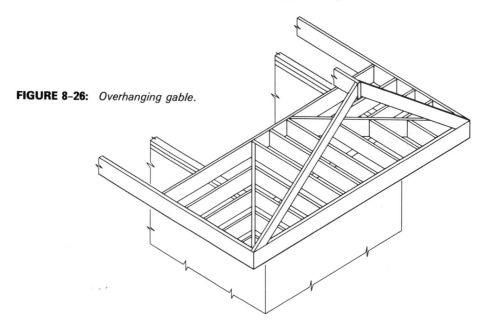

There are some advantages to this system for roofing that should not be overlooked. With the conventional type of roof, the wider the cornice is made, the lower the bottom of the cornice will be, and the height of windows is restricted. With the cornice formed by overhanging joists, windows may be carried as high as desired. For conventional type roofs, the width of the cornice is restricted and is limited to the practical height for the plancier. With overhanging joists, the width of the cornice is restricted only be the ability of the joists to carry the roof loads beyond the wall plates.

There are also disadvantages to this style of roof, the main one being inherent in any flat roof as well. It is that the full height of the wall is exposed in this type of construction. In some cases the result may be that the building looks out of proportion. If the building is small with regard to area, a roof which projects at the wall plate level will probably not be appropriate. Also, the wider the overhang, the deeper the joists will have to be to carry the load. Consequently, the width of the finish fascia is increased and may be out of proportion to the rest of the building. These are things which must be taken into consideration when the building is being planned.

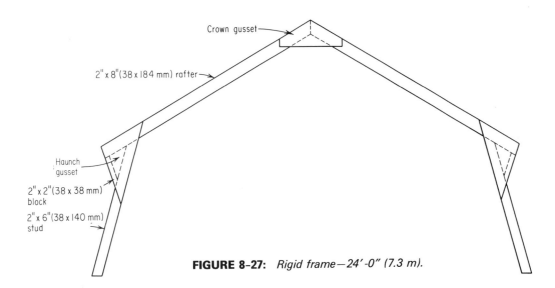

FIGURE 8–27: *Rigid frame—24′-0″ (7.3 m).*

RIGID FRAME CONSTRUCTION

Rigid frame is a relatively new concept in construction. As in the case of trusses, frames are made as complete units but with the difference that wall as well as roof framing is incorporated into one unit. A rigid frame is essentially an arch but is formed with four straight pieces of lumber held together by plywood gussets (see Fig. 8–27). The studs and rafters are 1½-in. (38-mm) material, varying in width from 4 to 12 in. (100–300 mm), depending on the span; the gussets are made from ¼- or ⅜-in. (7- or 9-mm) plywood. The shape which the frame will take is formed by laying

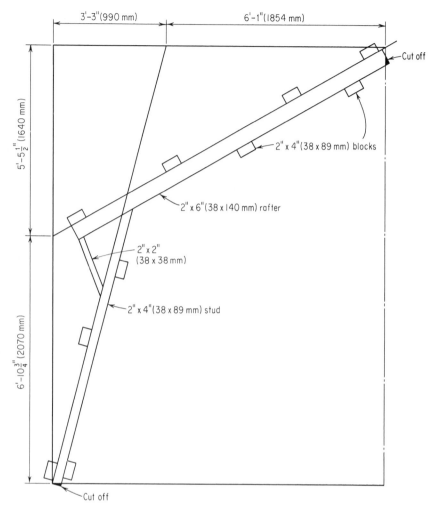

3'-3"(990 mm) 6'-1"(1854 mm)

5'-5½"(1640 mm)

6'-10¾"(2070 mm)

Cut off

2" x 4"(38 x 89 mm) blocks

2" x 6"(38 x 140 mm) rafter

2" x 2" (38 x 38 mm)

2" x 4"(38 x 89 mm) stud

Cut off

FIGURE 8-28: *Jig layout for 18'0" (5.4 m) span rigid frame.*

out a jig, as illustrated in Fig. 8–28. Jig layouts will be different for frames of different spans. The specifications for any particular span may be obtained by inquiring at a local lumber dealer's office or by writing to the nearest office of the Plywood Manufacturer's Association.

The proper size and length of material having been obtained from the jig layout plan, the next step is to cut the ends of the studs and rafters at the proper angles, which may also be obtained from the plan. The *haunch* and *crown gussets* are the next cut to size and shape. With the stud, rafter, and haunch block in position in the jig, nail the gussets into place, according to instructions, on both sides of the frame. The unit is now ready for erection.

Each frame is raised by itself and spaced usually on 24-in. (600-mm) centers. The bottom end will rest on a sill plate which is anchored to the top of a concrete foundation wall. The end is held in place by a metal shoe or angle. A number of styles of shoe are available, or a simple angle iron may be used, which is bolted to the sill plate and to the stud (see Fig. 8–29). The complete

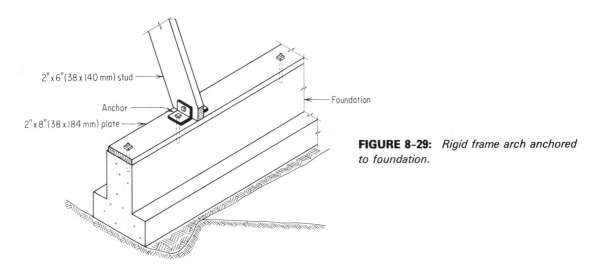

FIGURE 8-29: *Rigid frame arch anchored to foundation.*

skeleton is sheathed with 4 × 8 (1200 × 2400 mm) sheets of $\frac{5}{16}$- or $\frac{3}{8}$-in. (7- or 9-mm) plywood. These may be butt-jointed along their long edge, or they may be applied so that each horizontal row of sheets overlaps the one below by 3 or 4 in. (75 or 100 mm) (see Fig. 8-30). If the latter procedure is followed, the end joints may be caulked or strapped and the whole surface painted, with no further covering.

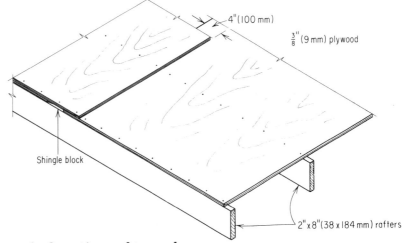

FIGURE 8-30: *Sheathing lapped on rafters.*

Windows are set between the haunched portions of two adjacent frames. They may be set on the slope of the stud, in which case they will be recessed between the haunches, or they may be set on the slope of the outside leg of the haunch. Modifications to either of these basic plans may quite easily be made (see Fig. 8-31).

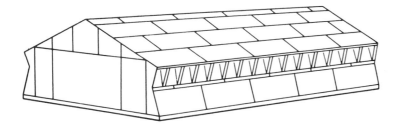

FIGURE 8-31: *Windows in rigid frame building.*

Rigid frame construction was originally designed for use in farm buildings, but today the design is being put to many other uses as well. Churches, community halls, summer cottages, garages, and warehouses are being built by this system. The whole idea is quite recent, and, no doubt, as time goes on it will be put to many more uses. Simplicity of construction, ease of erection, rigidity of the structure, and relative inexpensiveness are certainly factors in favor of its use.

GLUE-LAMINATED ARCHES

Glue-laminated arches are structural components made by gluing together thin pieces of lumber of any required width. In this way a member of any desired thickness and shape may be produced. Because of the wide scope in shape, size, and span, buildings using laminated arches for their frames are becoming increasingly popular. A great many are made commercially, some very large, with spans of 180 ft (60 m) or more. Great care is taken in the selection of lumber and the cutting of splices so that pieces fit perfectly, end to end, in the manufacture of these large arches. Large bandsaws and planers are used to cut and dress them to size. They are finally sanded, treated, and wrapped so that when they arrive on the job they are unmarked and, when in place, produce a finished appearance.

Smaller, lighter arches, in two or three standard shapes, with spans up to about 54 ft (18 m), are made with less sophisticated equipment for use in buildings such as that illustrated in Fig. 8-32.

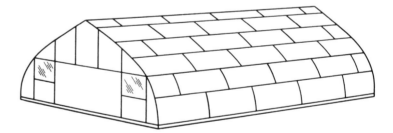

FIGURE 8-32: *Simple arched rafter building.*

Although many shapes are possible, probably the most often used ones are the semicircular arched, the parabolic arch, the gothic arch, and a shape, half curve, half straight line, known as the shed rafter (see Fig. 8-33). The semicircular and sometimes the parabolic arch are made in one piece, but in most other cases two separate halves of the complete arch are formed. The two halves are put together on the site, being secured together at the top by bolts or with steel plates acting as gussets.

Spacing of the arches varies greatly, depending on the design, the span, and the type of roof deck to be applied. In some instances the roof decking will be carried on purlins running across

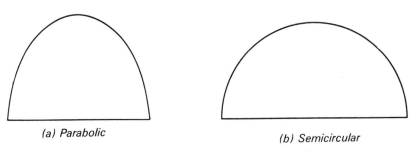

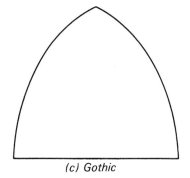

(c) Gothic

FIGURE 8–33: *Common rafter shapes.*

the arches or hung in between them. With other designs, the decking will be 3- or 4-in. (75- or 100-mm) tongue-and-grooved material, usually cedar, spanning directly from arch to arch.

A-FRAME

In *A-frame* construction, the wall and roof are combined in a single, plane surface. Straight members are joined together at the top and anchored to a floor at the bottom to form a tall, rigid triangle (see Fig. 8–34). The floor frame for an upper floor may be introduced at the appropriate level, to give the A-shape.

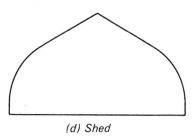

(d) Shed

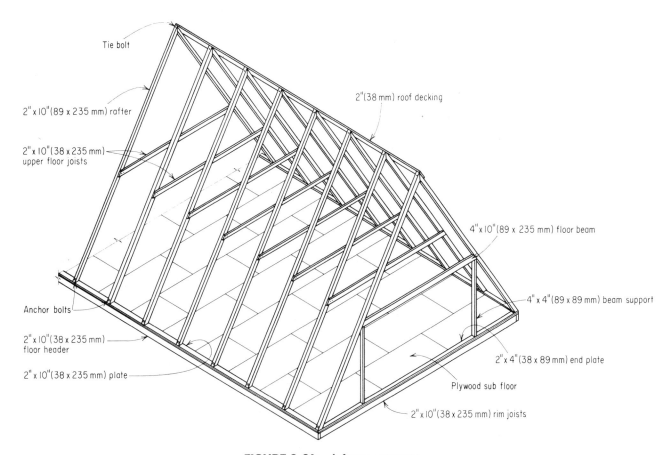

FIGURE 8–34: *A-frame structure.*

FIGURE 8-35: *A-frame with conventional framing. (Courtesy NRC)*

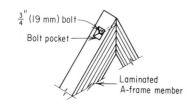

FIGURE 8-36: *Bolted connections for A-frame.*

The frame may consist of relatively large, laminated members [often 4 in. (100 mm) wide] on spacings of 4 ft (1200 mm) or more (see Fig. 8-34), or of conventional dimension material at normal spacing of 16 or 24 in. (400 or 600 mm) (see Fig. 8-35).

The larger members are usually bolted together at the top (see Fig. 8-36) and may be bolted through the sole plate at the bottom (see Fig. 8-36) or anchored to the plate with metal *shoes* or *angles.* Conventional framing members will be tied at the top with a gusset and toenailed to the plate at the floor.

Ends may be framed with conventional, 2 × 4 (38 × 89 mm) framing or with heavier members, more widely spaced, to provide large glass areas.

Sheathing will normally be 2-in. (38-mm) decking over widely spaced frames or plywood over conventional framing. Wood or asphalt shingles will complete the cover.

Stressed-Skin Roof Panels

A *stressed-skin* panel consists of a frame of dimension material [usually 2 × 4 (38 × 89 mm)] covered on one or both sides with a *plywood skin* (see Fig. 8-37). The plywood skin, in addition to providing a surface covering, acts with the framing members to form a complete structural member. To develop the full potential strength of the panel, the skins must be attached to the frame with glue and, in order to tie panels together effectively, the skin will project over the frame along one edge to overlap the adjoining panel.

Panels may be of various shapes but are most often either *rectangular* or *triangular.* Rectangular panels are supported top and bottom, while triangular ones often fit together to form a self-supporting structure.

The spacing of framing members for stressed-skin panels depends on the thickness of the plywood skin and the direction of the face grain in relation to the long dimension of the framing member.

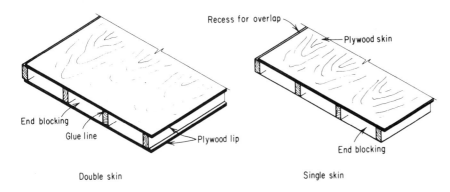

FIGURE 8-37: *Stressed skin panels.*

REVIEW QUESTIONS

8-1. Give two reasons for using blocking between joists at the wall plates in flat roof construction. Why should the blocks not be as wide as the joists?

8-2. Give two reasons for using flashing around the perimeter of a flat roof.

8-3. When insulating a flat roof, what type of insulation is placed over the roof deck?

8-4. (a) List the materials used in laying a built-up roof over a wood deck.

   (b) How much of each of these materials will be required for a flat roof over a building 24 × 30 ft (7200 × 9000 mm) with a 24-in. (600-mm) overhang on all sides?

8-5. Outline the main difference between *laminated* beams and *box* beams.

8-6. Explain the difference between a trussed rafter and a rigid frame unit.

8-7. List two major advantages of rigid frame and arch construction over conventional construction.

8-8. List three advantages of open web joists over conventional wood joists.

9

STAIR BUILDING

Stair building is virtually a trade in itself, and it would be impossible to cover the subject in all its facets in a single chapter. Nevertheless, a stair of one kind or another is still a necessity in many types of light construction, and we shall look into the basic elements of the subject to find out how to construct some of the most commonly required kinds of stairs.

TERMS USED IN STAIR BUILDING

Stairs lead from one floor level to another through a *stairwell opening* (see Fig. 4–34). It must be framed during the floor-framing stage of construction, so the length and width of the opening must be known. Stairs consist primarily of *risers* and *treads* carried on *stringers* (see Fig. 9–1). The height of each riser is called the *rise,* and the width of the tread (exclusive of nosing) is called the *run* (see Fig. 9–2). The sum of all the individual rises gives the *total rise,* and the sum of the runs gives the *total run* (see Fig. 9–1).

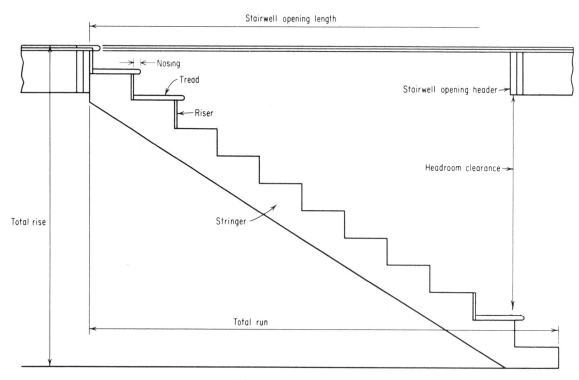

FIGURE 9-1: *Stair elevation.*

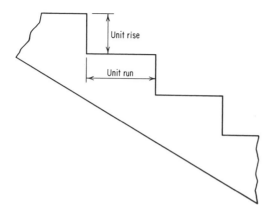

FIGURE 9-2: *Stair unit rise and unit run.*

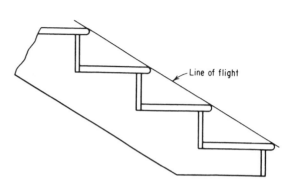

FIGURE 9-3: *Line of flight.*

Two important terms used in stair work are *line of flight* and *headroom clearance.* The line of flight is a line drawn through the extremities of the nosings (see Fig. 9–3). Headroom clearance is the vertical distance from the underside of the end of the stairwell opening to the line of flight (see Fig. 9–1).

STAIR DIMENSIONS

Most building codes provide guidelines for dimensions of stair parts. For example, interior stairs leading to unfinished basements, cellars, or attics are generally allowed a maximum riser height of 9 in. (225 mm), a minimum run of 8 in. (200 mm), and a minimum tread width of 9¼ in. (230 mm). Other interior stairs in residences and exterior stairs serving residences are allowed a maximum riser height of 8 in. (200 mm), a minimum run of 8¼ in. (210 mm), and a minimum tread width of 9¼ in. (235 mm). Interior stairs in buildings other than residences and exterior stairs, except those serving single residences, are restricted to a maximum riser height of 8 in. (200 mm), a minimum riser height of 5 in. (125 mm), a minimum run of 9¼ in. (230 mm), and a minimum tread width of 10 in. (250 mm). The product of the run times the rise for such stairs must not be less than 70 in. (45,000 mm) or more than 75 in. (48,500 mm).

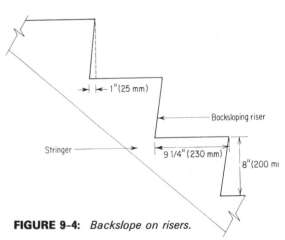

FIGURE 9-4: *Backslope on risers.*

If the run of a stair is less than 10 in. (250 mm), it is necessary to provide a *nosing* which will project at least 1 in. (25 mm) beyond the face of the riser (see Fig. 9–1) or to provide at least as much backslope to the risers (see Fig. 9–4).

Stairs should have a minimum width of 2 ft, 10 in. (900 mm) and a minimum headroom clearance of 6 ft, 6 in. (1.95 m) in residences (see Fig. 9–1). In other types of buildings, the minimum headroom clearance should be 6 ft, 10 in. (2.05 m). The stairwell opening length will depend on the steepness of the stair and must be calculated accordingly.

TYPES OF STAIRS

Stairs are classified by the *kind of stringer* used in their construction, by the way the complete *stair is fitted into the building,* and by the *shape* of the complete stair.

Stringer Types

One type of stringer used is the *open stringer* (see Fig. 9–5), in which pieces are cut from the stringer and risers and treads attached so that their ends are exposed. Another type is the *semihoused stringer,* in which a piece of ¾-in. (19-mm) material is cut out, as in Fig. 9–5, and then fastened to the face of a solid 1½-in. (38-mm) stringer, as illustrated in Fig. 9–6. The ends of the risers and treads are thus concealed in the finished stair. A third type is the *housed stringer,* which has *dadoes,* ½ in. (12 mm) deep, cut on its inner face to receive the ends of the risers and treads (see Fig. 9–7). The dadoes are tapered so that wedges can be driven below the tread and behind the riser to hold each tightly in place.

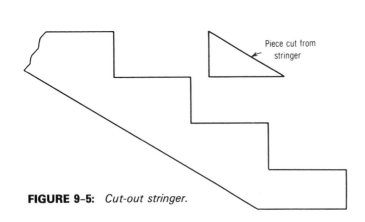

FIGURE 9–5: *Cut-out stringer.*

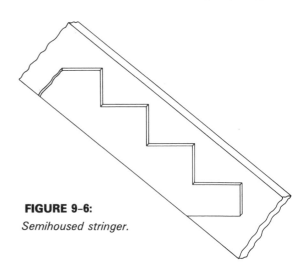

FIGURE 9–6:
Semihoused stringer.

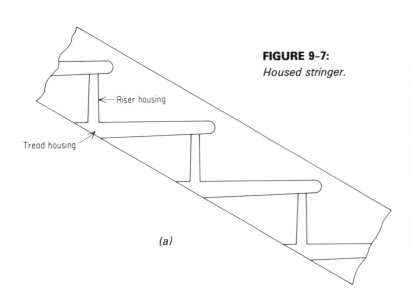

FIGURE 9–7:
Housed stringer.

(a)

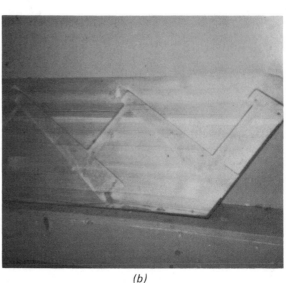

(b)

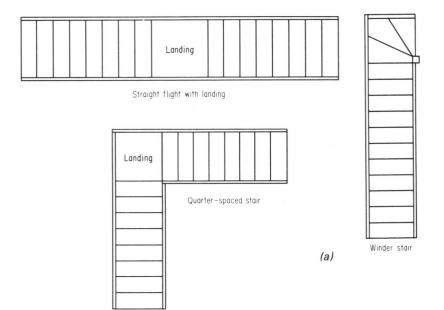

Straight flight with landing

Quarter-spaced stair

Winder stair

(a)

Straight flight

FIGURE 9-8: *a) Stair shapes.*
b) Prefab winders. c) U-stairs.

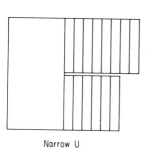

Narrow U

Wide U

(c)

(b)

Stair Types

A complete stair that has no wall on either side is called an *open stair*. If it has a wall on one side only, it is a *semihoused stair;* if it is built between two walls, it is a *housed stair.*

Stair Shapes

A stair that rises with uninterrupted steps from floor to floor is known as a *straight flight;* if it has a break between top and bottom, it is a *straight flight with landing.* A stair that makes a right-angle turn by means of a landing is called an *L stair,* and a stair that makes a right-angle turn step by step is called a *winder.* A stair that makes a 180° turn is called a *U stair* (see Fig. 9-8).

PARTS OF A STAIR

The parts of a stair include some already mentioned. The first is the *stringer,* the purpose of which is to carry the other members of the stair and to support the live load. Next is the *tread,* the surface on which to walk and then the *riser,* which covers the vertical spacings. A *nosing piece* is attached at the upper floor level, and a *molding* may cover the joint between riser and tread. *Wedges* hold the risers and treads tight in a housed stair. A semi-housed or open stair will have a *newel post* or newel posts at the bottom, which carries the bottom end of the *handrail.* Between handrail and treads there are *balusters;* the complete unit of handrail and balusters, with newel post, makes up a *balustrade* (see Fig. 9–9).

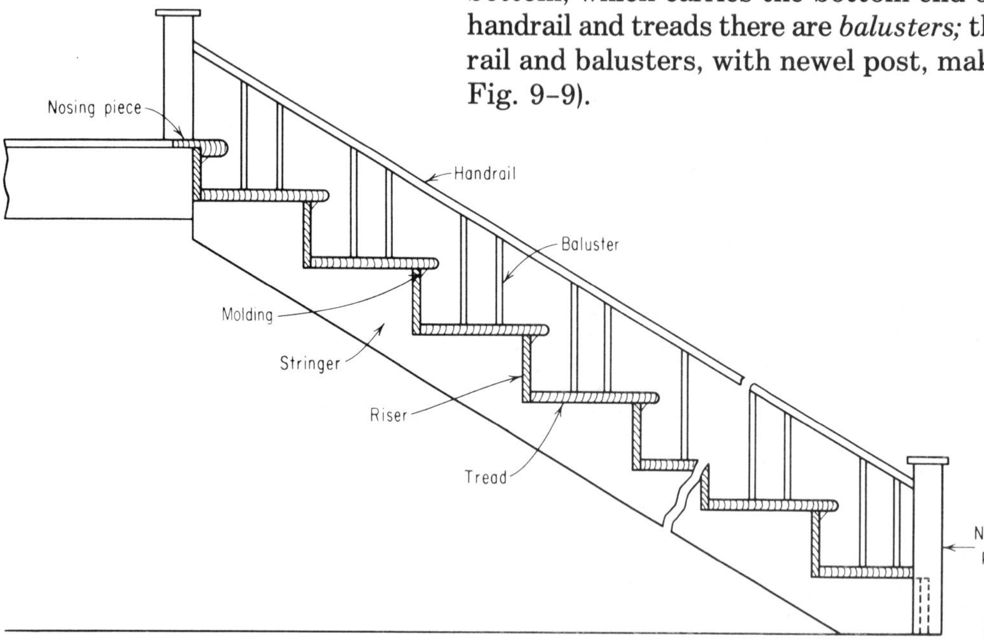

Nosing piece

Handrail

Baluster

Molding

Stringer

Riser

Tread

Newel post

FIGURE 9–9: *Parts of a stair.*

HOW TO BUILD AN OPEN STRINGER STAIR

The first calculation necessary in connection with a stair is that of finding the length of the stairwell opening. This actually must be done during the floor-framing stage, because the opening must be framed at that time.

Calculation of Stairwell Opening Length

A simple method of calculating the stairwell opening length is by comparison of two similar triangles involved in the stair layout. One contains unit rise and unit run as two sides of a triangle and the other the stairwell opening as one side and the headroom clearance plus thickness of ceiling finish, floor frame, subfloor, and finish floor as another. A portion of the line of flight is the third side of both.

As a specific example, suppose the finish floor to finish floor height is 108 in. (2.7 m), the headroom clearance required is 6 ft, 6 in. (1.95 m), and the floor assembly thickness is 11½ in. (240 mm). The unit rise is to be between 7 and 7½ in. (175 and 190 mm).

The product of the unit rise and the unit run should be 70–75 in. (45,000–48,500 mm). What length of stairwell opening is required?

Use the following procedure to determine stairwell opening length:

1. Determine the number of risers required by dividing the total rise by 7 (175):

$$^{108}\!/_7 = 15.4 = 15 \text{ risers}$$
$$(^{2700}\!/_{175} = 15.4 = 15 \text{ risers})$$

2. Determine the size of each riser by dividing the total rise by the number of risers:

$$^{108}\!/_{15} = 7.2 \text{ in.}$$
$$(^{2700}\!/_{15} = 180 \text{ mm})$$

3. Determine the unit run by dividing 72 (45,000) by the individual riser size:

$$^{72}\!/_{7.2} = 10 \text{ in.}$$
$$(^{45000}\!/_{180} = 250 \text{ mm})$$

4. Determine the total drop by adding the floor assembly thickness to the headroom (see Fig. 9-10):

6 ft, 6 in. + 11½ in. = 7 ft, 5½ in.   or   89.5 in.
(1950 mm + 240 mm = 2190 mm)

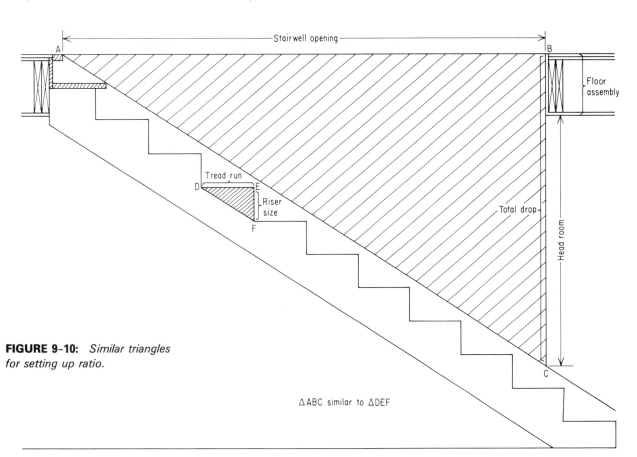

FIGURE 9-10: *Similar triangles for setting up ratio.*

△ABC similar to △DEF

5. Determine the stairwell opening by setting up a ratio with the common sides of the two triangles:

$$\frac{\text{unit rise}}{\text{unit run}} = \frac{\text{total drop}}{\text{stairwell opening}}$$

$$\frac{7.2}{10} = \frac{89.5}{\text{S.O.}}, \quad \text{S.O.} = \frac{89.5 \times 10}{7.2} = 124.3 \text{ in.}$$

$$= 10 \text{ ft, } 4\tfrac{5}{16} \text{ in.}$$

$$\left(\frac{180}{250} = \frac{2190}{\text{S.O.}}, \quad \text{S.O.} = \frac{2190 \times 250}{180} = 3042 \text{ mm}\right)$$

6. The rough stairwell opening is equal to the stairwell opening plus the riser thickness plus the nosing and finish on the stairwell header. Usually adding 3 in. (75 mm) to the opening will allow for these items.

$$\text{rough stairwell opening} = 10 \text{ ft, } 4\tfrac{5}{16} \text{ in.} + 3 \text{ in.}$$
$$= 10 \text{ ft, } 7\tfrac{5}{16} \text{ in.}$$

$$(3042 + 75 = 3117 \text{ mm})$$

Layout of Stair Stringer

Take the unit rise on the tongue of a framing square and the unit run on the blade and clamp a pair of stair gauges to the square at these points (see Fig. 9–11). Lay the square on the stringer stock, close to one end, as illustrated in Fig. 9–12, and carefully draw on the stock the rise and run. Slide the square along to the end of the run line and repeat. Continue until the required number of risers and treads have been laid out. The stock will now look like the illustration in Fig. 9–13.

Since the stair begins with a riser at the bottom, extend the last riser and tread lines to the back edge of the stock, as shown in Fig. 9–14. Cut the stringer off along these lines.

At the top end, extend the last riser and tread lines to the back edge of the stock and cut as shown in Fig. 9–14. Cut very carefully along all the rise and tread lines to produce a cutout stringer (see Fig. 9–15).

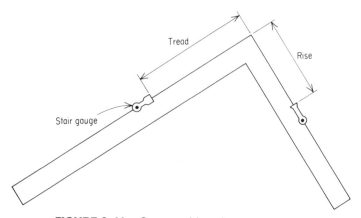

FIGURE 9-11: *Square with stair gauges attached.*

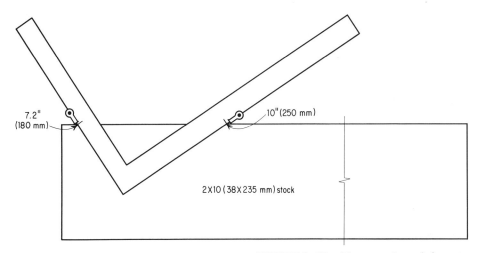

7.2"
(180 mm)

10"(250 mm)

2 X 10 (38 X 235 mm) stock

FIGURE 9-12: *First step in stair layout.*

FIGURE 9-13: *Stringer laid out.*

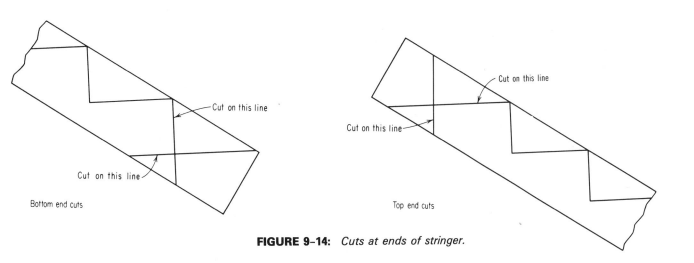

Cut on this line

Cut on this line

Cut on this line

Cut on this line

Bottom end cuts

Top end cuts

FIGURE 9-14: *Cuts at ends of stringer.*

FIGURE 9-15: *Stringer cut out.*

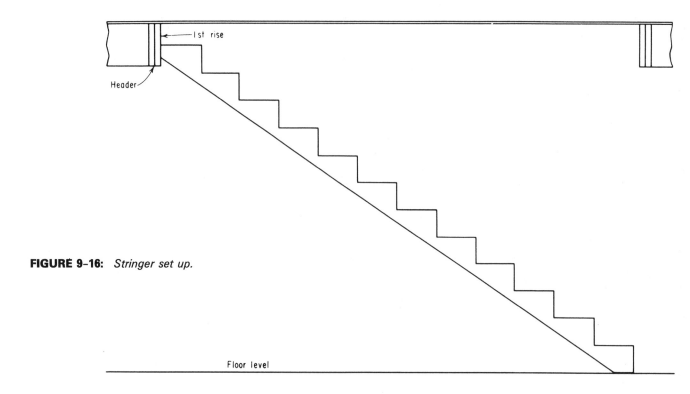

FIGURE 9-16: *Stringer set up.*

When this stringer is set up in place (see Fig. 9–16), it is seen that the face of the header becomes the first riser. But when the treads are set in place (see Fig. 9–17), the bottom step becomes higher than the rest, and the top step becomes shorter than the others, each by the thickness of the tread. To remedy this discrepancy, cut off from the bottom an amount equal to the thickness of the tread (see Fig. 9–18). Now when the stringer is set in place, the bottom rise is less than the rest, and the top is greater, but when the treads are in place, all the rises will be equal.

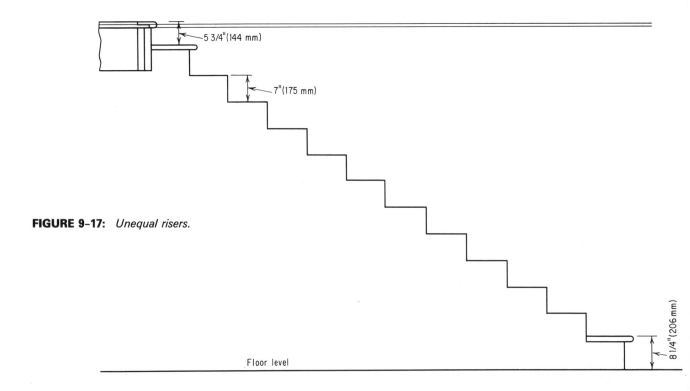

FIGURE 9-17: *Unequal risers.*

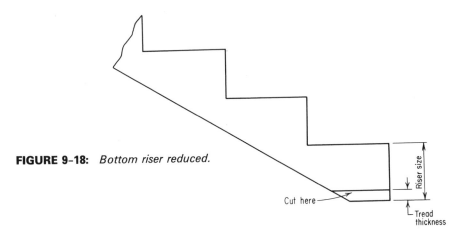

FIGURE 9-18: *Bottom riser reduced.*

Stair Assembly

At the top of the stair, a nosing piece is used (see Fig. 9-19). This produces the same tread overhang at the floor level as the other treads have.

Tread material should be clear, dry, edge grain (if the treads are exposed), 1⅛ or 1⅜ in. (28 or 32 mm) in thickness. The nosing is usually half-round, and the overhang should extend a distance equal to the thickness of the tread (see Fig. 9-20). If the tread nosing is to be capped with a metal molding, it should be shaped to fit that molding rather than being made half-round.

The risers are put in place first, nailed with 2½ in. 65-mm finishing nails. The treads are then put on, nailed in place, and from the back the bottom edges of the risers are nailed to the back edge of the treads. Cove molding may then be fitted under the overhang to hide the joint between riser and tread (see Fig. 9-20).

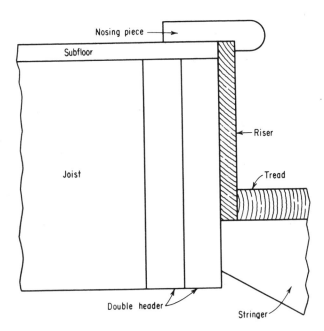

FIGURE 9-19: *Nosing piece in place.*

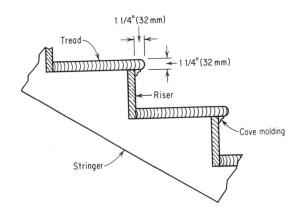

FIGURE 9-20: *Tread overhang.*

HOW TO BUILD A SEMIHOUSED STRINGER STAIR

When this type of stair is to be built, the ¾-in. (19-mm) cutout stringers are made in exactly the same way as described above. The ends of the solid, backup stringers will, however, be cut differently. At the bottom, the stringer will be allowed to run past the riser until the vertical cut is equal to the height of the baseboard, if any (see Fig. 9–21). At the top end, it will be cut as illustrated in Fig. 9–21, again so that the vertical cut will be the same height as the baseboard.

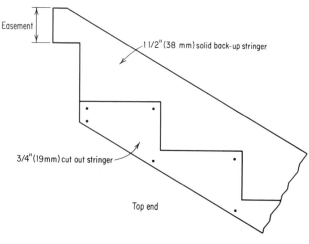

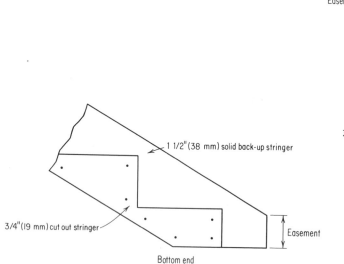

FIGURE 9-21: *Top and bottom ends of semihoused stringer.*

Glue and nail ¾-in. (19-mm) cutout pieces to the solid stringers so that the tread nosings will be the required distance from the upper edge of the stringer. Set the stringers in place, and set in the risers and treads. Care must be taken so that the ends fit snugly against the face of the stringer.

HOW TO LAY OUT AND CUT A HOUSED STRINGER

Calculations for riser heights are the same as for other types of stairs, but the method of layout is considerably different.

Tapered dadoes are cut in the stringer to receive the treads and risers; these can be drawn with a template but are more commonly cut out with the aid of a router. Adjustable templates are available that make the cutting of the dadoes a simple chore. These templates can be set to the required riser and tread size making a 1 in 16 taper to allow for installation of wedges to hold pieces in place (see Fig. 9–22). A simple template can be made from hardboard to be used with a router equipped with a collar (see Fig. 9–23). To lay out and cut the stringer, proceed as follows:

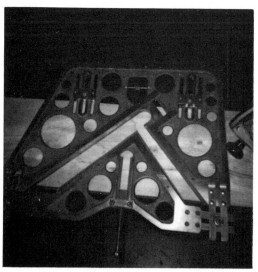

FIGURE 9-22: *Adjustable metal template.*

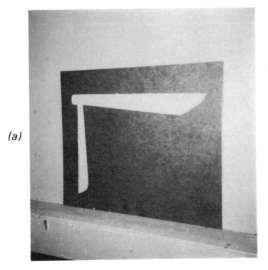
(a)

(b)

FIGURE 9-23: *a) Masonite template. b) Router, collar and bit.*

1. Joint the top edge of the stringer so that it is perfectly straight.

2. Using the required run and rise figures, draw the slope of the tread across the face of the stringer (see Fig. 9-24).

3. Lay the template to this line, making sure the nosing is the required distance from the top edge of the stringer. Draw a line through this point parallel to the top edge. This distance is usually at least 1 in. (25 mm), as shown in Fig. 9-25.

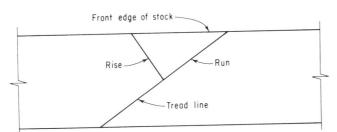

FIGURE 9-24: *Tread line drawn on stringer.*

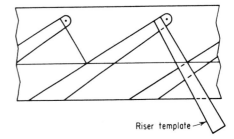

FIGURE 9-25: *Nosing line.*

4. Mark on the tread line the width of the tread, and through this point draw a line parallel to the front edge. This is the baseline from which the layout is made (see Fig. 9-26).

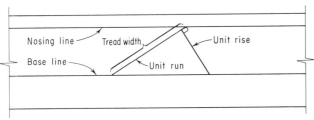

FIGURE 9-26: *Base line drawn.*

FIGURE 9-27:   *Template ready for routing.*

5. Fasten the template to the stringer in the correct location and cut out the dadoes for each step using the router (see Fig. 9-27).

6. At the bottom end, cut the stringer as illustrated in Fig. 9-28.

7. At the top end, cut the stringer as shown in Fig. 9-29, making sure to cut the dado for the nosing piece.

8. Chamfer the top inner edge of the stringer as required and sand the surface clean.

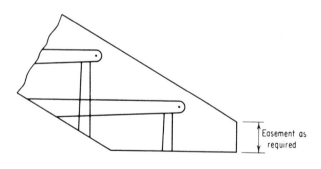

FIGURE 9-28:   *Bottom end cut.*

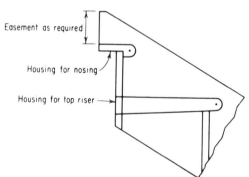

FIGURE 9-29:   *Top end cut.*

ASSEMBLING THE STAIR

Lay the stringers top edge down on the floor, far enough apart to take the treads and risers. Set all the treads in place first and hold the assembly together with bar clamps near the top and bottom. Check to see that the front edge of the groove in each tread lines up with the front edge of the riser dado. Set the nosing piece in place.

Now set all the risers in place. Check to see that the back edge of each tread fits snugly against the front face of its riser. The top riser cannot be wedged, since its dado is open at the back. Apply glue to the ends and nail it in place from the back.

The next step is to wedge the treads in place. Apply glue to the wedges and drive each firmly into place. The tread should thus be brought into a tight fit against the upper edge of the housing. Be sure that the wedges do not project beyond the back edge of the treads.

Now wedge the risers in place. Cut off the thin end of each wedge so that, when driven, it will not interfere with the tread above. Nail the bottom of each riser to the back edge of the tread, as illustrated in Fig. 9-30.

Nail and glue a cant strip into the junctions between risers and treads (see Fig. 9-31). The stair is now complete and ready to install.

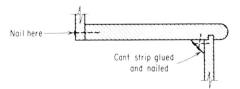

FIGURE 9-30: *Riser and tread nailed.*

FIGURE 9-31: *Glued and nailed strip at junction of step.*

REVIEW QUESTIONS

9-1. Fill in the blank or blanks in each sentence with the correct word or phrase:

(a) A stair begins at the bottom with a _____ .

(b) A stair leads to an upper floor through a _____ .

(c) A line drawn through the extremities of the nosings of stair treads is the _____ .

(d) Stair risers and treads are carried on _____ .

(e) The vertical distance from the underside of the stairwell opening header to the line of flight is the _____ .

(f) The vertical distance from one floor level to the next is the _____ of a stair.

(g) A stair tread is the width of the _____ wider than than the run.

(h) Stairs should have a slope of from _____ to _____ degrees.

(i) Minimum stair width should be _____ .

(j) Run multiplied by rise should equal approximately _____ .

9–2. Name three types of stringer used in stair building and indicate the differences between them.

9–3. Explain briefly the difference between an open, a semihoused, and a housed stair.

9–4. What is the purpose of

(a) Wedges in a stair?

(b) A newel post?

(c) A baluster?

(d) A stringer?

(e) A tread and riser template?

9–5. If the headroom clearance is to be 6 ft, 8 in. (2.0 m), the floor frame 2×10 (38×235 mm) material, the subfloor $\frac{5}{8}$-in. (15-mm) plywood, the finish floor $\frac{1}{8}$-in. (3-mm) inlaid, the underlay $\frac{3}{8}$-in. (10-mm) particle board, the ceiling and header finish $\frac{1}{2}$-in. (12-mm) gypsum wallboard, the risers $\frac{3}{4}$ in. (19 mm) thick, and the nosing projection $1\frac{1}{4}$ in. (32 mm),

(a) What should the length of the stairwell opening be if the tread run is 10 in. (250 mm)?

(b) How many risers will there be in the stair if the rise is to be kept as close to 7 in. (175 mm) as possible and the total rise between floors is 9 ft, $4\frac{1}{2}$ in. (2858 mm)?

(c) How many treads will there be in the stair?

9–6. By means of a diagram, illustrate how a nosing piece fits at the top of an open stair stringer.

9–7. Explain why no cutoff of the stringers at the bottom is necessary when a housed stringer stair is being built.

During the course of construction, a variety of equipment is often employed, in addition to the hand and power woodworking tools used in the actual erection of the building.

10

CONSTRUCTION EQUIPMENT

SOIL TESTING EQUIPMENT

In some cases it may be desirable to determine the types and depth of soil strata to be encountered before excavation takes place. This may be done by digging a *test hole* or by the use of electronic soil testing apparatus, such as a *refraction seismograph* or an earth-resistivity meter, such as that illustrated in Fig. 10–1.

EARTH-MOVING EQUIPMENT

Earth-moving equipment includes such machines as the *bulldozer* and the *backhoe* (see Chapter 3), used for excavating, and a *front-end loader* (see Fig. 10–2), used for moving and loading earth.

FIGURE 10–1: *Earth Resistivity Meter. (Courtesy Soiltest Inc.)*

(a)

(b)

FIGURE 10–2: *Tractor with bucket for moving earth.*

FIGURE 10-3: *Portable concrete mixer. (Courtesy Western Equipment Ltd.)*

CONCRETE EQUIPMENT

Conventional foundations require the use of some type of concrete mixer, varying from the small, *portable mixer* shown in Fig. 10-3 or a large, *stationary mixer* such as that illustrated in Fig. 10-4 to a *transit-mix truck*, like that illustrated in Fig. 10-5.

At the site, concrete may have to be transported from mixer to the forms, and this may be done by the use of wheelbarrows, by *power buggies*, or by lifting buckets (see Fig. 10-6), and concrete in the forms may be consolidated by the use of an *internal vibrator*, such as the one shown in Fig. 3-69, page 98. A number of tools are used in finishing concrete slabs, including a *hand float*, illustrated in Fig. 10-7, *hand trowels* and *edgers*, and a *power trowel*, such as that illustrated in Fig. 10-8.

FIGURE 10-4: *Stationary concrete mixer. (Courtesy London Concrete Machinery Co.)*

FIGURE 10-5: *Transit-mix truck.*

FIGURE 10-6: *Lifting bucket.*

FIGURE 10-7: *Concrete hand float. (Courtesy Portland Cement Ass'n)*

FIGURE 10-8: *Concrete power trowel. (Courtesy Master Builders Co.)*

SOIL BREAKING AND COMPACTION EQUIPMENT

At some sites, it may be necessary to break up the surface to facilitate excavation, and this may be done by the use of a *jack hammer*, illustrated in Fig. 10-9. A *rammer* (see Fig. 10-10) may be used to compact backfill and a *vibro-compactor* to compact a subgrade (see Fig. 10-11). Vibratory rollers may be used to compact and smooth earth and asphalt surfaces at the site (see Fig. 10-12).

FIGURE 10-9: *Jackhammer in use. (Courtesy Wacker Corp.)*

FIGURE 10-10: *Rammer compacting backfill. (Courtesy Wacker Corp.)*

FIGURE 10-11: *Vibro-compactor at work. (Courtesy Wacker Corp.)*

FIGURE 10-12: *Vibratory/static roller. (Courtesy Wacker Corp.)*

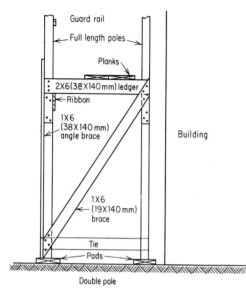

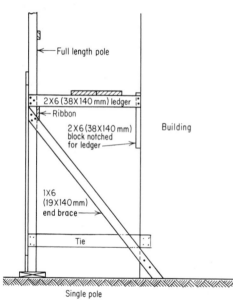

FIGURE 10-13: *Wooden pole scaffolds.*

SCAFFOLDS

As construction progresses, *scaffolding* or *staging* is required to reach work areas which are beyond the normal reach of a workman standing on the ground. Scaffolding consists of elevated platforms, resting on rigid supports and strong and stiff enough to support workmen, tools, and materials safely.

Scaffolding may be made of wood or metal, the wooden ones normally being made on the job, while metal scaffolds are manufactured articles, made in a variety of styles.

Pole Scaffolds

The most common wooden scaffolds are *pole scaffolds,* made in either *single-* or *double-pole* style scaffolds. Double-pole scaffolds have two legs per section (see Fig. 10-13), while the single-pole scaffold has but one. In both cases, 2 × 6 in. (38 × 140 mm) minimum *ledgers* support the platform, usually made up of two or more planks, at least 10 in. (235 mm) wide. 1 × 6 (19 × 140 mm) angle braces and a ribbon under the ledger join scaffold sections together. Figure 10-13 gives recommended minimum sizes of materials used in the construction of such scaffolds. The horizontal distance between scaffold sections should not exceed 10 ft (3 m) and the maximum height for such scaffolds should be limited to 20 ft (6 m). Bearing pads should always be placed under the bottom ends of the poles in order to obtain maximum stability, and double-pole scaffolds should be tied to the building at intervals for greater safety.

Manufactured Scaffolds

Manufactured scaffolding is easily assembled and dismantled. Adjustments can easily be made to facilitate different heights, and they are very mobile. Many types of scaffolds are available, including brackets, ladder jacks, metal pole, climbing, sectional, tower, and swing stage.

FIGURE 10-14: *Wall bracket.*

Brackets

Planks can be supported by special brackets which are fastened to the studs in wall construction and to the roofing members in roof installations (see Fig. 10-14). Brackets are very convenient as they can be quickly erected and they require little material. Brackets are also easily moved from one job site to another. Roof brackets are fastened with nails through the sheathing into the rafters (see Fig. 10-15). They can be removed without pulling the nails so damage does not happen to the roof covering.

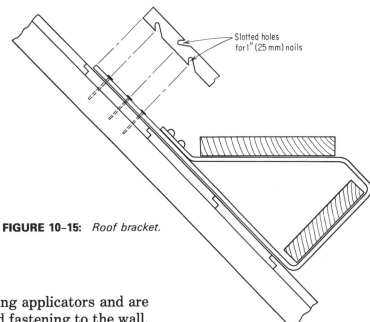

Slotted holes for 1" (25 mm) nails

FIGURE 10-15: *Roof bracket.*

Ladder Jacks

Ladder jacks are commonly used by siding applicators and are especially convenient as they do not need fastening to the wall. This allows for placing of finish with little or no interference (see Fig. 10-16). A ladder jack is a device for hanging a plank from a ladder. Ladder jacks are convenient when only one workman is involved. Two ladders and a single plank are all that is required.

FIGURE 10-16: *Ladder jack.*

FIGURE 10-17: *Metal pole scaffold.*

Metal Pole Scaffold

A metal pole scaffold is similar in design to a single-pole wooden scaffold except the poles are made of aluminum in most cases. This makes for a very light pole to which the metal ledger is clamped. The other end is nailed to the wall at the desired height. Metal poles are also used for the required bracing (see Fig. 10–17). A wooden pad is used under the leg so that it doesn't sink into the ground.

Climbing or Adjustable Scaffold

One type of adjustable scaffold looks similar to a single-pole metal scaffold but works on much the same principle as ladder jacks. A 2 × 4 (38 × 89 mm) is used as the vertical pole. It is fastened to the roof so that it does not rotate. Metal brackets are fastened to the support with horizontal ledgers for supporting the plank. These brackets are adjustable and can be moved up or down the support as the inner end is not fastened to the wall (see Fig. 10–18).

FIGURE 10-18: *Climbing or adjustable scaffold.*

FIGURE 10-19: *A workhorse.*

Another type of adjustable scaffold that has come on the market recently is a workhorse. It is supported by four legs and is easily moved on each of these legs individually. The height that it can be used is limited to about 6 ft (2 m) unless diagonal bracing is attached to the legs (see Fig. 10-19).

Sectional Scaffolds

Sectional scaffolding is made up of tubular material in double-pole style, as illustrated in Fig. 10-20. A standard unit consists of two end frames and two crossed braces with holes at each end [see Fig. 10-20(b)]. An end frame has threaded studs attached to the legs which fit into the holes in the brace ends. Braces are held in place by wing nuts so that the assembled unit appears as illustrated in Fig. 10-20(c)]. To increase the length of a scaffold, additional sets of end frames are added, using two additional crossed braces for each end frame used. To extend the height, one standard unit is mounted above another. Coupling pins [Fig.

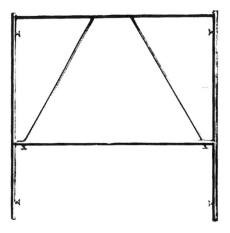

(a) Scaffold end frame

(b) Crossed brace

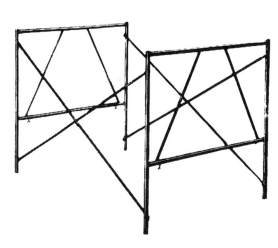

(c) Assembled unit

(e) Metal scaffold plank

FIGURE 10-20:
Tubular metal scaffolding.

(d) Assembled scaffolding

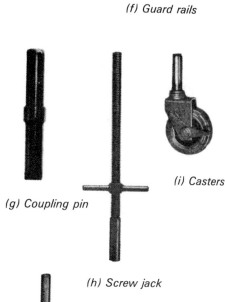

(f) Guard rails

(i) Casters

(g) Coupling pin

(h) Screw jack

(j) Base plate

FIGURE 10-20 *(continued)*

10-20(g)] are inserted into the top of the tubular legs of the lower unit, and the legs of additional end frames are slipped over the top half of the coupling. Additional crossed braces can be used horizontally on any unit to give it greater rigidity. By adding units as indicated, scaffolds of any length or height may be assembled [see Fig. 10-20(d)].

Nonskid metal planks, like the one shown in Fig. 10-20(e), which span a standard unit, are available with this type of scaffold, or wooden planks may be used if desired.

Base plates [see Fig. 10-20(j)] fit into the bottom of frame legs to form the footing for a scaffold, the short one for level surfaces and the long one to compensate for uneven ground. Casters [see Fig. 10-20(i)] fitted with safety brakes, may be used in place of base plates on hard surfaces when it is necessary to move the assembled scaffold from one position to another. *Screw jacks* [see Fig. 10-20(h)] are available for use with base plates or casters where floors are uneven or where a fine adjustment in height is required.

(k) Metal scaffolding.

Steel scaffolding of this type is made with the utmost consideration for safety, but of course care must be exercised in its erection and use. The following rules should be observed by erectors and workmen:

1. Provide sufficient sills or underpinning in addition to standard base plates on all scaffolds to be erected on fill or otherwise soft ground.

2. Compensate for unevenness of ground by using adjusting screws rather than blocking.

3. Be sure that all scaffolds are plumb and level at all times.

4. Anchor running scaffold to the wall approximately every 28 ft (8.5 m) of length and 18 ft (6 m) of height.

5. Adjust scaffold until braces fit into place with ease.

6. Use guard rails on all scaffolds, regardless of height.

7. Use ladders—not the cross braces—to climb scaffolds.

8. Tighten all bolts and wing nuts which are a part of the scaffold.

9. Use horizontal bracing to prevent racking of the structure.

10. All wood planking used on a scaffold should be of sound quality, straight grained, and free from knots.

11. When using steel planks, always fill the space because steel planks tend to skid sideways easily.

12. Handle all rolling scaffolds with additional care.

13. Do not extend adjusting screws to full extent.

14. Horizontal bracing should be used at the bottom, at the top, and at intermediate levels of 18 ft (6 m).

In addition to its many conventional uses, sectional scaffolding may be used as temporary *shoring* or support for various parts of a building under construction, as illustrated in Fig. 10-21.

For interior work a smaller sectional scaffolding can be obtained. It is equipped with casters and will fit through a standard doorway. Height adjustments are quickly made, making it convenient for different ceiling heights (see Fig. 10-22).

FIGURE 10-21: *Sectional scaffolding used as shoring.*

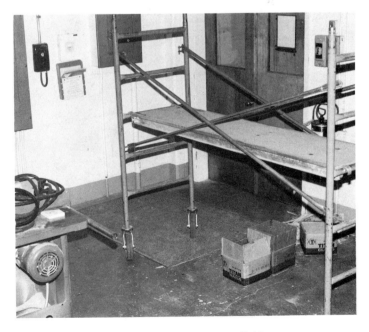

FIGURE 10-22: *Rolling scaffold.*

FIGURE 10-23: *Tower scaffold in use. (Courtesy Morgen Mfg. Co.)*

Tower Scaffold

Another commonly used type of metal scaffolding is the *tower* scaffold, shown in Fig. 10-23. A standard unit is made up of a pair of metal towers, tied together with both horizontal—*stringer*—braces and crossed braces (see Fig. 10-23). Pairs of towers are then tied together with stringer braces to form a continuous scaffold of any desired length. Each tower supports a *carriage*, on which rests a workman's plank platform or, in the case of scaffolding intended for use by masons, two platforms, one for the tradesmen and one for laborers and materials. A *winch* is attached to each tower to raise the platform to the required height.

The height of these scaffolds can be increased by adding sections to the ends of towers, thus producing scaffolds which will reach to any desired height.

The same rules of safety which apply to the erection and use of sectional scaffolding also apply to the use of tower scaffolds.

Swing Stage Scaffold

Swing stage scaffolds are platforms, complete with guardrail and toeboard, suspended from the roof of the building by a block-and-tackle system attached to each end. By this means, the workers on the platform can adjust the height of the scaffold as required (see Fig. 10-24). Such scaffolds are intended for workmen using only light equipment, working high above the ground.

FIGURE 10-24: *Swing stage scaffold.*

LADDERS

Several types of aluminum and wooden ladders are used on construction projects. They include *step ladders,* ranging in height from 4 to 18 ft (1.2–6 m); *extension ladders,* in lengths up to about 24 ft (8 m); *safety rolling ladders,* usually consisting of three or four steps mounted on rollers; and *single ladders,* usually available in lengths of from 8 to 24 ft (2.4 to 8 m).

A sturdy, reliable single ladder may be made on the job as follows:

1. Choose seasoned, straight-grained stock, preferably fir or Sitka spruce, which is free from knots and other defects which would impair the strength. For ladders up to 16 ft (4.8 m) in length, 2 × 4 (38 × 89 mm) stock is used; for ladders over that length, 2 × 6 (38 × 140 mm) material should be used.

2. Taper rails for shorter ladders from 3⅜ in. (85 mm) at the bottom to 2⅝ in. (65 mm) at the top and the larger ones from 5¼ in. (140 mm) at the bottom to 4 in. (100 mm) at the top. Dress all four sides and bullnose the edges.

3. Cut notches for the cleats on the top edges, ¾ in. (19 mm) deep, 3 in. (75 mm) wide for the shorter ladder, and 4 in. (100 mm) wide for the longer one [see Fig. 10–25(a)]. Notches must not be more than 12 in. (300 mm) o.c.

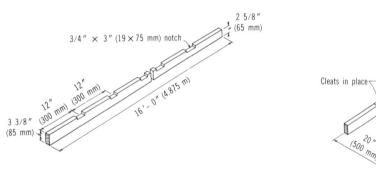

(a) Side rails.

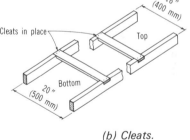

(b) Cleats.

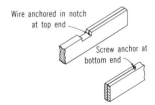

(c) Wire reinforcement.

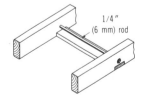

(d) Rod reinforcement.

FIGURE 10–25: *Single ladder details.*

4. Cut a cleat not less than 20 in. (500 mm) long for the bottom step. Cleats may be made successively shorter up the ladder so that the top is not more than 4 in. (100 mm) narrower than the base. Dress the cleats on all four sides and bullnose the top edges. Be sure that each cleat fits snugly into its notches.

5. Nail the cleats into place with two nails in each end (see Fig. 10–25b).

Ladders may be reinforced by running a heavy galvanized wire in a groove along the underside of each side rail. The ends must be securely anchored by bringing them over the ends of the rails and providing suitable anchorage on the top edges of the rails [see Fig. 10–25(c)]. Further reinforcement can be provided by running a ¼-in. (6-mm) rod with threaded ends behind each cleat through the side rails. Put a washer and nut on each end and tighten them down snugly [see Fig. 10–25(d)].

Safety Rules for Ladders

1. Always check a ladder for defects before use.

2. In use, place the ladder so that the distance from its bottom end to the wall is at least one-quarter the length of the ladder, unless the top end is secured.

3. Single and extension ladders should extend at least 3 ft. (1 m) above the landing against which the top end rests.

4. Check to see that both rails rest on solid, level footing.

5. Equip the bottom ends of the rails with *safety shoes* if the ladder is to be used on hard surfaces that may allow it to slip.

6. Lubricate locks and pulleys on extension ladders, keep fittings tight, and replace worn rope.

7. Do not allow paint, oil, or grease to accumulate on ladder rungs or rails.

MATERIALS HOIST

Some type of equipment is required to lift materials which are to be applied to a building from a scaffold or used on the roof, and normally a *hoist* of some sort will be employed for the purpose.

For relatively low lifts, a *fork lift* on a tractor or front-end loader may be satisfactory, or a lift of the type illustrated in Fig. 10–26 may be used. Such a hoist will lift about 500 lbs (225 kg) to a height of about 18 ft (6 m). For greater heights, an *elevator-type materials lift* (see Fig. 10–27), a mobile crane, or a tower crane may be necessary.

(a)

(b)

(c)

FIGURE 10-26: *Materials lift.*

REVIEW QUESTIONS

10-1.   Explain the basic difference between a double-pole and a single-pole scaffold.

10-2.   (a) Name three groups of workmen for whom "light trade" scaffolds are made.

   (b) Name three groups for whom "heavy trade" scaffolds are made.

10-3.   Carefully draw a diagram of a double-pole scaffold and name each part.

10-4.   What is the purpose of a *life line* on a swing stage scaffold?

10-5.   What species of lumber, from the standpoint of weight, is best for making ladders?

10-6.   (a) What size material should be used for ladders not over 16 ft (4.875 m) in length?

   (b) What is the accepted taper in the rails of ladders over 16 ft (4.875 m) in length?

   (c) What is the maximum spacing of cleats in a ladder?

   (d) What is the minimum distance that the bottom end of a free-standing ladder should be from the wall against which it rests?

   (e) By how much should a portable ladder project above a landing?

10-7.   List three advantages of metal scaffolding over wooden scaffolding.

10-8.   Explain why two platforms are often used on scaffolds being used by bricklayers or stone masons.

FIGURE 10-27: *Materials hoist.*

11

FINISHING EXTERIOR

ORDER OF OPERATIONS

Once the frame of the building has been completed, the next step is to apply the exterior finish. This involves installing window and door frames and outside casings, roofing, exterior paper, cornice work, and the application of whatever is to be used to cover the exterior.

The usual order in which these operations are carried out is as follows: (1) cornice work, (2) roofing, (3) fitting of door and window frames (4) application of exterior paper, (5) application of exterior finish.

CORNICE WORK

The *cornice,* or eave, is formed by the overhang of the rafters and may be *open* or *boxed.* In either case, the work of finishing the eave should be done as soon as the rafters are in place.

Open Eaves

With open eaves, the rafter tails and the underside of the roof sheathing are exposed from below, and consequently the roof sheathing over that section of the roof may be replaced by a better quality of material.

In addition, it is necessary to block off the openings between the rafters above the cap plate, and this is done by the use of *windblocks.* These are pieces of 1½-in. (38-mm) material cut to fit snugly between each pair of rafters (see Fig. 11–1).

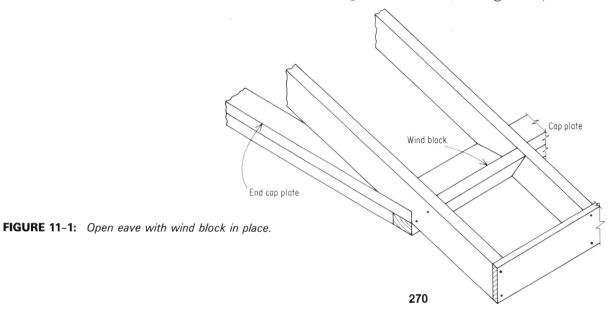

FIGURE 11–1: *Open eave with wind block in place.*

With this type of eave, a single fascia may be used over the ends of the rafter tails, as shown in Fig. 11–1.

Boxed Eaves

With boxed eaves, the underside of the roof overhang is enclosed. This may be done by applying a *soffit* to the bottom edge of the rafter tails, as illustrated in Fig. 11–2. In such a case, the *rough fascia*, which supports the outer edge of the soffit between rafters, has both top and bottom edges beveled to the slope of the roof.

(a)

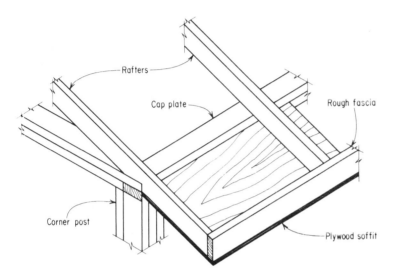

FIGURE 11–2: *Eave boxed on slope of roof.*

An alternative method is to build a frame to which a horizontal soffit may be attached to enclose the eave (see Fig. 11–3). Such a frame is made up of a *lookout ledger,* nailed to the wall, a *rough fascia* over the ends of the rafter tails, and *lookouts,* which span between the two and lie alongside each rafter tail (see Fig. 11–4). It may be fabricated on the ground and raised into place as a unit.

(b)

FIGURE 11–4: *Boxed eave framing.*

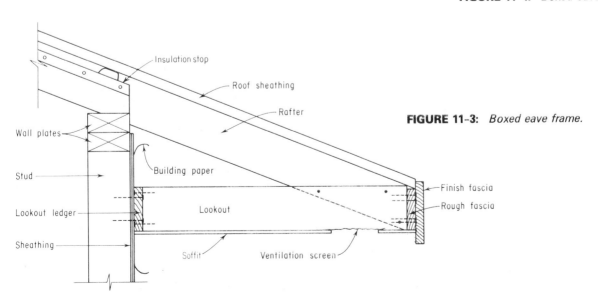

FIGURE 11–3: *Boxed eave frame.*

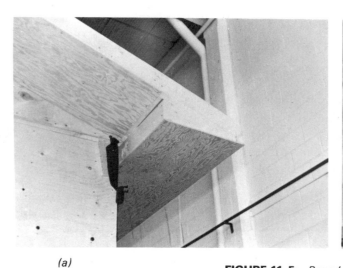

(a)   **FIGURE 11-5:** *Boxed eave with gable roof.*   *(b)*

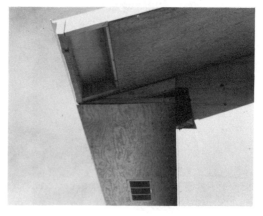

FIGURE 11-6: *Soffit stopped at corner of building.*

In a hip roof building, the end wall soffit frames are made to fit snugly between those of the side walls. When a gable roof is involved, the gable rafter tails fit outside the end lookouts, and a filler piece of 2×4 (38×89 mm) material is cut to fit between the end of the lookout ledger and the underside of the gable rafter, as illustrated in Fig. 11-5(a). The end of the eave is then finished as shown in Fig. 11-5(b). Figure 11-6 shows another method of finishing the eave with a gable roof. The soffit on the underside of the gable end overhang will run to the edge of the roof, and the horizontal soffit will stop at the corner of the building.

Another method of finishing the boxed eave at a gable end is illustrated in Fig. 11-7. This is known as *returning the cornice.* A set of rafter tails is nailed to the gable wall, producing the same overhang as the regular eave. The tops of the *returns* are shingled like the roof, and the soffit, carried around, finishes the underside.

Framing is placed between the gable end rafter and the verge rafter. Some of the lookouts are extended back to the next rafter to give support (see Fig. 11-8).

FIGURE 11-7: *Cornice returns.*

FIGURE 11-8: *Overhang support.*

Some finishing work must also be done at the gable ends before the roofing is applied. If there is a gable overhang, its underside is lined with ¼-in. (6-mm) plywood, and *barge boards,* at least 1 in. (25 mm) wider than the rafters, are used to cover the ends of the sheathing and to act as trim. Their lower ends should be mitered to meet the fascia.

If there is no gable overhang (see Fig. 11–9), *verge rafters,* 2 in. (50 mm) wider than the regular ones, should be nailed to the gable wall, covering the ends of the roof sheathing. A narrow mold may then be added, flush with the top edge of the verge rafters.

FIGURE 11–9: *Gable ends with no roof overhang.*

Also before the roofing is applied, provision must be made at the lower edges of the roof for the proper direction of rainwater into the eaves trough. One method of doing this is to fasten eave flashing, like that illustrated in Fig. 11–10, to the edge of the roof, under the roofing.

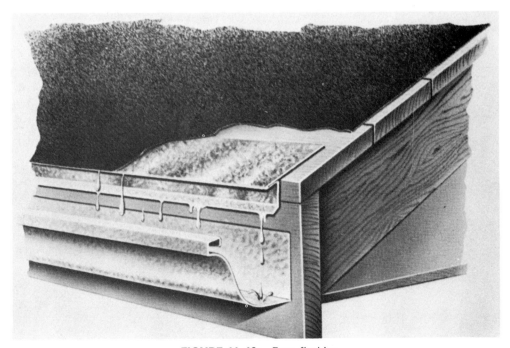

FIGURE 11–10: *Eave flashing.*

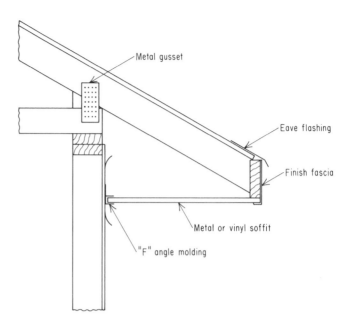

FIGURE 11-11: *Metal or vinyl soffit.*

Metal and Vinyl Soffits

Metal and vinyl soffits have become popular in recent years. They eliminate many framing members and also reduce the maintenance required. A 2 × 6 (38 × 140 mm) rough fascia is the only framing member required. The soffit material is supported on the inside by an F channel and on the outside end by the fascia (see Fig. 11-11). It is stapled or nailed to the bottom edge of the rough fascia. Soffit material is available in individual pieces and in rolls in some areas. A finish fascia of the same material is nailed to the rough fascia (see Fig. 11-12).

ROOFING

The most common type of roofing used on sloping roofs is shingles of one sort or another. These include wood (usually cedar), asphalt, and hardboard. Rolled roofing, roofing tile, and sheet metal roofing are also used on sloping roofs. Roofing tile is commonly made from concrete or metal. In recent years plastics have been used for roofing, particularly on roofs of unusual shape that do not lend themselves to the application of more conventional types of material. Built-up roofing is commonly used for flat or nearly flat roofs.

Wood Shingles

Wood shingles are commonly made from cedar, because it changes very little with atmospheric changes and withstands weathering better than most woods.

Cedar shingles are made in a number of grades, both edge and flat grain, as indicated in Table 11-1.

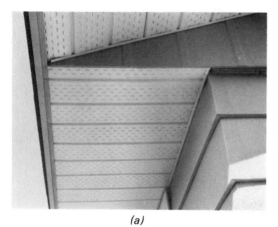

(a)

(b)

FIGURE 11-12: *Finished metal soffit.*

TABLE 11-1: *Western red cedar shingle grades*

Grade	Length	Thickness (at butt)	No. of courses per bundle	Description
No. 1: blue label	16 in. (400 mm) 18 in. (450 mm) 24 in. (600 mm)	$\frac{3}{8}$ in. (10 mm) $\frac{7}{16}$ in. (11 mm) $\frac{1}{2}$ in. (13 mm)	$^{20}/_{20}$ $^{18}/_{18}$ $^{13}/_{14}$	The premium grade of shingles for roofs and walls. These shingles are 100% heartwood, 100% clear, and 100% edge-grain.
No. 2: red label	16 in. (400 mm) 18 in. (450 mm) 24 in. (600 mm)	$\frac{3}{8}$ in. (10 mm) $\frac{7}{16}$ in. (11 mm) $\frac{1}{2}$ in. (13 mm)	$^{20}/_{20}$ $^{18}/_{18}$ $^{13}/_{14}$	A good grade for many applications. Not less than 10 in. (250 mm) clear on 16-in. (400-mm) shingles, $11\frac{3}{16}$ in. (280 mm) clear on 18-in. (450-mm) shingles, and 16 in. (400 mm) clear on 24-in (600-mm) shingles. Flat grain and limited sapwood are permitted in this grade. Reduced weather exposures recommended.
No. 3: black label	16 in. (400 mm) 18 in. (450 mm) 24 in. (600 mm)	$\frac{3}{8}$ in. (10 mm) $\frac{7}{16}$ in. (11 mm) $\frac{1}{2}$ in. (13 mm)	$^{20}/_{20}$ $^{18}/_{18}$ $^{13}/_{14}$	A utility grade for economy applications and secondary buildings. Not less than 6 in. (150 mm) clear on 16-in. (400-mm) and 18-in. (450-mm) shingles and 10 in. (250 mm) clear on 24-in. (600-mm) shingles. Reduced weather exposures recommended.
No. 4: under coursing	16 in. (400 mm) 18 in. (450 mm)	$\frac{3}{8}$ in. (10 mm) $\frac{7}{16}$ in. (11 mm)	$^{14}/_{14}$ or $^{20}/_{20}$ $^{14}/_{14}$ or $^{18}/_{18}$	A utility grade for undercoursing or double-coursed sidewall applications or for interior accent walls.
No. 1 or No. 2: rebutted-rejointed	16 in. (400 mm) 18 in. (450 mm) 24 in. (600 mm)	$\frac{3}{8}$ in. (10 mm) $\frac{7}{16}$ in. (11 mm) $\frac{1}{2}$ in. (13 mm)	$^{33}/_{33}$ $^{28}/_{28}$ $^{13}/_{14}$	Same specifications as above but machine-trimmed for exactly parallel edges with butts sawn at precise right angles. Used for sidewall application where tightly fitting joints between shingles are desired. Also available with smooth sanded face.

The portion of shingle which should be exposed to the weather will depend on the steepness of the roof slope and the grade and length of shingles used. Table 11-2 gives recommended maximum exposure, depending on the slopes.

TABLE 11-2: *Shingle exposure*

Roof slope	Maximum shingle exposure [in. (mm)]					
	No. 1 grade, length of shingle			No. 2 grade, length of shingle		
	16 in. (400 mm)	18 in. (450 mm)	24 in. (600 mm)	16 in. (400 mm)	18 in. (450 mm)	24 in. (600 mm)
1 in 3 or less	$3\frac{3}{4}$ (95)	$4\frac{1}{8}$ (105)	$5\frac{3}{4}$ (145)	$3\frac{1}{2}$ (90)	$3\frac{7}{8}$ (100)	$5\frac{1}{2}$ (140)
Over 1 in 3	$4\frac{7}{8}$ (125)	$5\frac{1}{2}$ (140)	$7\frac{1}{2}$ (190)	$3\frac{7}{8}$ (100)	$4\frac{1}{2}$ (115)	$6\frac{1}{2}$ (165)

(Courtesy Council of Forest Industries of B.C.)

The area of roof covered by one bundle of shingles will depend on the shingle exposure, and Tables 11–3 give the approximate area covered by one bundle of shingles for various exposures.

TABLE 11–3a: *Shingle coverage*

	Approximate coverage in ft² of one bundle based on the following weather exposures								
Length and thickness	*3½ in.*	*4 in.*	*4½ in.*	*5 in.*	*5½ in.*	*6 in.*	*6½ in.*	*7 in.*	*7½ in.*
16 × ⅜	18.2	20.2	23.2	25.3*	28.3	30.3	33.4	36.4	38.4‡
18 × 7/16	16.4	18.2	21.0	22.7	25.5*	27.3	30.0	32.7	34.5
24 × ½	12.3	13.7	15.7	17.1	19.2	20.4	22.5	24.5	25.9*

NOTES: *Maximum exposure recommended for roofs.
†Maximum exposure for double-coursing No. 1 grade on sidewalls.
‡Maximum exposure recommended for single-coursing No. 1 and No. 2 grades on sidewalls.

Approximate coverage in ft² of one bundle based on the following weather exposures										
8 in.	*8½ in.*	*9 in.*	*9½ in.*	*10 in.*	*10½ in.*	*11 in.*	*11½ in.*	*12 in.*	*12½ in.*	*13 in.*
40.5	43.5	45.5	48.5	50.6	53.6	56.6	58.6	61.7†	—	—
36.4	28.4‡	41.0	44.0	45.5	48.2	51.0	52.7	55.5	57.4	60.0
27.3	29.4	30.7	32.7	34.1	36.2	38.2	39.6‡	41.6	43.0	45.1

Approximate coverage in ft² of one bundle based on the following weather exposures					
13½ in.	*14 in.*	*14½ in.*	*15 in.*	*15½ in.*	*16 in.*
—	—	—	—	—	—
61.9	64.6†	—	—	—	—
46.4	48.4	49.8	51.9	53.3	55.3‡

(Courtesy Council of Forest Industries of B.C.)

TABLE 11–3b: *Shingle coverage*

	Approximate coverage in m² of one bundle based on the following weather exposures												
Length and Thickness	*90 mm*	*100 mm*	*115 mm*	*125 mm*	*140 mm*	*150 mm*	*165 mm*	*180 mm*	*190 mm*	*200 mm*	*215 mm*	*225 mm*
400 mm × 10 mm	1.69	1.88	2.16	2.35*	2.63	2.82	3.10	3.38	3.57‡	3.76	4.04	4.23
450 mm × 11 mm	1.52	1.69	1.95	2.11	2.37*	2.54	2.79	3.04	3.21	3.38	2.64‡	3.81
600 mm × 13 mm	1.14	1.27	1.46	1.59	1.78	1.90	2.09	2.28	2.41*	2.54	2.73	2.85

NOTES: *Maximum exposure recommended for roofs.
†Maximum exposure for double-coursing No. 1 grade on sidewalls.
‡Maximum exposure recommended for single-coursing No. 1 and No. 2 grades on sidewalls.

Approximate coverage in m² of one bundle based on the following weather exposures													
240 mm	*250 mm*	*265 mm*	*280 mm*	*290 mm*	*305 mm*	*315 mm*	*330 mm*	*340 mm*	*355 mm*	*365 mm*	*380 mm*	*390 mm*	*405 mm*
4.51	4.70	4.98	5.26	5.45	5.73†	—	—	—	—	—	—	—	—
4.06	4.23	4.48	4.74	4.90	5.16	5.33	5.58	5.75	6.00†	—	—	—	—
3.04	3.17	3.36	3.55	3.68‡	3.87	4.00	4.19	4.31	4.50	4.63	4.82	4.95	5.14‡

(Courtesy Council of Forest Industries of B.C.)

Decking for Cedar Shingles

Roof decking for wood shingles may be solid or spaced. If the deck is solid, it may be made of shiplap, common boards, or plywood. When plywood is used, it should be $\frac{5}{16}$ in. (7 mm) in thickness when the rafters are 16 in. (400 mm) o.c. and $\frac{3}{8}$ in. (9 mm) on rafters spaced 24 in. (600 mm) o.c.

Spaced sheathing is made up of 1×8 (19 × 184 mm) boards, spaced on centers equal to the exposure of shingles.

Paper for Cedar Shingles

Asphalt-impregnated paper should not be used under cedar shingles. It prevents the completed roof from "breathing," that is, from allowing the moisture vapor under the roof to escape. If paper is required, plain or asbestos paper is recommended.

Application of Wood Shingles

Shingles should extend beyond the finish fascia about 1 in. (25 mm) to provide sufficient watershed (see Fig. 11–13). Tack a narrow piece of board 1 in. (25 mm) thick lightly to the fascia as a guide.

The first course of shingles at the eave must be doubled or, better still, made up of three layers (see Fig. 11–14). Lay the first row with shingles $\frac{1}{4}$ in. (6 mm) apart, butts to the guide strip.

FIGURE 11-13: *Shingle overhang for watershed.*

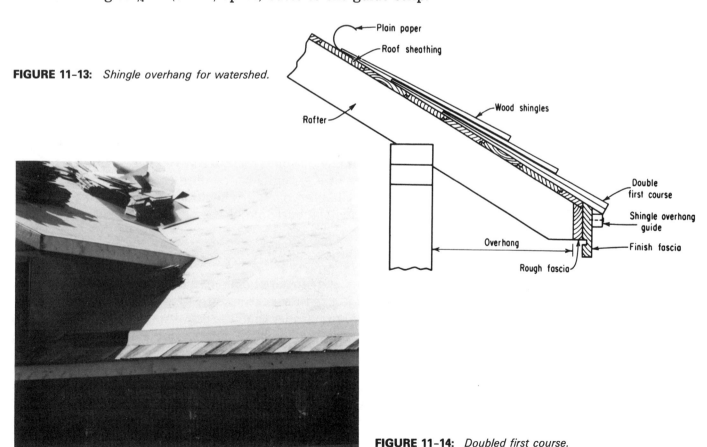

FIGURE 11-14: *Doubled first course.*

Use *two nails only* for each shingle, regardless of its width; nails should not be more than ¾ in. (20 mm) from the edge of the single. The second layer of the first course is laid directly over the first, shingles also spaced ¼ in. (6 mm) apart. Be sure that a joint between shingles does not come closer than 1½ in. (38 mm) to a joint in the layer below (see Fig. 11–15). If a third layer is used, it too must have joints offset 1½ in. (40 mm) from those underneath.

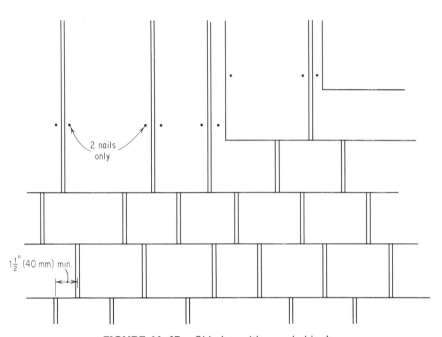

FIGURE 11-15: *Side lap with wood shingles.*

The second and succeeding courses are laid, with the specified exposure, measured from the butts of the course below. A strip of lumber may be used as a straightedge against which to lay the shingles; a chalk line may also be used as a guide, or a shingling hatchet is sometimes used by expert shinglers to keep the shingles in a straight course (see the section on shingling hatchets in Chapter 1). Nails should be placed ¾ to 1½ in. (19 to 38 mm) above the butt line of the next course.

Shingling Ridges and Hips

When ridges and hips are being shingled, the so-called modified *Boston lap* is used for best results. Use edge grain shingles at least 1 in. (25 mm) wider than the exposure used and select them all the same width. For hips, cut the butts at an angle so that they will be parallel to the butt lines of the regular courses (see Fig. 11–16).

Run a chalk line up the roof on each side of the hip centerline at a distance from the centerline equal to the exposure. Lay the first shingle to one of the chalk lines and bevel the top edge parallel to the plane of the adjacent roof surface. Now apply a

FIGURE 11-16: *Shingling a hip.*

shingle on the opposite side in the same way, beveling its top edge, as shown in Fig. 11-16. Place the nails so that they will be covered by the course above. The next pair of shingles is applied in *reverse order,* and each pair is alternated up the hip.

Ridges are done in much the same way. Start at the ends and work toward the center of the ridge, the last pair of shingles being cut to exposure length.

Shingling Valleys

Valleys have to be *flashed* before they are shingled. A strip of sheet metal—galvanized iron or aluminum—is laid in the valley, extending from 8 to 12 in. (200–300 mm) on each side of the valley centerline, depending on the steepness of the rise. One long edge of each shingle has to be cut at an angle so that it is parallel to the valley centerline (see Fig. 11-17). The open part of the valley is usually 4 in. (100 mm) wide and will increase in size toward the lower end at a rate of 1:100. Underlayment of No. 30 asphalt-saturated felt should be installed under metal valley sheets.

Two layers of roll roofing can also be used to create a water-tight valley. The bottom layer shall consist of an 18-in. (450-mm) strip of type S smooth surface roll roofing or type M mineral surface roll roofing applied with the mineral surface down. A 36-in. (900-mm) strip is placed over the first strip with the mineral side up. Type M mineral roll roofing is applied over a 4-in. (100-mm) wide strip of cement along each edge of the bottom layer and fastened with enough nails to hold until shingles are applied.

Roof Flashing

Flashing has to be placed around chimneys, dormers, or other structures projecting above the roof surface. Along the sides which are perpendicular to the shingle courses, individual flashings must be used, one for each course of shingles. Each is bent at a right angle, so that part of the flashing lies flat on the roof, on top of the shingle, while the other half rests against the side of the chimney or dormer, with the upper edge embedded in a mortar joint or covered by dormer finish.

The upper and lower sides are flashed and counter-flashed—one flashing running under the shingles and the other lying on top of them (see Fig. 11-18).

Chimney Saddle

To prevent water from collecting on the *up-roof* side of a chimney which projects through the roof, a *saddle* should be built behind it to shed the water. It should be the same width as the chimney, built like a miniature gable roof [see Figs. 11-19 and 11-12(b)] with flashing at the chimney and in the valleys.

FIGURE 11-17: *Shingling a valley.*

FIGURE 11-18: *Chimney flashing.*

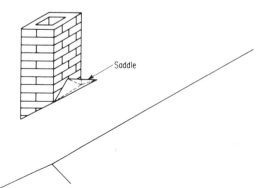

FIGURE 11-19: *Chimney saddle.*

FIGURE 11-20: *Shake roof.*

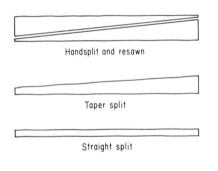

FIGURE 11-21: *Wood shake types.*

Wood Shakes

Wood shakes are made of the same material that is used for machine-cut wood shingles. Shakes, however, are longer and thicker, and they have a textured finish due to the splitting action used in the manufacture (see Fig. 11-20). There are three types of wood shakes generally available: handsplit and resawn, taper split, and straight split (see Fig. 11-21). Shakes are generally produced in 18- or 24-in. (450- or 600-mm) lengths with butt thicknesses varying from $\frac{3}{8}$ to $1\frac{1}{4}$ in. (9–32 mm). The maximum exposure recommended for double coverage on a roof is 10 in. (250 mm) for 24-in. (600-mm) shakes and $7\frac{1}{2}$ in. (190) for 18-in. (450-mm) shakes. Shakes are recommended on slopes of 1 in 3 or steeper, and spacing should conform to limits set for wood shingles.

Application of Shakes

Apply a starter strip of No. 30 asphalt-saturated felt underlayment 36 in. (900 mm) wide along the eaves and 12 in. (300 mm) along hips and ridges. The first course is doubled, just like regular wood shingles. After each course of shakes is applied, an 18-in. (450-mm) strip of No. 30 asphalt-saturated felt is applied over the top portion of the shakes, extending onto the sheathing (see Fig. 11-22). The bottom edge of the underlayment should be positioned above the butt a distance equal to twice the exposure (see Fig. 11-23). Individual shakes should be spaced from $\frac{1}{4}$ to $\frac{3}{8}$ in. (6–10 mm) to allow for expansion. The joints in each row should be offset $1\frac{1}{2}$ in. (40 mm) from the joints in the previous row.

Two nails are used in each shake, regardless of width. The nails used are rust resistant, normally 2 in. (50 mm) long placed approximately 1 in. (25 mm) from each edge and 1 to 2 in. (25–50

FIGURE 11-22: *Wood shake roof.*

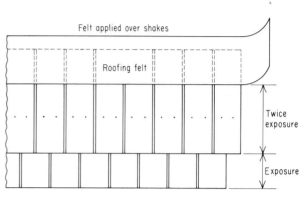

FIGURE 11-23: *Positioning of felt.*

FIGURE 11-24: *Stapling shakes.*

mm) above the butt line of the following course. Staples are commonly used to fasten shingles and shakes with the advent of air- and electric-powered staplers (see Fig. 11-24).

Eave Protection

Eave protection shall be provided on all shingle and shake roofs, extending from the edge of the roof to a line up the roof slope not less than 12 in. (300 mm) inside the inner face of the wall. This protection is not necessary when the building is not likely to have ice forming along the eaves, causing a backup of water (see Fig. 11-25). Eave protection is not required over unheated garages, carports, and porches, or where the roof overhang excedes 36 in. (900 mm) measured along the roof slope, or where low-slope shingles are used.

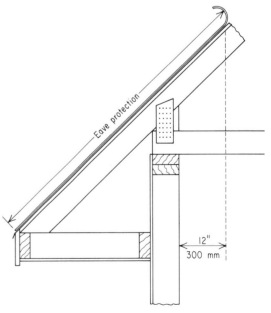

FIGURE 11-25: *Eave protection.*

FIGURE 11-27: *Roof shingled with interlocking shingles.*

FIGURE 11-28: *Starter strip.*

FIGURE 11-29: *Strip along rake.*

Application of Asphalt Shingles

So-called *asphalt shingles* are made by impregnating heavy felt paper with hot asphalt and covering the upper surface with finely crushed, colored slate. Many types are available, one of the most common being what is known as a *triple-tab* shingle, actually three singles in one (see Fig. 11–26). Interlocking asphalt shingles are also available, and they are very good for reroofing jobs (see Fig. 11–27).

FIGURE 11-26: *Roof shingled with triple-tab shingles.*

Laying starts at the bottom edge of the roof, and on some occasions a chalk line is snapped down the center of the roof to act as a guide for vertical alignment. The starter course along the lower edge of the roof is applied over the eave protection. The first course is doubled, with the first layer being placed with the tabs pointing up the roof slope (see Fig. 11–28). This row extends over the edge of the roof to provide a watershed. A strip of shingles is also nailed up the rake to provide a straight edge and to divert moisture away from this edge (see Fig. 11–29). The first layer should be nailed near the eave edge, making sure the nails are not exposed in the slots in the next layer. The second layer of the first row is staggered so the joints do not line up with the joints in the first layer, and each succeeding row is either offset one-half or one-third of a tab to provide a watertight roof [see Fig. 11–30(a)]. The shingles are fastened by nailing or stapling about ½ in. (12 mm) above the top of the slot [see Fig. 11–30(b)]. To ensure straight vertical alignment of slots (see Fig. 11–31), some shinglers will build a pyramid up the roof slope rather than in horizontal rows (see Fig. 11–32). They will also use all of the shingles from one bundle before opening another, as shingle lengths between bundles can vary, causing crooked slot lines.

Most triple-tab shingles have a strip of adhesive along the face of the shingle to cement the tabs down. Interlocking shingles are manufactured so the bottom edge hooks in and is very good in windy regions (see Fig. 11–26). A starter strip is needed on

(a)   *(b)*

FIGURE 11-30: *a) Staggering slots. b) Using tacker for applying shingles. (Courtesy Bostitch Textron)*

FIGURE 11-31: *Alignment of slots.*

FIGURE 11-32: *Building a pyramid.*

FIGURE 11-33: *Open valley.*

FIGURE 11-34: *Closed valley.*

the bottom edge to provide a double layer over the entire roof, through individual shingles with the bottom portion cut off can be used. Open valleys are formed (see Fig. 11–33) using asphalt roofing felt, as discussed under wood shingles. A closed valley is often used (see Fig. 11–34). The shingles from one side are placed over a valley liner of roll roofing and up the opposite slope at least 12 in. (300 mm). The shingles on the other slope will be cut along the valley line and embedded in a continuous strip of roofing cement.

FIGURE 11-35: *Ridge cap.*

Individual shingles are used to cap hips and ridges (see Fig. 11-35). Fold the shingle over the ridge or the hip and nail so that the nails will be covered by the succeeding shingles. On the ridge they are placed in a direction so the prevailing winds do not lift the edges of the tab.

Low-slope shingles are larger so that three layers are on the slope. The shingles have two tabs instead of three. A continuous band of roofing cement is applied so the width of the band is equal to the exposure plus 2 in. (50 mm) located 2 in. (50 mm) above the butt of the overlying shingle. Low-slope shingles can be used on a minimum slope of 1 in 6, while normal asphalt shingles can be placed on a minimum slope of 1 in 3.

Hardboard Shingles

Hardboard shingles are made of wood fibers, are resin treated, and are then pressed into a dense mat. The mat has a cedar shake pattern embossed into the fibers (see Fig. 11-36). The individual shingles are 12 × 48 in. (300 × 1200 mm) and are manufactured with a self-aligning exposure line and a nailing line (see Fig. 11-37).

FIGURE 11-36: *Hardboard shingled roof.*

Nailing line   →
Alignment line   →
Embossed pattern   →

FIGURE 11-37: *Hardboard shingle.*

FIGURE 11-38: *Valley treatment. (Courtesy Masonite Canada Inc.)*

On the job, conventional tools are used in the application of this roofing material. Precut or field-cut hip and ridge caps are available, and valley treatment follows the conventional open valley practice (see Fig. 11–38).

Built-up Roofing

A description of the application of built-up roofing has already been given in Chapter 8. It should be noted, however, that if there is to be traffic on the roof, preparation must be made for this during the application of the roofing.

While the liquid roofing is still hot, a walk base composed of two treated 2 × 4 (38 × 89 mm) pieces laid on the flat, 18 in. (450 mm) apart, is laid along the path intended for traffic. After the gravel has been applied, ¾-in. (19-mm) slats are nailed to the base pieces to form a walk.

Rolled Roofing

Rolled roofing is made from very heavy felt paper impregnated with asphalt. One side is coated with finely crushed slate. This type of material is useful for sloped roofs of the more or less temporary variety but is not recommended for permanent installation.

This roofing comes in rolls, usually 36 in. (900 mm) wide, and is applied horizontally. Start at one end and apply one width across the roof. Allow the lower edge to fold over the fascia. Nail the top and bottom edges with broad-headed roofing nails, spaced about 6 in. (150 mm) apart. Then apply a band of liquid asphalt roofing gum about 3 in. (75 mm) wide along the top edge and lay the next strip, overlapping the first by at least 3 in. (75 mm). Nail both edges as before, the lower one being bedded into the asphalt adhesive.

Work up to the ridge from both sides and place the last strip so that it is evenly folded over the ridge, overlapping the last strip on both sides. Any exposed nail heads should be coated with asphalt gum.

Roofing Tile

A number of types of tile are available as roof covering. They include concrete tile and metal roofing tile. Clay tile has been in use for a long time and is still used in some areas. Concrete tile is a more recent product, and for the most part, shapes resemble clay tiles (see Fig. 11–39). Concrete roof tiles have a nominal size of $16\frac{1}{2} \times 13$ in. (420×330 mm) with an interlocking side lap of $1\frac{3}{16}$ in. (30 mm). Metal roofing tiles are made of a lightweight galvanized steel, shaped to resemble clay or concrete tiles. The surface has a coating of stone particles to give good protection from weather.

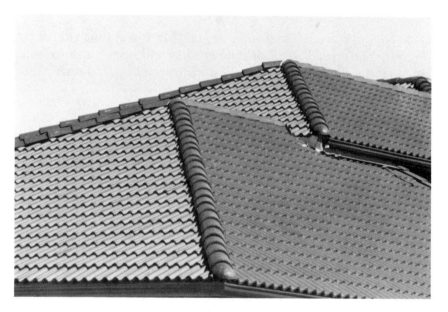

FIGURE 11–39: *Concrete roofing tiles.*

FIGURE 11–40: *Roofing felt and strapping application.*

(a)

(b)

FIGURE 11–41: *a) Roof ready for shingling—note raised barge boards.*
b) Rake tiles installed.

Roofing Tile Application

Concrete and clay tiles are applied over solid sheathing such as plywood. Spaced sheathing may be used with thermo-ply underlayment. Fifteen-pound (6.8-kg) asphalt-saturated felt or thermo-ply roofing tile underlayment is placed on top of the sheathing prior to the application of the strapping (see Fig. 11–39). 1 × 2 strapping spaced 12 in. (300 mm) is used with solid sheathing, 1 × 4 (19 × 89 mm) is used with spaced sheathing over trusses spaced 16 in. (400 mm) o.c., and 1 × 6 (19 × 140 mm) is used over trusses spaced 24 in. (600 mm) o.c. using 2½-in. (63-mm) nails at each support. Tiles are predrilled to allow nailing near the top edge with corrosion-resistant roofing or shingle nails. Fascia is raised 1 in. (25 mm) above sheathing when rake tiles are used along a roof edge [see Fig. 11–41(b)].

Strapping is built up under ridges and hips to provide support under caps (see Figs. 11–40 and 11–42). Joints should be sealed with caulking or a putty made with masonry mortar [see Fig. 11–42(b)]. Open valleys are used and installed with metal flashing, using the method discussed under wood shingles [see Fig. 11–42(c)].

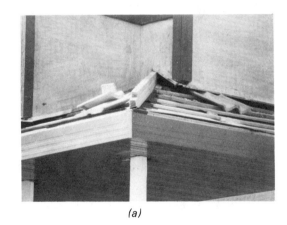

(a)

FIGURE 11–42: *a) Strapping built up for hip.*
b) Hip completed including sealing.
c) Valley construction.

(b)

(c)

a) Metal roofing tiles.

b) Metal rake in place.

c) Forming a valley.

d) Applying metal roofing tiles.

FIGURE 11–43:

Metal roofing tiles are available in Mediterranean or Spanish roofing styles. The tiles are interlocking with a natural stone finish on the galvanized tiles (see Fig. 11–43). Strapping and asphalt-saturated paper are used under the tiles similar to method used with concrete tiles [see Fig. 11–41(a)]. Strapping must also be built up under hips and ridges, as illustrated in Fig. 11–42(a). The main difference is that these tiles are started at the top of the roof working down the slope using rust-resistant screws along the bottom edge to fasten the tiles to the strapping (see Fig. 11–43). Open valleys are used, similar to the style used with wooden shingles (see Fig. 11–43).

WINDOWS

Windows form an important part of a building exterior, and a great variety of styles and sizes is manufactured to suit every conceivable purpose. Basically, they consist of one or more panes of *glass* surrounded by a *sash,* which is set into a *frame.* The sash may be wood, aluminum, vinyl or steel, and the frame also may be made of any one of these four materials, the most common one being wood.

Window Styles

Common window styles including *gliding* windows, in which the sash slide horizontally past one another (see Fig. 11–44), and *double hung,* in which the sash slide vertically (see Fig. 11–45). Another style is the *swinging* window, in which the sash are hinged at either top or bottom (see Fig. 11–46). If the sash swings inward, it is called a *hopper* type; if it swings outward, it is an *awning* type. *Casement* windows have the sash hinged at the side (see Fig. 11–47), and in *fixed* windows, the sash does not move in the frame (see Fig. 11–48).

FIGURE 11–44: *Horizontal sliding windows.*

FIGURE 11-45: *Vertical sliders.*

FIGURE 11-46: *Awning windows.*

FIGURE 11-47: *Casement windows.*

FIGURE 11-48: *Fixed sash.*

(a)

(b)

FIGURE 11-49: *Picture windows.*

Many windows are made with no sash and consist of a single sheet of glass or two sheets sealed together, set directly into a window frame. The glass may be fixed, or two single sheets may glide horizontally. When a large glass unit is fixed in a frame, it is commonly known as a *picture* window (see Fig. 11-49). If the glass reaches from floor to ceiling, the result is a *window wall* (Fig. 11-50).

FIGURE 11-50: *Window wall. (Courtesy Libby-Owens-Ford)*

Window Manufacture

Modern windows are made in a factory—a millwork plant—and shipped to the construction site, either ready for assembly or, in most cases, already assembled and ready for installation in the openings. Windows with a wood frame will usually have the outside casing in place and be provided with braces or battens to keep them square.

FIGURE 11-52: *Sealed-unit and multi-pane casement.*

FIGURE 11-51: *One- and two-sash casement.*

Windows may be made with a variety of combinations of panes in a sash. For example, Figs. 11-51 and 11-52 illustrate various combinations available in single- and two-sash casement window styles.

Window Details

Frames for the various window styles vary, depending on the manufacturer, the kind of sash being used, and the type of wall into which the window will be installed.

For example, Fig. 11-53 indicates one manufacturer's details for a single, fixed wood sash, intended to be installed in a 2 × 4 (38 × 89 mm) stud wall, with 5/16-in. (8-mm) plywood exterior sheathing and 1/2-in. (12-mm) plasterboard interior finish. In Fig. 11-54, another manufacturer's frame details are shown for a casement window like that illustrated in Fig. 11-52, which is intended to be installed in a 2 × 4 (38 × 89 mm) stud wall, with 3/8-in. (10-mm) exterior sheathing and 3/4-in. (19-mm) laminated gypsum board inside finish.

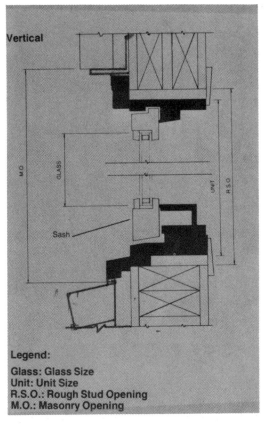

Legend:
Glass: Glass Size
Unit: Unit Size
R.S.O.: Rough Stud Opening
M.O.: Masonry Opening

FIGURE 11-53: *Fixed wood sash.*

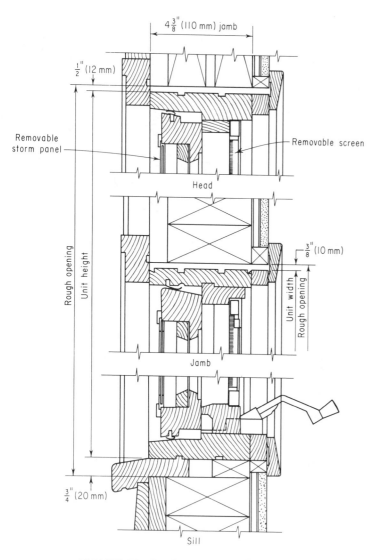

FIGURE 11-54: *Casement window details.*

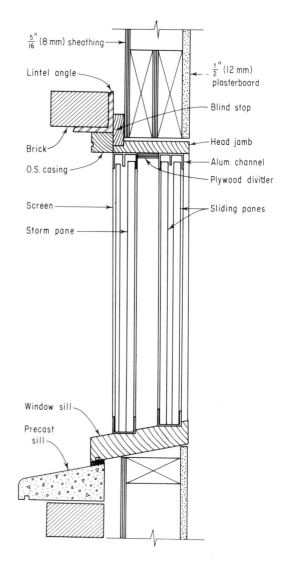

FIGURE 11-55: *Details for sliding aluminum sash.*

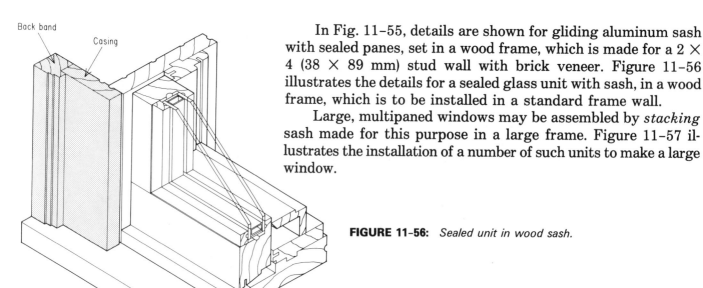

In Fig. 11-55, details are shown for gliding aluminum sash with sealed panes, set in a wood frame, which is made for a 2 × 4 (38 × 89 mm) stud wall with brick veneer. Figure 11-56 illustrates the details for a sealed glass unit with sash, in a wood frame, which is to be installed in a standard frame wall.

Large, multipaned windows may be assembled by *stacking* sash made for this purpose in a large frame. Figure 11-57 illustrates the installation of a number of such units to make a large window.

FIGURE 11-56: *Sealed unit in wood sash.*

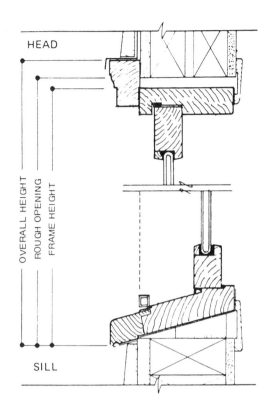

FIGURE 11-57: *a) Horizontal combination of casement and fixed units. b) Vertical and horizontal combination of units. (Courtesy Lock-Wood Ltd.)*

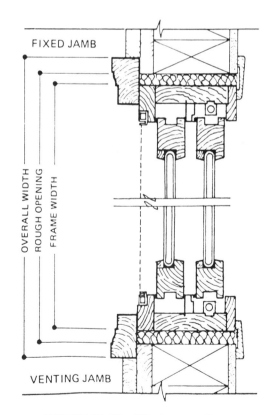

FIGURE 11-58: *Window openings. (Courtesy Mason Windows)*

Window Openings

The openings in the wall frame for the installation of windows must be large enough that the frame can be leveled and plumbed. It is therefore necessary to ascertain from the manufacturer either the size of rough opening that is required for the particular style, the size and combination of window chosen, or the dimensions of the frame (see Fig. 11–58). If the frame dimensions are given, it is common practice to allow at least ⅜ in. (10 mm) *on each side* of the frame and ¾ in. (20 mm) *above* the head for this purpose.

In residential construction, it is standard practice to frame the top of the rough opening so that the height of window and door openings will normally be the same. The height of the bottom of the opening above the floor will depend on the location of the window. In a living room, for example, a common distance is 12 in. (300 mm) above the floor, while in a dining room, it will normally be 30 in. (50 mm). Kitchen windows are usually over a counter, and the standard height for the bottom of the opening for such windows is $42\frac{1}{2}$ in. (1065 mm). In other rooms the height is optional and will depend on the size of windows to be used.

Window Installation

If the rough opening is the proper size and is level and plumb, it is a relatively easy matter to install windows. The first step is to tack a 12-in. (300-mm) wide strip of exterior sheathing paper around the opening on the outside (see Fig. 11–52). Then place the window, with the outside casing attached, into the opening from the outside and secure it temporarily, after having closed the sash and locked them in place.

Wedge blocks are placed under the sill and can also be adjusted so that the sill is perfectly level. Long sills should have three or more wedges under them to prevent them from sagging in the middle. Nail the lower end of the side casings to secure the bottom of the frame in place.

Plumb the side jambs with a level and check the corners of the window frame with a framing square. Nail the top end of the side casings temporarily and check to see that the sash slides properly and that the window hardware operates as it should.

Finally, nail the window permanently in place with noncorrosive nails, long enough so that they will reach well into the building frame.

EXTERIOR DOORS

Exterior doors may be made of wood, glass, metal, or combinations of these, such as wood and glass or metal and glass.

Wooden doors vary from a plain slab to ornate paneled doors, either of which may have glass inserts or a full glass panel. In some cases, *sidelights* may be introduced on one or both sides of the door (see Fig. 11–59).

Wooden doors and their frames are made in a millwork plant and may come to the construction site as a complete unit, with the door hinged in the frame and the outside casing attached.

Glass doors are usually sliding doors, mounted in an aluminum track, top and bottom, and traveling on rollers. They are frequently used on conjunction with a window wall, as illustrated in Fig. 11–60.

FIGURE 11–59: *Exterior door with sidelights.*

(a)

(b)

Metal doors have become very popular in residential construction. The door is made with two steel faces, separated by wooden stiles and a polyurethane foam core. This makes the door more energy efficient and resistant to warping. The door surface is usually embossed with a series of patterns, and often glass panels are included in the design (see Fig. 11–61).

FIGURE 11-60: a) Sliding patio door detail. b) Sliding glass doors. (Courtesy Pella Windows)

Door Sizes

The standard size of exterior door used in residential construction is 2 ft, 8 in. (810 mm) wide by 6 ft, 8 in. (2030 mm) high by 1¾ in. (45 mm) thick. Door widths of 2 ft, 10 in. and 3 ft (860 and 910 mm) are used when larger openings are required.

FIGURE 11-61: a) Insulated metal door detail: (1) Galvanized steel treated with rust inhibitive prime finish; (2) Polyurethane foam core; (3) Wood stiles and rails act as a thermal break; (4) Adjustable sill.
b) Decorative insulated metal door.

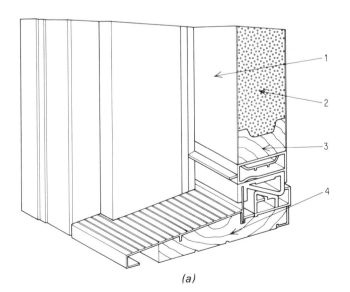

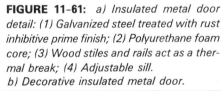

(a)

(b)

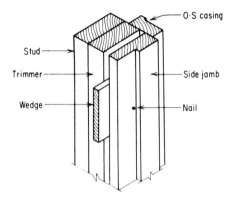

FIGURE 11-62: *Nailing side jamb wedge.*

Door Frame Installation

If the sole plate is still in the opening, it must be cut out, flush with the trimmers. A filler piece may be required under the door sill to raise the top of the sill a sufficient distance above the finish floor to allow for a mat in wet weather.

If a prehung door unit is being used, the door is left in the frame during installation. Staple a strip of paper around the outside of the opening, as was done with windows. Then apply two or three beads of caulking along the bottom of the opening to seal the crack under the sill. Place the frame into the opening from the outside, center it, and fasten it temporarily. Check the hinge side of the frame for plumbness and shim it to hold it in the correct position and then nail the bottom ends of the side casings to secure the bottom of the frame. Place additional wedges between the trimmer and the jamb at two evenly spaced locations to make sure the jamb is straight. Secure the wedges by driving a nail through the jamb and the wedges into the trimmer, as illustrated in Fig. 11-62. Wedge the head jamb in a similar fashion to make sure it is straight and level. Finally, nail the rest of the casings to secure the frame. A temporary cover should be placed over the sill during construction to protect it from damage.

If a prehung unit is not being used, the *hinge gains* should be cut in one side jamb before the frame is assembled. Check to see whether the door is right- or left-hand (see Chapter 12 for hand of doors) and mark the positions of the hinges on the *hinge jamb*. A common position for hinges in residential construction is 7 in. (175 mm) from the top of the door and 11 in. (275 mm) from the bottom, with the third hinge, when required, located midway between the other two. The hinge gain should be cut the exact length of the hinge leaf and normally $1\frac{1}{4}$ in. (30 mm) wide and $\frac{1}{8}$ in. (3 mm) deep for $3\frac{1}{2} \times 3\frac{1}{2}$ (89×89 mm) butt hinges (see Fig. 11-63).

There should be about $\frac{1}{16}$-in. (1.5-mm) clearance between the door edges and the jambs, and the lock edge should be beveled about 5° toward the closing edge.

Set the door in the opening, block it in place at the correct height, and mark on it the position of the hinge gains. Take the door down, lay out the hinge gains, and cut them exactly the same size as those in the jamb. Separate the leaves of each hinge and install one-half of each in the jamb gains and the other half on the door. Hang the door and check to see that it swings freely and opens and closes easily without binding or rubbing and will remain in any open position in which it is placed. (For complete details on cutting hinge gains and installing hinges, see Chapter 12).

If a threshold is used, the door is shortened to allow room for the threshold. The door should fit snugly against the threshold to seal the space between the bottom of the door and the door sill (see Fig. 11-64).

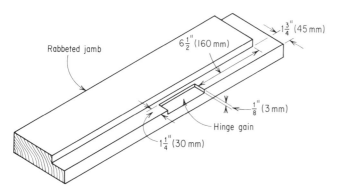

FIGURE 11-63: *Top end of hinge jamb.*

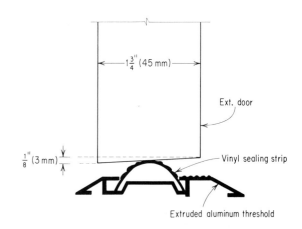

FIGURE 11-64: *Aluminum threshold.*

Lock Installation

Locks for exterior doors are often more elaborate than interior locks, and the installation instructions, contained in the package with the lock, should be followed carefully.

Open the door to any convenient ajar position and fix it there with a wedge placed between it and the floor. Measure up from the floor a distance of 36 in. (900 mm) (optional) and mark a horizontal line on both faces and across the edge of the door, which will be on a level with the center of the knob or thumb latch. Now, following the instructions and using the template provided, mark the centers of the holes required on the faces and edge of the door and drill holes of the proper size. Holes in the face of the door should be bored from both sides to prevent any splintering. In place of the template, a *boring jig* may be used to locate holes in the proper position. Drill holes in the jamb about ⅝ in. (15 mm) deep at the location of the mortise for the latch bolt and the dead bolt and square the holes out with a chisel.

Lay out the position of the rebates for the lock anchor plate and the striker plate, one on the edge of the door and the second on the jamb, over the mortise just completed and cut them out with a chisel. A *mortise marker* may be used to mark out the rebates. (For details on lock installation, see Chapter 12).

EXTERIOR PAPER

Building paper is applied to the exterior of a frame building for two reasons. One is to provide a moisture-proof barrier on the outside and the other is to give some extra insulation. Need for the latter will, of course, depend on the climate.

A moisture barrier paper is one which has been impregnated with asphalt and is reasonably hard and tough. A good insulating paper, on the other hand, must be soft and porous. This means that it will have little resistance to weathering. If both types are to be used, the insulating paper must be applied first and immediately covered with the asphalted paper.

Apply both horizontally, allowing at least 3 in. (75 mm) of lap. Be sure that the top edge of the last course is tucked under the paper hanging below the cornice. At door and window frames a good lap should be provided over the paper projecting from under the casings.

One point about which you must be particularly careful is that the asphalted paper you use is not also a vapor barrier. Some asphalted papers are also coated with wax to render them vapor proof. You *do not* want a vapor barrier on the outside of the building (see Chapter 12 on vapor barriers).

TYPES OF EXTERIOR WALL FINISH

A great many exterior finishes are used today, particularly for a wood frame. They include *stucco, cedar siding, shake and shingle siding, boards and battens, hardboard siding, aluminum siding, vinyl siding, vertical wood, plywood, and masonry finishes.*

Stucco. Stucco finish can be applied directly over a masonry wall, but when a sheathed wood frame is to be stuccoed, wire backing must be applied first. The wire should be of a type that has a relatively small mesh—about 2 in. (50 mm) is satisfactory—and is formed as illustrated in Fig. 11-65. When nailed in place, the main body of the wire stands away from the wall surface, allowing plaster to form all around it.

The wire should be stapled, with the curves to the wall, at least every 8 in. (200 mm), and therefore the sheathing must be a solid material that will allow such nailing between studs. Figure 11-66 illustrates typical application of a stucco finish.

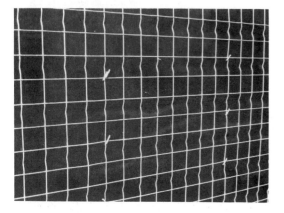

FIGURE 11-65: *Stucco wire.*

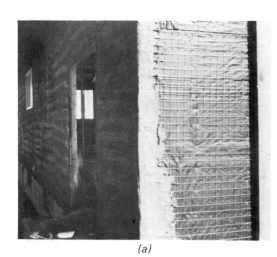

(a)

(b)

FIGURE 11-66: *a) First coat—wire only filled. b) Finish coat—Is often placed at the same time as second coat.*

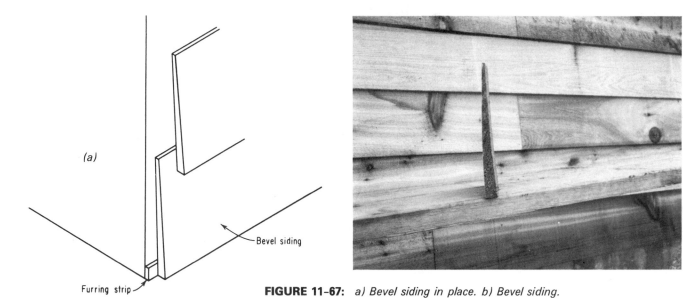

FIGURE 11-67: *a) Bevel siding in place. b) Bevel siding.*

Cedar siding. A number of styles of cedar siding are manufactured, but probably the most popular is bungalow siding, available in 6-, 8- and 10-in. (140-, 184-, and 235-mm) widths. Each course should overlap the one below from 1 to 1½-in. (25–38 mm), the exact lap and exposure depending on the width of board being used and the space to be covered.

Under the bottom edge of the first course, nail a furring strip about 1 in. (25 mm) wide and the thickness of the siding at the point of overlap. This furring strip will allow the surface of the first course to have the same slope as that of succeeding courses (see Fig. 11-67).

There are three common methods of fitting bungalow siding at external corners. One is to miter the two meeting ends. Another is to butt-join and cover the ends with a metal cap. The third method is to butt the siding ends against corner boards (see Fig. 11-68).

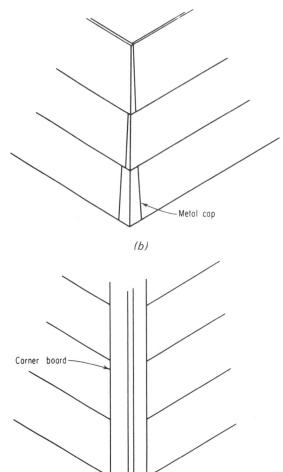

FIGURE 11-68: *a) Mitred corner. b) and c) Bevel siding at external corners.*

(a) Angle cut at butt joint.

FIGURE 11-69:

(b) Applying bevel siding.

At internal corners, nail a 1 × 1 (25 × 25 mm) strip into the corner and butt the siding ends to it from both sides.

Plan to have butt joints between boards in a course fall on a stud. Be very careful that the ends meet in a perfect joint (see Fig. 11–69). Use one 2½-in. (65-mm) siding nail per board at each stud, just above the lap. This system of nailing will allow seasonal contraction and expansion without interference. Do not drive the nails so hard as to crack the siding.

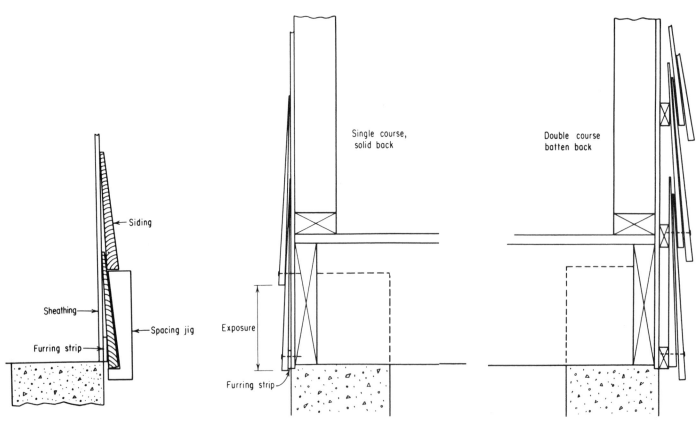

FIGURE 11-70: *Jig for spacing siding.*   **FIGURE 11-71:** *Sidewall shingles.*   **FIGURE 11-72:** *Sidewall shingles.*

To get the proper exposure on each course, make a jig as illustrated in Fig. 11–70. The last course must be ripped to the exact width of the exposure.

Shake and shingle siding. Shakes and shingles are applied to sidewalls in either single or double course. They may be applied directly to walls with solid sheathing, but in other cases, furring strips must first be nailed to the wall, spaced the required amount of exposure (see Figs. 11–71 and 11–72).

Boards and battens. This is the name given to a type of finish in which square-edged boards are placed vertically on the wall, and the joints between them are covered with narrow strips about 2½ in. (64 mm) wide called battens. Boards may be all the same width, two different widths placed alternately, or random widths. Cedar is the best material, either dressed or rough-sawn.

For this type of finish, nail in two rows of blocking, evenly spaced, between the top and bottom plates. Boards can then be nailed at four points. Allow approximately ⅛ in. (3 mm) between boards.

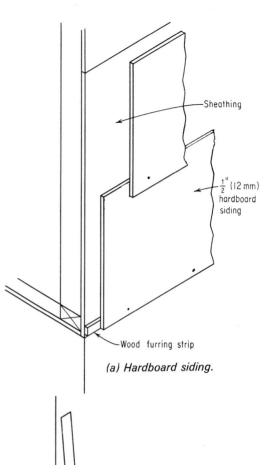

(a) Hardboard siding.

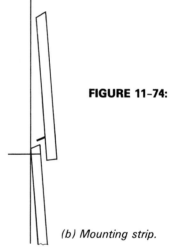

FIGURE 11–74:

(b) Mounting strip.

FIGURE 11–73: *Hardboard siding.*

Hardboard siding. This type of siding is made from wood fibers pressed into a hard, thin sheet, ½ in. (12 mm) thick and used as a lap siding or used in full-sized panels. They are impregnated with a baked-on tempering compound providing a tough facing (see Fig. 11–73). When used as bevel siding, a wood furring strip is needed under the first row to maintain the bevel. A special strip is mounted on the back of siding to provide attachment without requiring face nailing (see Fig. 11–74). Special vinyl or aluminum strips are used for butt joints (see Fig. 11–75), and exterior and interior corner pieces are used to finish ends of siding pieces.

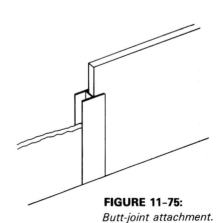

FIGURE 11–75:
Butt-joint attachment.

(a) Horizontal.

FIGURE 11-76: *Aluminum siding.*

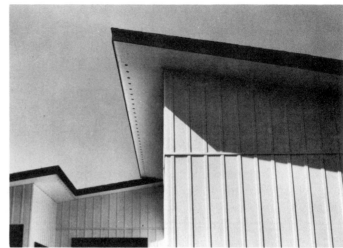

(b) Vertical.

Aluminum siding. Aluminum siding is available in horizontal and vertical applications (see Fig. 11-76). Metal siding has a baked-on finish that requires little maintenance or care. Textures are applied to some surfaces to simulate wood siding, and insulation-backed siding manufactured by some companies provides greater insulative value to the building (see Fig. 11-77).

Horizontal application starts with a starter strip at the bottom of the wall (see Fig. 11-78). The pieces are interlocking along the bottom edge (see Fig. 11-78) and are nailed in the slots at the top edge with rust-resistant nails. The nail should not be driven home so that movement is possible without buckling of the face. Special molding is used at window and door frames (see Fig. 11-78), and either individual or full-length corner pieces are used to finish exterior corners.

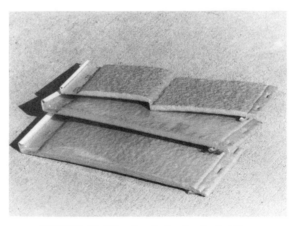

FIGURE 11-77: *Insulative-backed siding.*

(a) Starter strip.

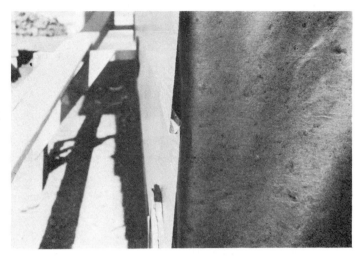

(b) Interlocking effect.

(c) J mold.

(d) Interior corner.

(e) Application.

(f) Exterior corner.

FIGURE 11-78: *Horizontal metal siding application.*

FIGURE 11-79: *Application of vertical metal siding.*

Vertical application siding has the interlocking feature on one side, and the strips are nailed or stapled along the other side (see Fig. 11–79). Similar moldings are used to finish edges and corners, though on some occasions the pieces are bent around a corner [(see Fig. 11–76b)].

Vinyl siding. Vinyl siding is a low-maintenance product and is the same color all the way through so it is not affected by scratches as is metal siding. The application is similar to metal siding and is illustrated in (Fig. 11–80). Siding is available in single or double wall, the double being more rigid and having a higher insulative value.

FIGURE 11-80:

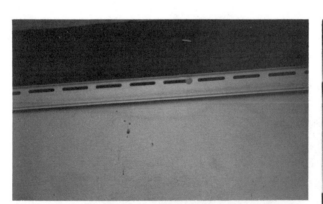

(a) Nailing slot.

(b) "J" mold adjacent to frame.

(c) Exterior corner mold.

Vertical wood siding. This type of exterior finish is similar in style to boards and battens. However, the materials have tongue-and-groove or shiplap edges so that battens are not needed to cover the joints (see Fig. 11–81). Horizontal joints in vertical boards should be sloped to the outside to shed water (see Fig. 11–82). A light layer of caulking in these joints will ensure a weather-tight joint. On some occasions, to provide variety, this siding is placed at an angle, as illustrated in Fig. 11–83.

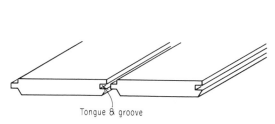

Tongue & groove

a) Tongue and groove joint.

FIGURE 11–81:

b) Vertical wood siding.

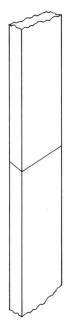

FIGURE 11–82: *Horizontal joints in vertical boards.*

FIGURE 11–83: *Angle application of tongue and groove siding.*

Plywood siding. Any exterior grade of plywood is suitable for siding purposes. It may be applied vertically in 4 × 8 ft (1200 × 2400 mm) sheets, with the joints covered by battens, or it may be cut in strips and applied horizontally like bevel siding. However, several types of plywood are made particularly for exterior finishing. These include one with a striated surface which is usually applied horizontally in 16-in. (400-mm) wide, overlapping strips, with the striations being vertical (see Fig. 11–84). Another is made from ⅜-in. (10-mm) plywood with one surface coated with a smooth plastic coating. The plastic coating provides a very smooth paint surface through which the grain of the wood will not show.

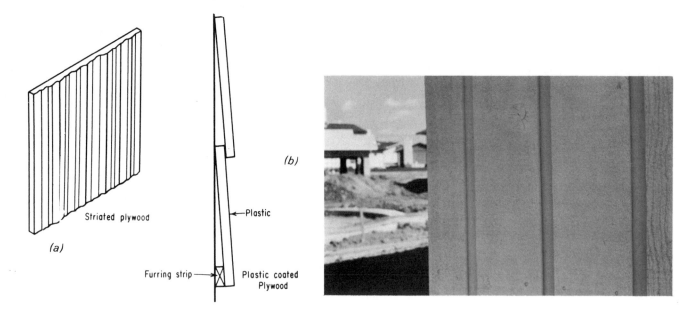

Striated plywood

(a)

(b)

←Plastic

Furring strip→ Plastic coated Plywood

FIGURE 11–84: *Plywood siding.*

Masonry finishes. Masonry finishes which are commonly used over a wood frame include *brick* and *stone veneer*.

Brick veneering is done in two ways: (1) by facing a wall with a single wythe of brick, nominally 4 in. (100 mm) in thickness and (2) by facing it with a thin layer of brick material approximately ½ in. (12 mm) in thickness.

In the first case, a single wythe of brick is built up outside the sheathed framework; preferably, about 1 in. (25 mm) of space is left between the sheathing and the brick. The brick should rest on the concrete foundation, which must be extended beyond the floor frame to provide a shelf from which to start the brick (see Fig. 11–85). The brick facing is held to the wall by metal ties which have one end nailed to the sheathing and the other laid in a mortar joint between two courses of brick. These should be spaced about 24 in. (600 mm) apart horizontally every six or seven courses. The space between sheathing and brick can be partially filled with a rigid type of insulation. The brick can also be set on an angle iron that has been fastened to the wall (see Fig. 11–86).

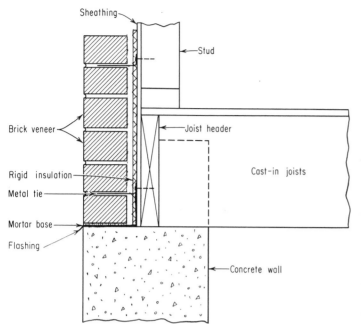

FIGURE 11-85: *4" (100 mm) brick veneer.*

FIGURE 11-86: *Angle iron support for veneer.*

The second type of brick facing is set into a mortar base. The wall is prepared as if a stucco finish were to be used. A base coat of plaster is applied over stucco wire and allowed to harden. As the second coat is applied, the brick facing pieces are set into it in the same positions as regular brick would be (see Fig. 11-87), and the joints are dressed after the mortar has partially hardened.

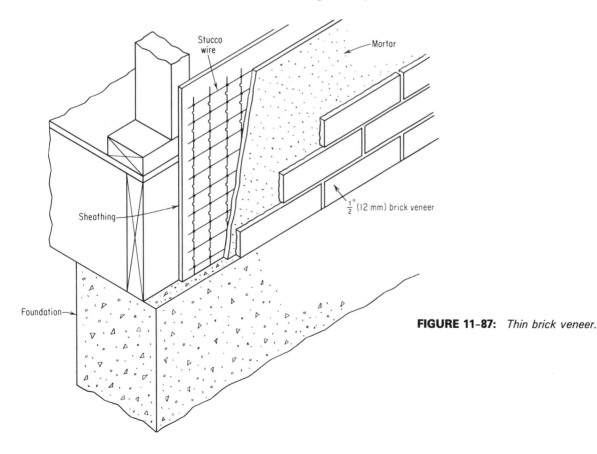

FIGURE 11-87: *Thin brick veneer.*

(a)   **FIGURE 11-88:** *a) Stone veneer. b) Stone facing set in mortar.*   *(b)*

Stone, like brick, can be used in two ways as a veneer over a wood frame. In one, stones, not over 4 in. (100 mm) in thickness, can be laid up in the same way as a single wythe of brick, carried on the concrete foundation, and tied to the wall with metal ties. In the other, stone which is available in thin sections, usually not over ¾ in. (19 mm), notably *slate, galena,* or *argillite,* is laid in a mortar bed in the same way as the thin brick veneer is applied, except that the pattern will usually be random rubble (see Fig. 11-88).

Sometimes two materials are used together to finish an exterior. For example, siding may be used part way up the wall and the remainder finished with stucco. In such a case it is necessary to use a *drip cap* to divide the two. Be sure that the drip cap is flashed (see Fig. 11-89) before the stucco wire is applied.

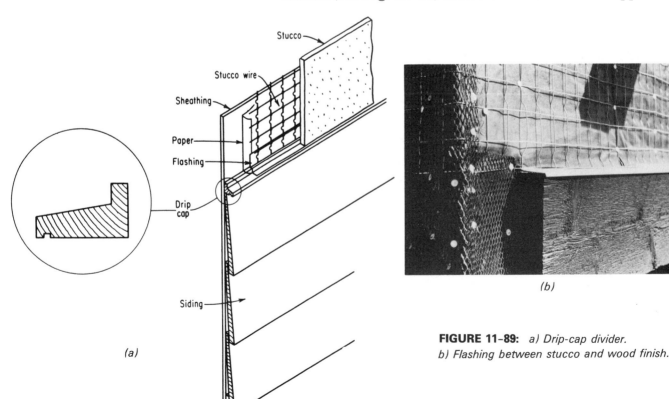

(a)

FIGURE 11-89: *a) Drip-cap divider.*
b) Flashing between stucco and wood finish.

Brick veneer and bungalow siding may be used together. In this case, one material will be used to finish all or part of one wall from top to bottom. The other material then is used either on the remainder of that wall or on an adjoinining one. The two materials should meet at an external corner or at a vertical dividing piece of the same size as the outside casing (see Fig. 11-90).

(a)   **FIGURE 11-90:** *a) Brick veneer—wood siding finish. b) Brick veneer.*   *(b)*

REVIEW QUESTIONS

11-1. List five building operations that may be classed under the general heading of "exterior finishing."

11-2. (a) What are *wind blocks?*

(b) Under what circumstances are they used?

11-3. Where is each of the following located:

(a) Barge board

(b) Verge rafter

(c) Finish fascia

11-4. Name five different types of roofing that may be applied to light construction roofs.

11-5. Why should vapor barrier paper not be used under wood shingles?

11-6. Fill in the blanks in each statement below:

(a) Joints in successive rows of cedar shingles should be not less than _____ mm apart.

(b) Each shingle should be fastened with _____ nails only.

(c) Standard exposure for 16-in. (400-mm) #1 shingles is _____ in. (mm).

(d) Proper spacing of shingles in a course is _____ in. (mm).

11-7. What is the primary advantage of concrete tiles?

11-8. What is meant by *triple-tab* asphalt shingles?

11-9. By means of a diagram, illustrate how concrete tiles are placed on a roof.

11-10. Describe briefly the chief distinguishing characteristic of each of the following window styles:

(a) Slider

(b) Casement

(c) Awning

(d) Hopper

11-11. What is the purpose of

(a) Wedges under the sill of an exterior door frame?

(b) A drip groove in a door sill?

(c) The door rabbet in the jamb?

11-12. (a) What should be the minimum overlap when applying bevel siding?

(b) What is the purpose of a furring strip under the first course bevel siding?

(c) List three ways of fitting bevel siding at outside corners.

11-13. Outline two advantages of vinyl siding.

11-14. (a) What is meant by *brick veneer?*

(b) Name two types of brick veneer used in exterior finish.

(c) Explain how a brick course is tied to a wood sheathed wall.

Upon completion of the exterior of the building, attention may then be given to the interior. This part of the job can be done at any time of the year, and it is wise to plan so that interior finishing can be done when the weather does not permit outside work.

First, arrangements must be made to have the wiring, plumbing, heating, and air conditioning installed. This is necessary for two reasons. One is that a considerable portion of these services will be situated in the walls and partitions and above the ceiling, and the work must be done while the space is open. The other is that heat and power particularly may be required during the finishing operation.

INSULATION

Once these services have been installed, the next step is the placing of insulation in the outside walls and ceiling. Two distinct types of insulation are used, one to insulate against heat loss by conduction and convection and the other to insulate against heat loss by radiation. For the former purpose, materials are used which contain large numbers of trapped air spaces and are themselves poor conductors of heat. For insulation against radiant heat losses, a material is required that has a smooth, shiny surface—a material which is, in fact, a reflector.

Modern insulators against conduction and convection are made in the form of batts, blankets, rigid slabs, or loose fill, whereas radiant insulators are usually aluminum or copper foil.

Batts are commonly used for insulating in framed walls. They are usually friction-fit batts that are stuffed into the stud space and are held there by the friction of the batts on the studs. Glass fiber rolls with one surface which is a vapor barrier are available in some areas. They have flaps along the side for easy attachment to the studs (see Fig. 12-1). The application of a polyethylene vapor barrier over the insulation is recommended to achieve an effective seal (see Fig. 12-2).

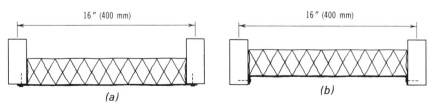

FIGURE 12-1: *Two methods of attaching batts.*

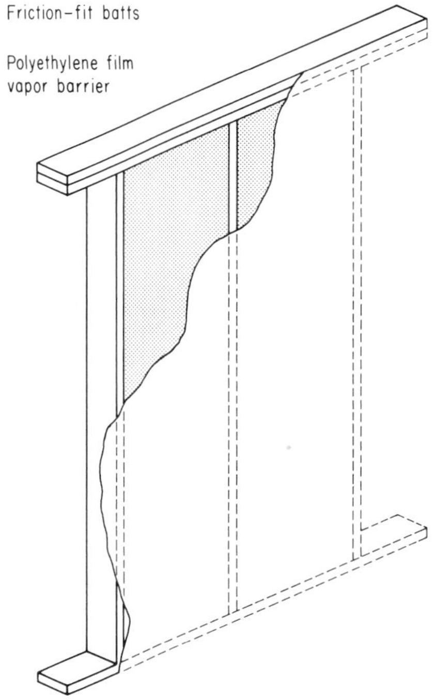

Friction-fit batts

Polyethylene film vapor barrier

FIGURE 12-2: *Vapor barrier over friction-fit batts.*

12

INSULATION AND INTERIOR FINISHING

Batts may also be used between ceiling joists, but it is often more convenient to use loose fill for ceiling insulation. It must be placed after the ceiling has been installed.

Rigid insulation is best suited for application against flat surfaces, such as concrete or masonry walls or a roof deck. It can, however, be cut to fit between studs or applied over the studs, as described in Chapter 5.

Over concrete or masonry walls, rigid insulation is applied with an asphalt adhesive, while over a wood deck either adhesive, nails, or staples may be used.

Reflective insulation is applied in vertical strips. The roll is usually 36 in. (900 mm) wide, so a strip will span two stud spaces. Do not stretch the foil tightly across the studs but rather allow it to sag back between them. Lap the strips at studs so that the finish material will seal the two pieces together (see Fig. 12-3).

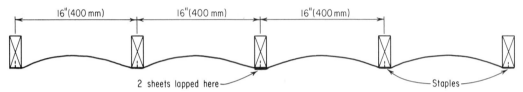

FIGURE 12-3: *Reflective insulation draped between studs.*

Be sure that when insulation has to be cut to fit around electrical outlet boxes, etc., it fits snugly. Open spaces will allow heat to escape.

Vapor Barriers

A *vapor barrier* is an essential component of a building wherever there is a considerable difference between inside and outside temperature and where the inside air has a high moisture content. The moisture vapor, under relatively high pressure, tries to escape to the outside where pressure is lower. If the temperature *in* the wall is low, the vapor will condense there to form water or ice. If the insulation in the wall becomes damp as a result of this condensation, its effectiveness is reduced. In addition, trapped moisture may cause the wood frame to deteriorate.

Vapor barriers must be materials which are impervious to the passage of water vapor. Common ones are polyethylene film and aluminum and copper foil. The latter, properly used, may serve two purposes.

Vapor barriers are applied in horizontal strips, lapped on the edge of studs. No *draping* between studs is necessary. Great care must be taken to see that the cover is as complete as possible. Where openings do occur, the vapor barrier must be made to fit tightly around them.

INTERIOR WALL FINISHES

In the past, *lath and plaster* was the most popular inside wall and ceiling finishing material. Today, *dry-wall* (gypsum board) finish is replacing lath and plaster to a large extent, and, in addition, *plywood, hardboard, insulating fiberboard, plastic laminates, tile, masonry finishes, wall fabrics,* and *wallpaper* finishes are common.

Lath and Plaster

In modern practice, gypsum or metal lath is used as a base for plaster, gypsum lath being commonly used over wood frame. However, before lath is applied, plaster grounds must be placed. These wooden grounds, ¾ in. (19 mm) thick, are nailed along the bottom of the walls and around door openings. They act as a guide for the plasterer in getting the plaster straight and of even thickness. Along the wall, they also provide a base to which to nail baseboard. Window and door frames in outside walls act as grounds for those openings (see Fig. 12–4), and door frames in inside walls may be used for the same purpose. A metal molding may also be used as a ground around door and window openings when there is to be no inside casing (see Fig. 12–5).

Gypsum lath is 16 in. (400 mm) wide, 48 in. (1200 mm) long, and ⅜ in. (9 mm) thick, with a plain or perforated surface. Apply the lath in horizontal courses, staggering the joints in each succeeding course. It is not necessary to leave a space between them, either at ends or sides. Use 1½-in. (40-mm) broad-headed nails, five to each stud. Be careful not to break the paper when driving the nails (see Fig. 12–6).

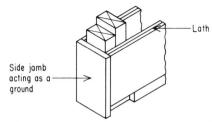
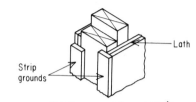

(a) Door jamb acting as plaster ground

(b) Narrow strip grounds

FIGURE 12–4: *Plaster grounds.*

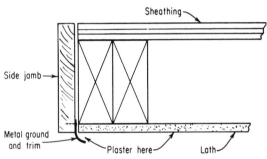

FIGURE 12–5: *Metal plaster ground.*

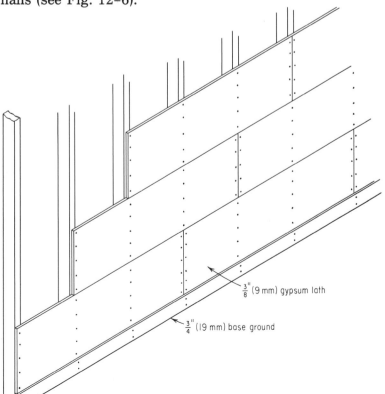

⅜" (9 mm) gypsum lath

¾" (19 mm) base ground

FIGURE 12–6: *Gypsum lath in place.*

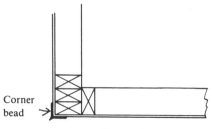

FIGURE 12-7: *Reinforcing gypsum lath at corners.*

FIGURE 12-9: *Double nailing system.*

FIGURE 12-10: *Gypsum board joints taped and filled. (Courtesy Westroc Industries Ltd.)*

When lathing is complete, it is necessary to add metal lath reinforcement at points where cracking is likely to occur. These are at internal corners where two walls meet, at the upper corners of windows, and at any point where there is a wide gap between two laths. For the corners, fold a 6-in. (150-mm) strip of metal lath 96 in. (2400 mm) long to a right angle and nail it into the corner. At window corners, cut strips of lath about 4 in. (100 mm) wide and 12 in. (300 mm) long and nail them diagonally across the corners of window or door openings. Nail strips of lath over wide spaces between laths. At external corners, use metal *corner bead* over the corner. It has a raised ridge at the angle which acts as a plaster ground as well as protecting the corner against damage (see Fig. 12-7).

Expanded metal lath is used as a plaster base over solid backing in the same way as stucco wire is used on an exterior. Ribbed lath (see Fig. 12-8) is applied over strapping nailed to a solid surface.

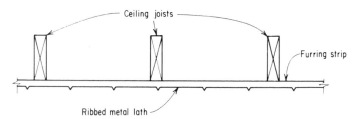

FIGURE 12-8: *Ribbed lath over furring.*

Dry-Wall Finish

Dry wall is a term used to describe a finish produced by applying gypsum board to the inside walls and ceiling. This material is made in sheets 4 ft (1200 mm) wide, from 8 to 16 ft (2.4–4.8 m) in length and in thicknesses of ⅜, ½, ⅝, and 1 in. (9, 12, 16, and 25 mm). The ⅜- and ½-in. (9- and 12-mm) thicknesses are most commonly used for interior finishing, but the ⅝ and 1-in. (16- and 25-mm) thicknesses are used in some applications and do offer some advantages. The board may be applied in single or double thickness, the latter being known as *laminated dry wall.*

In single application, ½ (12-mm) board is generally used, and sheets may be applied horizontally or vertically. In many cases, horizontal application of long sheets will result in a lesser amount of joints to be filled than with vertical application. Broad-headed, cadmium-coated, spiral, or ringed nails 1¼ to 1½ in. (30 to 40 mm) long are used. They should be spaced 6 to 8 in. (150–200 mm) o.c. around the outside edges of the board and on intermediate studs. Use enough force when driving the nails so that a slight depression is made in the surface when the nail is fully driven, but, at the same time, care must be taken not to break the paper.

The two-nail system is often used with success. After each nail is driven, a second one is driven about 2 in. (50 mm) away.

This serves to draw the board more tightly against the studs or joists and helps to prevent *nail popping*. *Dry-wall screws* may be used in place of nails (see Fig. 12-9).

The long edges of the board are depressed where two sheets meet so that a recess appears on the surface that must be filled to hide the joint and to produce a flat surface. To do this, gypsum joint filler and paper tape are used. The tape is about 2 in. (50 mm) wide, has *feathered* edges, and is perforated to allow filler to come through. The joint filler is mixed with water, allowed to stand for at least one-half hour, and applied with a broad spatula or trowel. The tape is pressed into the first layer of filler applied. Add a little filler over the tape, smooth off the joint, and let it dry. Now give it a light sanding and apply a wider coat, being careful to thin out the edges. Let dry, sand off, and apply a third, still wider coat which, when dry and sanded, should be ready for sizing and painting (see Fig. 12-10).

At internal horizontal and vertical corners, a strip of paper tape is folded into a right angle and set into the angle in a bed of filler. Second and third coats are added and sanded as above (see Fig. 12-11). At external corners a paper-covered metal angle is used (see Fig. 12-12). It is set over the corner in a bed of joint filler, the paper providing the binding surface. The corner is then treated in the same way as other joints.

Much of the taping and joint filling for dry-wall finish is now done by machine, although, in many cases, nail heads are covered (spotted) by hand (see Fig. 12-10).

At door and window openings, the gypsum board may simply be stopped against the frame if casings are to be used (see Fig. 12-13). When the opening is to be trimmed without casing, again paper-covered metal angles are used. A thin saw kerf is cut into the edges of the frames, and one leg of the angle, with the paper folded around it, is pressed full depth into the kerf. The other leg covers the edge of the gypsum board, and the paper flap is bedded in filler. Finishing is similar to other joints (see Fig. 12-14).

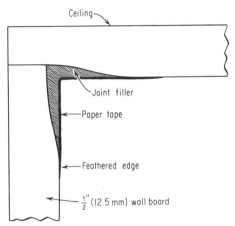

FIGURE 12-11: *Filling and taping internal corners.*

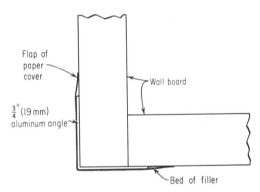

FIGURE 12-12: *Reinforcing external corners.*

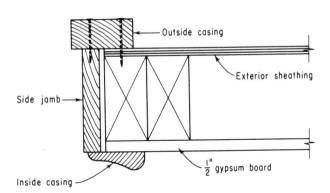

FIGURE 12-13: *Opening trimmed with casing.*

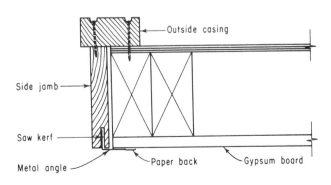

FIGURE 12-14: *Opening trimmed with metal angle.*

Two-ply laminated dry wall consists of two layers of gypsum board, each usually ⅜ in. (9 mm) thick. The first is applied vertically over studs, nailed as described above. The second layer is applied horizontally. Sheets should be long enough to reach completely across the wall if at all possible. They are fixed in place by a coating of gypsum cement (joint filler) or contact cement between the sheets.

The filler should be mixed into a creamy paste and allowed to stand for at least half an hour. It is then remixed, and a coating is spread over the back of the sheet to be applied using a broad, toothed spreader that will give an even distribution of cement. The first coated sheet is then applied to the upper half of the wall, with the upper edge up snug against the ceiling. The board is held in place with double-headed nails, one at each edge and two across the board, at each stud. Later, when the cement is dry, these nails are pulled out and the holes filled. Joints and corners are finished as described for single application. Taping and filling may be done by machine if desired. Also, the junction of wall and ceiling may be finished with a paper-covered gypsum cove molding. It is held in place with gypsum cement and nails.

With either single or double application, board should first be applied to the ceiling, and, in either case, only a single application is required. Board may be applied directly to ceiling joists spaced 16 in. (600 mm) o.c. or less, but if spacing is greater than that, the ceiling frame should be *strapped* at right angles to the framing members ¾-in. (19 mm) strapping, at least 1½ in. (38 mm) wide, and the board applied over the strapping. When two boards butt together at their ends, a small space (about ⅛ in. or 3 mm) should be left between them so that some joint filler may be forced into the crevice during the taping and filling process to act as a key for the joint.

Plywood Finishes

A great variety of plywoods is available for interior finishing, both in hardwood and softwood. They may be obtained in plain sheets, in a number of decorative faces, such as simulated driftwood or pressed patterns, and in sheets which have been scored to imitate plank or tile.

Joints between sheets can be treated in several ways. One is to make the edges meet as tightly as possible in an attempt to hide the joint. This is relatively easy with patterned plywoods, but it is difficult with plain sheets, particularly those in light colors. In such a case it is better to chamfer the meeting edges so as to accentuate the joint. Another method of finishing is to use battens over the joints.

Plywoods are nailed with 1¼- to 1½-in. (30- or 40-mm) finishing nails or glued in place with a panel adhesive. If the gluing process is being used, nailing is used at the top and bottom of the sheet to hold it in position.

Hardboards

Hardboards are produced in a great variety of face patterns for inside finishing. There are tile and plank effects, wood grain patterns, plastic-covered faces, and boards with baked enamel surface, among others. A very important consideration in applying any of the hardboards, other than those with plastic or enameled face, is that of pre-expansion. Since these are wood fiber products, they will expand and contract with changes in humidity. Therefore, it is desirable that they be applied while at their maximum size. Otherwise, any expansion on the wall would cause buckling between studs. To pre-expand hardboards, wet them on the back or screen side and stack them flat, back to back, for 24 hours. Most of the hardboards can be fastened in place with finishing nails; those with plastic faces must be held by metal moldings.

Insulating Boards

Insulating boards for interior finish are made in regular 48 × 96 in. (1200 × 2400 mm) sheets and in various smaller sizes, particularly for use on ceilings. These boards are made from wood fiber, cane fiber, and asbestos fiber. Some are perforated to improve their acoustical qualities.

Colored panel nails are commonly used for fastening these boards because of the ease of hiding the nail head. In wall applications, the nails should be driven at an angle of about 30° to the horizontal. For ceiling application, using the smaller units, cement is commonly used to hold them. Metal channels may also be employed to hold ceiling tile.

Plastic Laminates

Plastic laminates are hard, synthetic materials, commonly made in sheets 48 ft × 96 in. (1200 × 2400 mm), and in thickness of $\frac{1}{16}$ in. (1.5 mm). They are durable and wear-resistant but because of the thickness must be bonded to other backing materials, such as tempered hardboard, particle board, or plywood. Bonding is done with a *contact cement* that is applied to both the back of the plastic laminate and the face of the backing and allowed to dry before the sheet is placed in position. Bonding is instantaneous, and care must be taken to ensure that the sheet is in its proper location before the two coated surfaces are allowed to come into contact. A sheet of paper can be placed over the coating on the backing after it is dry to allow some adjustment to be made in the position of the plastic laminate sheet and then withdrawn to allow the surfaces to come into contact.

For wall application, the material is often prebonded to a backing such as particle board, which may have tongue-and-grooved edges so that the units can be blind-nailed into place. In other cases, metal moldings nailed to the sheathing are used to secure laminate-faced panels to the wall (see Fig. 12–15).

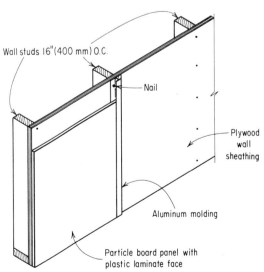

FIGURE 12–15: *Metal channel to hold plastic laminate panel.*

Tile

For interior finish, tiles in plastic, steel, ceramic, or glass are in common use, particularly in bathrooms and kitchens. Plywood or gypsum board forms a suitable base for the tiles, which are set in a prepared tile cement. The joints are filled (pointed) after the tiles are all in place. A grouting compound especially made for this purpose is used for pointing.

Masonry Finishes

Masonry products have an important place as interior finishes. In addition to brick, tile, concrete block, and stone of various types, thin veneers of brick and tile, both real and artificial, are available. These may be fixed to a solid backing by plastic adhesives, the joints being filled with grout after the units are in place.

FLOORING

Flooring materials include *hardwood; resilient tile* such as linoleum, asphalt, vinyl, vinyl-asbestos, cork, and rubber; *sheet flooring; clay tile;* and *carpet.* Each has some particular advantages, and many require some special preparation.

Hardwood Strip Flooring

Hardwood strip floors are a popular choice for many buildings, with *red and white oak, birch, beech,* and *maple* all being used. All are produced in both edge grain and flat grain—*quarter sawn* and *plain sawn.*

Flooring is milled in a number of widths and thicknesses with tongue-and-grooved edges and ends. Standard face widths run from 1½ to 3½ in. (38–89 mm) and thicknesses include from ⅜ to 1 in (9–25 mm).

The tongue and groove is placed below the center of the piece to allow for more wear, and the bottom surface is hollowed for a tighter fit against the subfloor and for greater resilience (see Fig. 12–16).

Most flooring is made in four grades, determined by the range of color, similarity of grain, and the number of defects in the pieces. The finished product is packaged in bundles made up of various lengths from 2 ft (0.6 m) up to the designated length of the bundle.

FIGURE 12-16: *Section through hardwood flooring strip.*

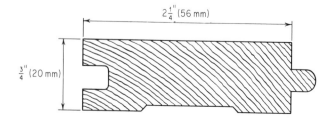

Hardwood flooring should be laid only after the humidity in the building has been brought to the normal range. The material should be stored in the place for several days before laying so that it may adjust to the atmospheric conditions. Heat and ventilation are necessary for satisfactory results.

Following is the procedure for laying strip flooring over wood subfloor:

1. Plan to lay the floor the long dimension of the building, if possible, and, in any case, lay it at right angles to the floor frame.

2. If the subfloor is shiplap, center match, or common boards, it should run diagonally across the floor frame. See that it is adequately nailed.

3. Lay a good-quality 15-lb, (0.75 kg/m²), asphalt-saturated felt paper over the subfloor with 4-in. (100-mm) laps. If the space beneath the floor is cold, paper with vapor-barrier qualities is required.

4. Start the first strip of flooring against an outside wall, unless the area to be covered is large. Place the groove edge to the wall and leave a ½-in. (12-mm) space between floor and wall to allow for expansion (see Fig. 12–17). Make sure that the strip is perfectly straight and then nail it with finishing nails.

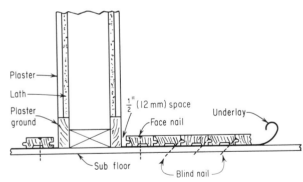

FIGURE 12-17: *Starting flooring at wall.*

5. Succeeding strips are *blind-nailed* (see Fig. 12–18) with flooring nails. The length and spacing of nails depend on the thickness of the flooring. Be sure that the strip is snug against the one behind it and that end joints are tight and square.

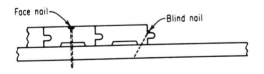

FIGURE 12-18: *Blind nailing.*

6. Use the piece cut off at the end of one strip of flooring to start the next strip, wherever possible. Watch that end joints in successive strips are at least 6 in. (150 mm) apart.

7. After four or five strips have been laid, place a piece of hardwood against the outside edge and strike firmly against it with the hammer to draw up the flooring.

8. Try to arrange the pieces in the floor so that there is as wide a separation of end joints as possible and there is as smooth a blending as possible of color and grain variation from piece to piece.

Sanding and finishing hardwood floors are usually done by qualified finishers. They use a drum-type power sander, starting with coarse (No. 2) sandpaper, progressing to medium (No. ½), and finishing with fine (No. 00). Then they use paste filler, rubbing first across the grain and then with it with a large piece of burlap. Excess filler must be wiped off immediately before it hardens. Oil, shellac, plastic finish, or floor varnish is applied in three coats, and finally, after drying, the surface is waxed.

Wood Floor over Concrete Base

Wood floors may be laid over a concrete subfloor, although the procedure will vary somewhat depending on whether the concrete slab is suspended or is resting on the earth. If the slab is suspended, with an airspace beneath, a moisture barrier is usually not needed, but for a slab-on-grade floor, a moisture barrier is required between the concrete and the flooring.

Wood strips—*sleepers*—are attached to the concrete slab with asphalt adhesive and nails, and they act as a nailing base for the strip flooring.

A satisfactory method of laying strip flooring over a concrete slab-on-grade may be carried out as follows:

1. Clean and prime the floor with asphalt primer.

2. Snap chalk lines down the length of the floor at 16-in. (400-mm) centers.

3. Apply ribbons of the special adhesive required to bond wood to concrete along the chalk lines.

4. Embed random length, $1 \times \frac{1}{2}$-in. (19×38 mm) or wider, treated wood strips in the adhesive and fasten them in place with $1\frac{1}{2}$-in. (40-mm) concrete nails about 24 in. (600 mm) apart or with power-actuated pins similarly spaced.

5. Lay 4-mil polyethylene film over the sleepers. Joints in the film must be made at a sleeper, and the material should overlap at least 4 in. (100 mm) on each side of the sleeper (see Fig. 12–19).

6. Nail another layer of wood strips of the same width over the first, on top of the polyethylene, using $1\frac{1}{2}$-in. (38-mm) spiral nails, spaced about 16 in. (400 mm) apart.

7. Apply strip flooring as previously described. If flooring ends are tongue-and-grooved, end joints may fall between sleepers, but joints in the succeeding strip must not fall between the same pair of sleepers.

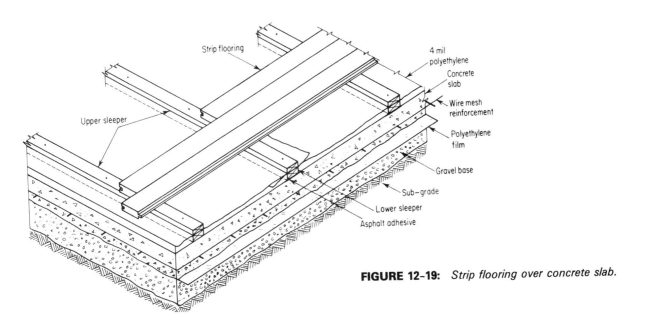

FIGURE 12-19: *Strip flooring over concrete slab.*

Hardwood Block Flooring

Block flooring consists of small sections of hardwood made either by gluing several short strips of hardwood flooring together—*parquet flooring* (see Fig. 12-20)—by bonding three thin layers of hardwood together—*laminated flooring*—to form small rectangles or squares with tongue-and-grooved edges, with a maximum size of about 12 × 12 (300 × 300 mm).

Both types are laid by the same general methods, although in the case of parquet flooring, allowance must be made for expansion by leaving at least a ¾-in. (20-mm) space on all sides between flooring and wall. Both may be blind-nailed in the same way as strip flooring, and both may be laid in mastic.

With this method, a thin coat of mastic is first spread over a smooth, level, dry base, to which is applied a layer of asphalt-impregnated felt paper. A second coat of mastic about ⅛-in. (2 mm) thick is spread over the paper, and the blocks are laid in this top coat. Laying patterns may be square or diagonal.

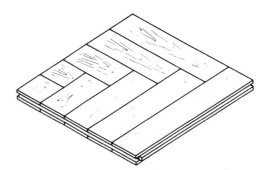

FIGURE 12-20: *Parquet flooring.*

Resilient Flooring

Resilient floor tiles of all kinds require a smooth, regular surface in order to give satisfactory service and to maintain a good appearance. In most cases the conventional subfloor does not provide a surface which is smooth and even enough, and some type of *underlayment*, such as plywood, particle board, or hardboard, must be applied over the subfloor.

Large defects in the subfloor, such as knotholes, should be patched before hardboard is used, and the material should be allowed to stand unwrapped in the room for at least 24 hours to adjust to the prevailing humidity conditions. A space of approximately ¹⁄₁₆ in. (1 mm) should be left between sheets when they are laid to allow for expansion. The end joints in the panels should be staggered, and the continuous joints should be at right angles to those in the subfloor.

Plywood and particle board are dimensionally stable products, and sheets may be butted against one another when they are laid. They are also rigid enough that they will bridge most defects in the subfloor. Joints should be staggered in the same manner as with hardboard.

Ring-grooved or *cement-coated, tapered-head* nails or divergent staples are used as fasteners for all these underlayment materials, with spacings not over 6 in. (150 mm) on the interior of the panel and 4 in. (100 mm) around the edges for hardboard and up to 8 in. (200 mm) for the interiors and 6 in. (150 mm) around the edges for plywood and particle board.

Following is the procedure for laying resilient tile:

1. Clean the surface thoroughly and check to see that it is smooth and the joints are level. Remove any rough edges with sandpaper or plane.

2. Snap a chalk line down the center of the room in the direction of the long dimension.

3. Lay out another centerline at right angles to the main one, using a framing square to get the chalk line in its proper alignment.

4. Spread adhesive over one-quarter of the total area, carrying it up to, but not over, the chalk lines. Use the type of spreader recommended by the manufacturer of the adhesive.

5. Allow the adhesive to acquire an initial set. It should be slightly tacky but not sticky, and the length of time required to achieve this condition will depend on the type of adhesive.

6. Lay the first tile at the center of the room, with two edges to the two chalk lines (see Fig. 12–21).

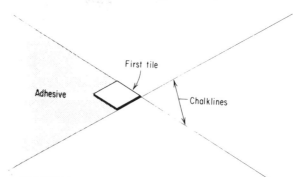

FIGURE 12-21: *First tile laid to center lines.*

7. Lay a row of tiles to both chalk lines, being careful to keep the butt joints tight and the corners in line. Lay each tile in position—do not slide it into place.

8. Cut the last tile in each row to fit against the wall, with the cut edge to the wall.

9. Complete the tile laying over that quadrant of the floor and roll the tile if the manufacturer recommends it.

10. Lay the opposite quadrant next and then the other two, in exactly the same manner as the first.

Sheet Flooring

Sheet flooring includes materials such as linoleum and a variety of synthetic products consisting of vinyl or other plastics, sometimes in combination with cork, asbestos, or other fibers and various resins. They are produced in rolls from 6 to 12 ft (1.8 to 3.6 m) wide, in thicknesses of ⅛ to ¼ in. (3–6 mm), with plain and patterned surfaces.

Some must be cemented to the floor, while others will hold their position without cement, due to the texture of the material on the underside. All require a smooth base, like that produced by the use of underlayment.

In the case of linoleum, a cushion is required under the material in the form of a layer of soft felt paper. The underlayment is covered with a coating of linoleum cement, into which the felt paper is laid and rolled smooth. Another coating of cement is applied over the paper, and the linoleum is laid in it and rolled down.

Clay Tile Flooring

Clay floor tiles are burned clay products, similar to brick, made in various dimensions and in thicknesses of from ¼ to 1 (6–24 mm), intended for use over a concrete base floor. The procedure for laying clay floor tile is as follows:

1. Wash the concrete base and saturate it with water.

2. Snap two chalk lines across the floor at right angles to one another, so that the floor is divided into four equal parts.

3. Mix the bedding mortar in the proportion of one part cement to three parts plaster sand, with enough water to make a plastic, workable mixture.

4. Starting at the center of the room, at the intersection of the two lines, apply a layer of mortar the area of one tile and about ½ in. (12 mm) thick and lay the first tile to both lines. Bed the tile firmly and remove any excess mortar around its edges.

5. Continue to lay a row of tile along one of the lines, bedding each one firmly and leaving a ⅜-in. (10-mm) space between tiles. Remove excess mortar around exposed edges.

6. Lay a row of tile along the second line in the same manner.

7. Lay the remainder of the tile over that section of the floor, keeping the tiles as level as possible and maintaining an even spacing between them.

8. Lay the opposite quarter and then the two remaining ones, in the same manner.

9. After the bedding mortar has set, prepare a grouting mix, using the same proportions of cement and sand but a little more water, for a more fluid mix. Pour the grout over a section of the tiled surface and rub it into the spaces between the tile with a rubber float.

10. Rub off the excess grout, and when it has set sufficiently, so as not to pull out of the spaces, clean the surface of the tile thoroughly.

Carpets

Carpets do not usually require underlayment because normally a rubberized underpad is laid under the carpet as part of the installation. However, the subfloor should be flat and even and cracks and knotholes filled with a reliable crackfiller. Actual carpet installation is normally done by a carpet layer, using specialized equipment for stretching the carpet and holding it in place.

INSIDE DOOR FRAMES

Several different kinds of jamb are used to make inside door frames, as indicated in Fig. 12-22. The flat jamb, ¾ in. (19 mm) thick, is often saw-kerfed on the back side to minimize cupping. Rabbeted jamb has one edge ploughed out ½ in. (12 mm) deep by 1⅜ or 1¾ in. (35 or 45 mm) wide, depending on the thickness of the door to be used. Hardwood jambs are made in a number of styles, one of which is illustrated in Fig. 12-22(e). Steel jambs are made for use in masonry walls [see Fig. 12-22(f)].

In cases where no casing is to be used around the door frame, a galvanized metal bead is available which can be nailed to the back of the jambs to accommodate lath and plaster wall finish [see Fig. 12-22(c)]. A similar type of trim is used with dry-wall finish, consisting of a paper-covered aluminum angle, with one leg set into a saw kerf cut in the edges of the jamb [see Fig. 12-22(d)].

Installing Inside Door Frames

In some cases, door frames are cut, sanded, and fitted in a millwork plant and come to the job ready for assembly. In others, sets of jamb material are supplied, consisting of two side jamb pieces about 7 ft (2100 mm) long and a head jamb piece 3 ft (900 mm) long.

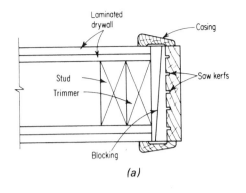

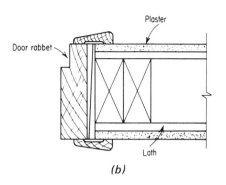

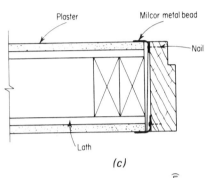

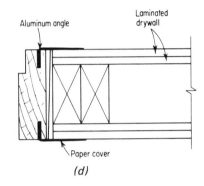

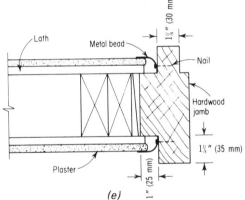

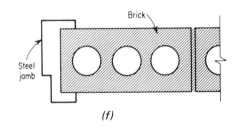

FIGURE 12-22:   *Inside door frames.*

The procedure for installing inside door frames, using precut jambs is as follows:

1. Check the length of the side and head jamb pieces to make sure that they are correct for the opening.

2. Nail the jambs together, using 2½-in. (65-mm) finish or casing nails.

3. Place the frame in the opening with the ends of the side jambs resting on the finish floor and check to see that the head jamb is level. If it is not, trim the bottom end of the side jamb that is too high.

4. Tack a 1 × 4 (19 × 89 mm) *spreader*, exactly the same length as the head jamb, across the bottom of the frame, as illustrated in Fig. 12-23.

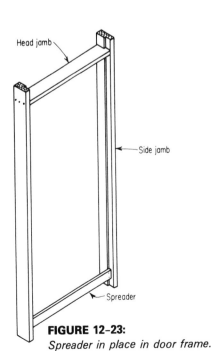

FIGURE 12-23:
Spreader in place in door frame.

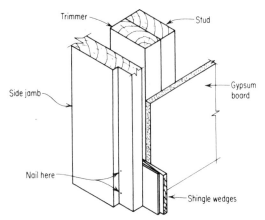

FIGURE 12-24: *Nailing side jamb wedges.*

5. Center the frame in the opening and wedge it in position with double shingle wedges at the top and bottom on both sides. Secure the wedges by nailing through the frame into the trimmer on either side, as illustrated in Fig. 12-24.

6. Complete the blocking by placing pairs of shingle wedges at each hinge location on one jamb, with a third midway between them and one at the lock position and at least one other midway between it and the top on the opposite side (see Fig. 12-25).

7. Adjust the pairs of wedges until the jambs are straight, using a long straightedge to check the straightness, and nail them in place as illustrated.

If the jambs are not precut, they will have to be cut to length and dadoed for the head jamb and the hinge gains cut before the frame is assembled. Proceed as follows:

1. Lay out the two pieces of side jamb material as a pair and check the bottom ends to make sure that they are square.

2. Measure up from the bottom of one side jamb the given height of the door plus ¼ in. (6 mm) and lay out and cut a dado across the jamb 1½ in. (38) mm wide (see Fig. 12-26). Cut the jamb to length above the dado.

3. Cut the opposite jamb in exactly the same way.

4. Cut the head jamb to the given door width plus ¼ in. (6 mm).

5. Hinge gains may be located as illustrated in Fig. 12-26 and cut out by hand, using a chisel and mallet, or a *hinge template* may be used to locate them and the cutting done with a hand electric router.

6. Assemble the frame and attach the spreader at the bottom as previously described.

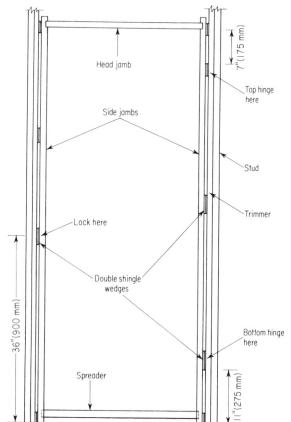

FIGURE 12-25: *Door frame wedges.*

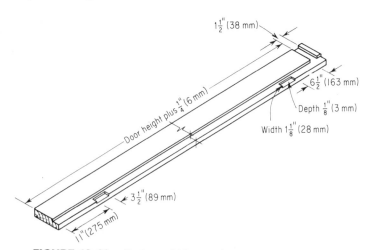

FIGURE 12-26: *Dado and hinge gains cut into side jamb.*

How to Hang an Inside Door

After the inside door frames have been installed, the doors must be *hung*, if prehung units are not being used. Hanging a door involves *trimming* it to fit the opening, *attaching hinges*, and *installing a lockset*. During this operation it is necessary to support the door firmly on edge, and this may be done with the aid of a woodworker's vise (see Fig. 12–27) clamped to a sawhorse.

FIGURE 12-27: *Sawhorse vise.*

Trimming the Door

1. Use only a door made to fit the size of the opening provided. Doors are made to fit standard openings, and a door of any given size should not be cut down to fit a smaller opening.

2. Trim the door to length. Be sure that the top edge is square with the sides and then trim enough from the bottom to make the door the right length. Inside doors should have about ⅛-in. (3-mm) clearance at the top and ¾-in. (20-mm) clearance at the bottom.

3. Dress the door to width, using a jointer or a hand electric plane to keep the edge straight. There should be about ⅛-in. (3-mm) clearance on the lock side and a little less on the hinge side. The lock edge should be beveled about 5° toward the closing side.

Attaching Hinges

The first step in attaching hinges is to provide hinge gains on the edge of the door. If a prehung unit is being used, this will have been done in the shop. If a hinge template has been used for the jamb hinge gains, it will also be used for those on the door. In other cases, the gains will be cut by hand.

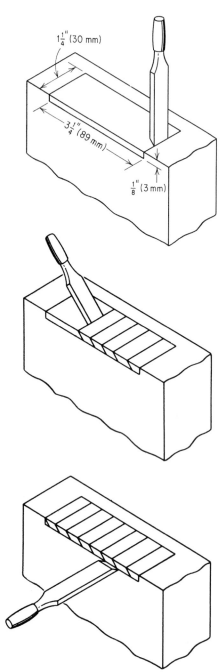

FIGURE 12-28: *Cutting a hinge gain.*

1. Set the door in the opening and block it to the right height.

2. Mark on the door the position of the hinge gains from those on the jamb.

3. Take the door down, secure it firmly on edge, and lay out the positions of the hinge gains on the door edge.

4. Use a chisel and mallet to cut out the gains as illustrated in Fig. 12-28.

5. When gains are complete, set one hinge leaf in each and mark the center of the screw holes with a self-centering punch (see Fig. 12-29).

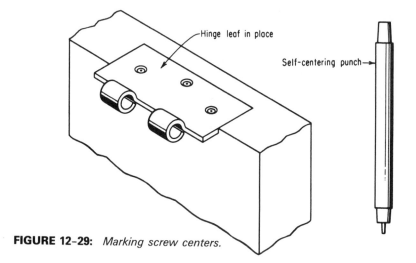

FIGURE 12-29: *Marking screw centers.*

6. Drill pilot holes and attach the hinge leaf by driving the screws supplied with the hinge.

7. Attach the corresponding leaf to a hinge gain in the door jamb.

8. Hang the door by its hinges. It should swing freely, close without rubbing or binding, and stand in any position in which it is placed. Failure to do any of these things means that some adjustment of the hinges or slight dressing off of the closing edge of the door is necessary.

Installing a Lockset

When the door is hanging properly, it is ready for the lockset. Two types are in common use, mortise locks and cylinder locks. The first is mounted through the edge of the door and the latter through the face.

Installing a Mortise Lock

1. Measure up 36 in. (900 mm) from the floor and mark that height on the face and edge of the door (see Fig. 12-30).

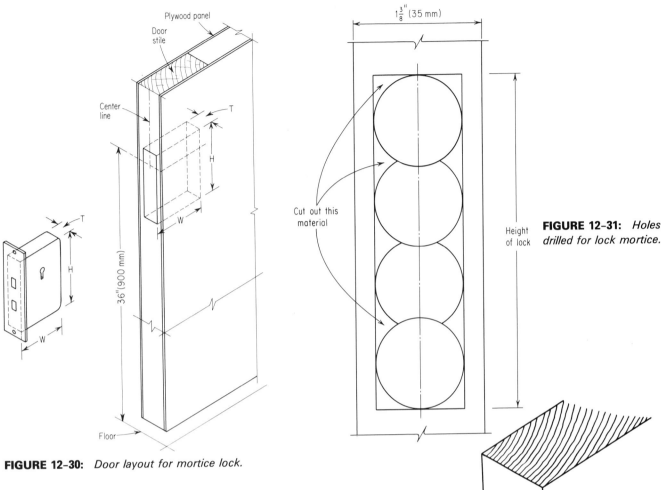

Plywood panel

Door stile

Center line

T

H

W

36" (900 mm)

Floor

T

H

W

FIGURE 12-30: *Door layout for mortice lock.*

$1\frac{3}{8}$" (35 mm)

Cut out this material

Height of lock

FIGURE 12-31: *Holes drilled for lock mortice.*

2. Measure the thickness and height of the *lock case* and lay out a mortise on the edge of the door with dimensions slightly larger than the height and thickness. Make sure that the layout is centered on the edge of the door and is located so that the center of the knob will fall on the *height-above-floor* line.

3. Select a wood bit with the same diameter as the thickness of the mortise layout and, from the centerline, drill a series of holes into the edge of the door ¼ in. (6 mm) deeper than the width of the lockcase (see Fig. 12-31).

4. With a sharp chisel, remove the remaining wood around the holes to form a rectangular mortise. Check to see that the lock will fit in the space.

5. Slip the lock into the mortise, and with a marking knife, lay out the outline of the lock-mounting plate on the door edge. Cut a gain for the plate as shown in Fig. 12-32.

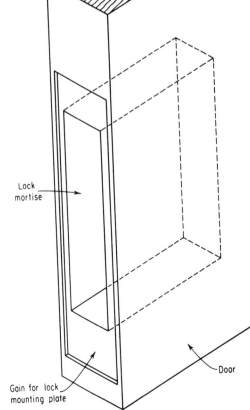

Lock mortise

Door

FIGURE 12-32: *Lock mortice complete.*

Gain for lock mounting plate

6. Measure the distance from the face of the mounting plate to the center of the knob shank hole and the keyhole. Lay out these points on the face of the door and drill holes of the size required.

7. Fit the lock into the mortise, drill pilot holes from the lock-mounting screws, and fasten the lock in place.

8. Install knob shank and knobs according to instructions included with the lockset.

9. Mark the location of the center of the latch bolt pocket on the jamb and fit the strike plate accordingly. A gain must be cut into which the strike plate fits and pockets for the latch bolt and dead bolt drilled and squared out (see Fig. 12–33). The strike plate must be positioned laterally so that the latch bolt will just engage when the door is closed and hold the door snugly in the closed position.

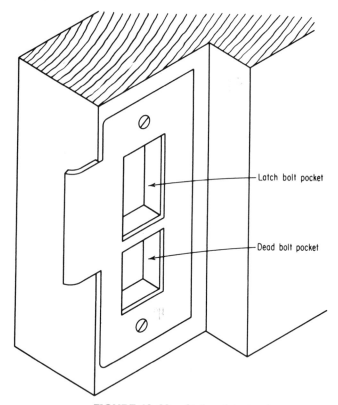

Latch bolt pocket

Dead bolt pocket

FIGURE 12-33: *Strike plate in place.*

Installing a Cylinder Lock

1. Mark the height of the lock above the floor on the edge and face of the door as before.

2. Use the template usually supplied with this type of lock and, laying it on the line drawn in step 1, mark the centers of holes to be drilled in the face and edge of the door (see Fig. 12–34).

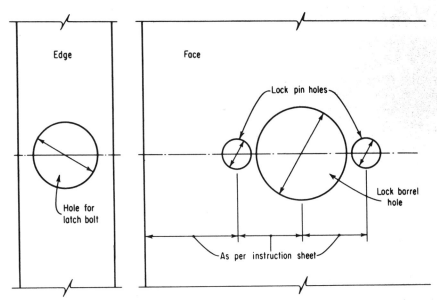

FIGURE 12-34: *Holes for cylinder lock.*

3. Drill the holes through the face of the door and then the one through the edge to receive the latch bolt. It should be slightly deeper than the length of the bolt.

4. Cut a gain for the latch bolt mounting plate, as shown in Fig. 12-35, and install the latch unit.

5. Install exterior knob as described in Fig. 12-36.

6. Next, install interior knob (see Fig. 12-37).

7. Find the position of the strike plate and install it in the jamb, as described for mortise locks.

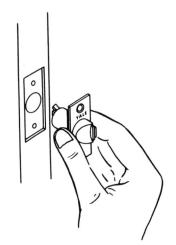

FIGURE 12-35: *Installing latch unit.*

Place exterior rosette with spindle into latch as shown below. Depress latch, position spindle and rosette stems correctly. Pass spindle and stems through holes in latch.

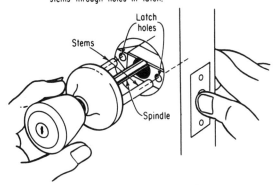

FIGURE 12-36: *Installing exterior knob.*

After exterior knob and rosette is placed, install interior knob and rosette as shown below and push rosettes tight against door. Line up screw holes with stems, insert screws and tighten until lockset is firm.

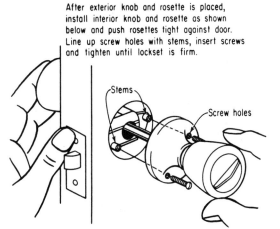

FIGURE 12-37: *Installing interior knob.*

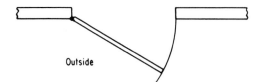

FIGURE 12-38: *Left-hand reverse door.*

The Hand of Doors

The term to describe the direction in which a door is to swing and the side from which it is to be hung is the *hand* of the door. The hand is determined from the outside.

The outside is the street side for entrance doors; it is the corridor side for doors leading from corridors to rooms; it is the room side for doors from rooms to closets; and it is the stop side—the side from which the butts cannot be seen—for doors between rooms.

Stand outside the door. If the hinges are on your right, it is a right-hand door, and if they are on your left, a left-hand door. If the door swings away from you, it is a regular, and if toward you, a reverse (see Fig. 12-38).

Sliding and Folding Doors

In addition to conventional hinged doors, a number of other types are commonly used in modern construction. They include *pocket-type sliding* doors, *bypass sliding* doors, and *folding* doors.

Pocket-type sliding door. This type of door may be considered to be a space-saver, since it opens by sliding into an opening in the partition. See Chapter 5 for a description of framing for such a door.

The door frame consists of one solid and one split side jamb and a solid head jamb with an apron along each edge to conceal the track. The track is an extruded aluminum product, and the rollers are usually nylon for longer wear and silent operations (see Fig. 12-39).

FIGURE 12-39: *Pocket-type sliding door head jamb details.*

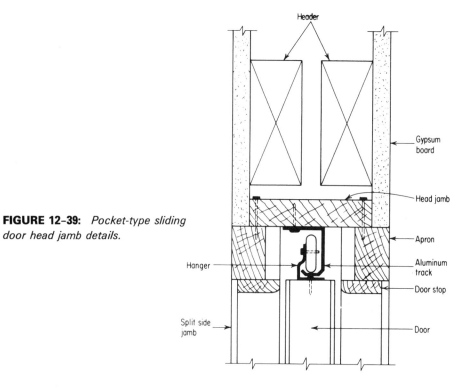

Bypass-type sliding doors. Two doors sliding past one another in an opening are used in this case, so that only half the width of the opening can be utilized at a time. A standard type of door frame may be used, with a head jamb long enough to accommodate two doors. The track may be mounted on the underside of the head jamb (see Fig. 12-40), or a split head jamb may be used to recess the track to permit the door to ride flush with the underside of the jamb (see Fig. 12-41).

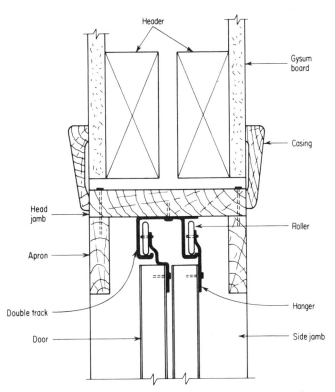

FIGURE 12-40: *Track mounted on underside of jamb.*

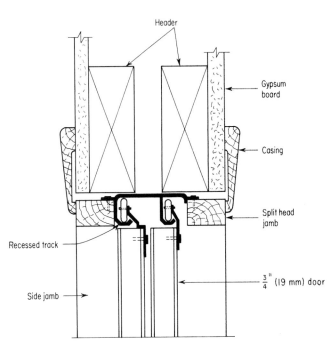

FIGURE 12-41: *Recessed track for sliding doors.*

Folding doors. A folding door unit consists of one or more pairs of doors hinged together at the center and pivoted top and bottom at the outer edge. The pivots fit into *pivot brackets* which are adjustable to provide the proper clearance between the edge of the door and the jamb.

The folding action is provided by the hinges joining pairs of doors together and is guided by a horizontal nylon roller pinned to the traveling edge of the door and enclosed in an overhead track.

WINDOW TRIM

Most of the windows used in modern construction come to the job as a complete unit—frames with the outside casing attached and the sash or glass already installed (see Fig. 11-52). All that is normally required is that the window be *trimmed* on the inside.

In the case of large picture windows, the frame, made for the purpose, is installed first, and the glass is then set in place. Depending on the length of the glass, three or more lead, neoprene, or rubber blocks about ¼ in. (6mm) thick are set on the sill and the glass set on them and tilted into place. Temporary stops may be used to hold the glass in place until the permanent ones have been fitted.

Moldings and Trim

These names designate the materials that are used to finish around the base of walls and around windows, doors, and other openings. Those with relatively small cross sections are *moldings*, while the larger ones are *trim materials*. They include *window and door casing, baseboard, window stop, cove mold, quarter-round, carpet strip,* and *astragal* (see Fig. 12–42).

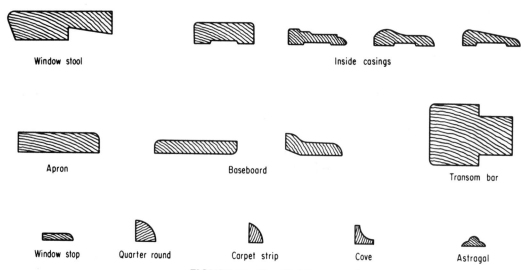

Window stool   Inside casings

Apron   Baseboard   Transom bar

Window stop   Quarter round   Carpet strip   Cove   Astragal

FIGURE 12–42: *Molding and trim sections.*

Great care must be taken in cutting these materials and in fitting them into place. A *miter box, coping saw, sharp panel saw, combination square, smooth plane, chisel,* and *marking knife* are essential tools for applying trim. *Butt, miter,* and *coped* joints are all used, and they must be made to fit perfectly.

Door Trim

Casings (see Fig. 12–42) are applied to both sides of interior doors to cover the space between the frames and the wall, to secure the frames to the wall, and to hold the jambs in a rigid position and must therefore be nailed to both the jambs and the wall frame. The procedure is as follows:

1. Select the necessary pieces and make sure that they are all the same pattern and width.

2. Draw a light pencil line on the jambs ³⁄₁₆ in. (5 mm) from the inner edge (see Fig. 12–43).

3. Check the bottom end of each side casing to see that it is square and will sit flat on the finish floor.

4. Set one side casing in place, with its inner edge to the line drawn in step 2 and mark the position of the miter joint at the top (see Fig. 12–43).

5. Use a miter box to make an accurate miter cut through that point.

6. Measure and cut the second side casing in exactly the same way.

7. Nail the first side casing temporarily in place.

8. Cut a miter on one end of the head casing to match the one on the side casing already in place.

9. Hold the head casing in place with the two miter cuts together and check to see that they fit perfectly. Then mark the position of the miter cut on the opposite end of the head casing and cut the miter accordingly.

10. Nail the second side casing in place and set the head casing in position. Make sure that both miter joints are close fits; glue before nailing.

11. Nail the casing securely with 1½-in. (35-mm) finishing nails into the edge of the jambs and 2½-in. (65-mm) nails into the wall frame (see Fig. 12–44).

12. Sand the external corners where side and head casings meet lightly and set the nails.

Window Trim

When windows are cased on all four sides, apply the bottom casing first, mitered on both ends, and then proceed with the side casings and the top (see Fig. 12–45).

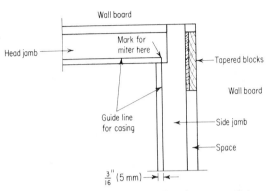

FIGURE 12–43: *Upper right-hand corner of door frame with guide line for casing.*

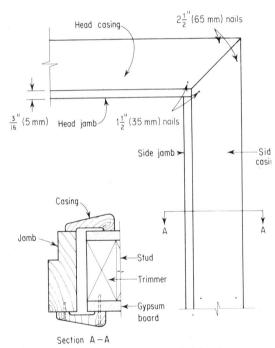

FIGURE 12–44: *Casing nailed in place.*

FIGURE 12–45: *Window cased on four sides.*

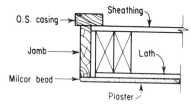

FIGURE 12-46: *Metal bead as trim.*

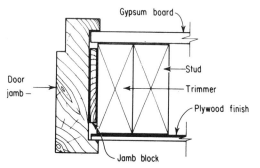

FIGURE 12-47: *Self-trimmed door jamb.*

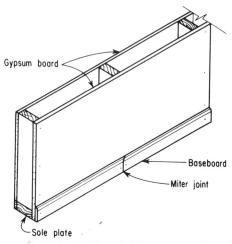

FIGURE 12-48: *Joining two pieces of baseboard.*

Many windows are trimmed without the use of casing. This is commonly done by using a paper-backed metal angle set into a saw kerf in the jamb [see Fig. 12-22(d)] when a drywall finish is applied or a curved metal bead which is nailed over the lath when walls are plastered (see Fig. 12-46).

Another method of trimming door and window openings is to use the edge of the door or window frame as the trim. This is best suited to dry types of finish. Frames are made (Fig. 12-47) and are installed in the same way as other types, and the edges of the wall finish are slipped into place as shown.

BASEBOARD, QUARTER-ROUND, AND CARPET STRIP

When the floors have been laid and the doors trimmed, the baseboard and carpet strip, if required, can be installed. Two types of baseboard are illustrated in Fig. 12-42—one plain and the other molded; the same basic procedures are used in applying any type.

The first step is to find and mark the location of the wall studs so that the baseboard may be nailed to them. Then try to select pieces which will be long enough to reach across the spans involved, if possible. Otherwise it is necessary to join two pieces together with a miter joint (see Fig. 12-48). The joint must come on a stud and should be made with precision so that the joint will be as inconspicuous as possible.

At door casings, butt the end of the baseboard to the casing with a tight, square joint. At inside corners, butt the end of the baseboard on one wall against the adjoining wall. The end of the meeting piece must be coped to fit against the other (see Fig. 12-49). To get a coping line, trace the shape of the baseboard on either the front or the back of the piece to be coped. Use a coping saw and undercut slightly from front to back in order to get a tight fit at the outside face. Pieces meet at outside corners with a 45° miter joint. Carefully mark the length to the corner of the wall on the inside edge of the baseboard and cut the miter *out* from this point (see Fig. 12-50).

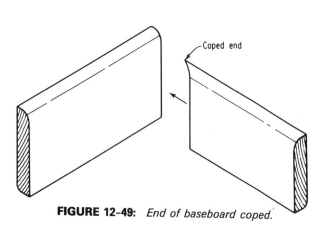

FIGURE 12-49: *End of baseboard coped.*

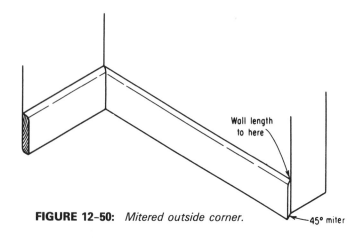

FIGURE 12-50: *Mitered outside corner.*

When plain baseboard is used, carpet strip may be added at the bottom. At the end where baseboard meets a door casing, the end of the carpet strip is mitered, as shown in Fig. 12–51. At an inside corner, the end of one piece is butted against the baseboard, and the end of the meeting piece is *coped* to fit over it (see Fig. 12–52). To cope the end, first cut a 45° miter and then follow the curved line of the miter cut with a coping saw, undercutting slightly to produce a tight fit. At outside corners, the ends of the two meeting pieces are mitered at 45°.

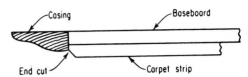

FIGURE 12–51: *Carpet strip meets casing.*

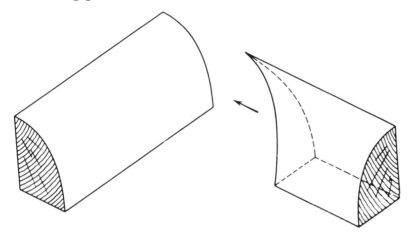

FIGURE 12–52: *End of carpet strip coped.*

CABINETWORK

The final finishing work usually consists of installing kitchen and bathroom cabinets; closets and wardrobes; built-in features such as dressing tables or desks (see Fig. 12–53) in bathroom, bedroom, or library; mantels; room dividers; counters; display racks; booths; etc. Such items may be *custom-built* in a cabinet shop for a specific installation, *mass-produced* in components in a millwork factory, or *built on the job* by the carpenter.

FIGURE 12–53: *Built-in library desk.*

Cabinetwork Drawings

The building plans usually include some details of the built-in cabinetwork. The floor plan will show the location of the units, while elevations, drawn to a larger scale, will provide detailed dimensions. Figure 12–54 illustrates a typical drawing of a kitchen cabinet, upper and lower sections. Detailed sizes of drawers and doors, facing widths, and overhangs may be scaled from the drawing.

When cabinets are to be built on the job, it is usually helpful to draw a *full-scale layout* of each of the various sections involved on plywood or heavy paper. These will be *section views*, both horizontal and vertical, in which each member and the various clearances that may be required are shown full size. These layouts are useful when cutting end or partition panels or doors to size, locating joints, and determining the size and location of drawer parts, etc., which are not included in the original drawings.

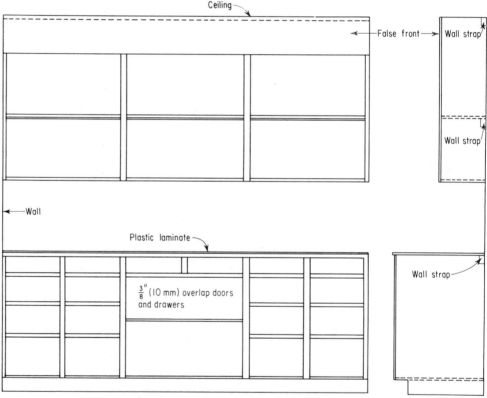

FIGURE 12–54: *Kitchen cabinet drawing.*

Standard Sizes

A number of dimensions and material sizes common in cabinetwork are relatively standard, and the finish carpenter should be familiar with them.

The counter section of a kitchen cabinet is normally 36 in. (900 mm) high and 24 in. (600 mm) deep, with a countertop overhang of about 1 in. (25 mm). The toe space at the bottom will be approximately 3 × 3½ in. (75 × 89 mm) (see Fig. 12-54). The vertical distance between the counter and the *wall* (upper) section will usually be 16 in. (400 mm) unless the wall section is located over a sink or the cooking surface of a stove, in which case the minimum distance is 30 in. (750 mm). This allows room for a hood over the range. The usual overall depth of the wall section is 12 in. (300 mm).

In bathrooms and dressing rooms, the top of a built-in vanity or dressing table should be 30–31 in. (750–775 mm) high, with the depth depending on the type of wash basin fixture, where applicable. Otherwise, the depth is normally 21–22 in. (525–550 mm).

In bedrooms and dressing rooms, the minimum clear depth for clothes closets will be 24 in. (600 mm), and shelves in such closets should be 66 in. (1650 mm) above the floor in order to allow a clothes bar to hang 60 in. (1500 mm) above the floor.

Drawers in kitchen or other cabinets should not usually extend 18 in. (450 mm) in width, with depths ranging from 3 to 12 in. (75–300 mm) in most cases. Drawer length will be about 2 in. (50 mm) less than the width of the counter top.

Doors in wall sections of cabinets should normally not exceed 18 in. (450 mm) in width, while those in lower sections may be up to 24 in. (600 mm) wide. Doors exceeding these widths will usually prove to be inconvenient in the open position.

Basic Framing

When cabinets are being built in place, there are several basic framing procedures to be followed:

1. A 1 × 2 (19 × 38 mm) strap is attached to the wall in a level position, ¾ in. (19 mm) below the level of the counter top, with its length equal to the *inside length* dimension of the cabinet. Draw a plumb line from the ends of the strap to the floor. These lines represent the inside of the end panels.

2. A cabinet base is built, usually from 2 × 4 (38 × 89 mm) material, with its length equal to the length of the wall strap and its width 4¼ in. (106 mm) less than the width of the counter top. It is nailed in place between the plumb lines drawn in step 1 (see Fig. 12-55).

3. The end panels are cut next, the length equal to the height of the top of the wall strap and the width 1¾ in. (44 mm) less than the counter top width. The bottom front corners must be notched to match the width and depth of the toe space (see Fig. 12-55).

4. The base cover can now be applied. It will be made from ½-in. (12-mm) plywood, the same width as the end panels and cut to fit snugly between them.

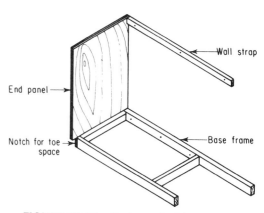

FIGURE 12-55: *End panel with toe-space notch cut in.*

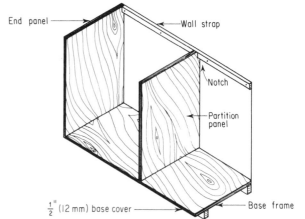

FIGURE 12-56:   *Partition panel with notch cut for wall strap.*

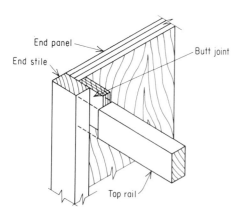

FIGURE 12-57:   *Top-rail-to-stile joint.*

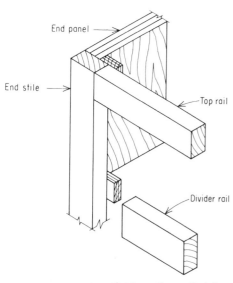

FIGURE 12-58:   *Divider-rail-to-stile joint.*

5. Partition panels are cut to size and installed. The length will be the distance from the top surface of the base cover to the top of the wall strap and the width equal to that of the end panels. The top back corner must be notched to fit over the wall strap (see Fig. 12-56).

Facings

With the base frame members in place, the facing strips are applied to the front edges of the basic frame. The horizontal members are called *rails* and the vertical members, *stiles*.

If the ends of the cabinet are not exposed, the top rail can be applied first, flush with the top end of the panels, its length equal to the distance from outside to outside of the panels. If the ends are exposed, the two outside stiles should be attached first and the top rail cut to fit between them. One method of making the joint between top rail and stile is illustrated in Fig. 12-57. Intermediate stiles are applied next, and then the rails which form the dividers between drawers are cut to fit between them. One method of making the joint between divider rail and stiles is shown in Fig. 12-58.

Drawers

Drawers are made with two general types of front—*flush* and *overlapping*. Flush drawers close so that the drawer front is flush with the facings around it and must be carefully fitted for good appearance. Overlapping drawers are made with a ⅜-in. (10-mm) lip around the edge of the drawer front, so that when the drawer is closed, the lip overlaps the edge of the facings around the opening, allowing more freedom in fitting (see Fig. 12-53).

Various construction techniques are used in making drawers, the choice depending on (1) whether the drawer is flush or overlapping, (2) the type of drawer guide to be used, and (3) the carpenter's perference as to the type of joints to be used.

The front will normally be ¾-in. (19-mm) plywood, the sides and back, ½-in. (12-mm) material, and the bottom ¼-in. (6-mm) plywood or hardboard. Figure 12-59 illustrates some drawer joint and assembly details and Fig. 12-60, an assembled overlapping front drawer.

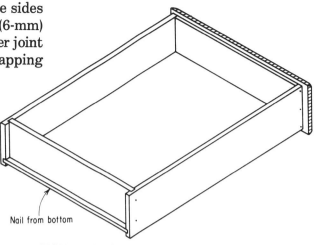

FIGURE 12-60: *Assembled overlapping drawer.*

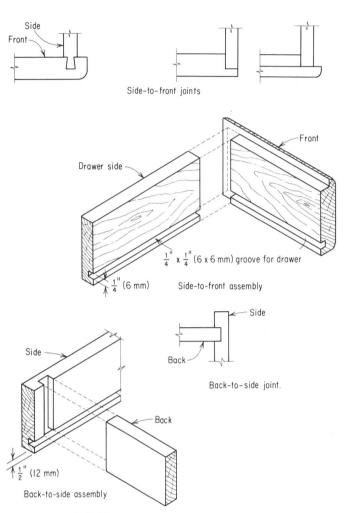

FIGURE 12-59: *Cabinet drawer detail.*

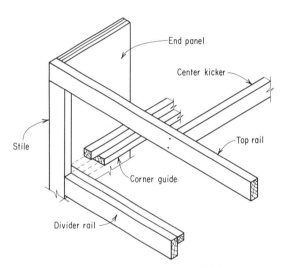

FIGURE 12-61: *Corner guide and kicker detail.*

Drawer Guides

Drawers must be guided during opening and closing and held in an approximately horizontal position when open. Types of drawer guide include *corner* guide, *side* guide, and *center* guide. When corner guides are used, an additional member, a *kicker*, must be added above the drawer to keep it level when it is open. Figure 12-61 illustrates a typical corner guide assembly and kicker location.

Side guides consist of narrow strips approximately ⅜ × ½ in. (10 × 12 mm), fixed to the side of the drawer opening, which fit into grooves cut in the sides of the drawer. They not only guide the drawer but hold it in a level position as well (see Fig. 12-62).

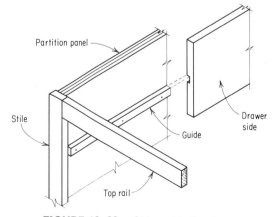

FIGURE 12-62: *Side guide for drawer.*

A center guide is located on the bottom of the drawer and consists of a grooved *guide bar* attached to the drawer bottom and a *runner* attached to the dividing rail. A kicker is also necessary with this type of guide (see Fig. 12–63). Center guide drawer hardware is also available for the same purpose.

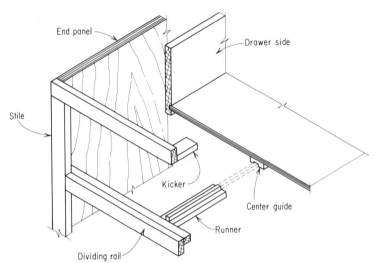

FIGURE 12-63: *Drawer center guide.*

Wall Cabinets

The construction of wall cabinets is a relatively simple procedure. Side panels of ¾-in. (19-mm) plywood are cut to size, rabbeted at both ends for the bottom shelf and cabinet top, and dadoed to receive the ends of intermediate shelves (see Fig. 12–64). Narrow straps glued under the back edge of the top and one intermediate shelf are used to fasten the cabinet to the wall. After assembly, the cabinet is secured in place and the facings installed.

Revolving Shelves

In the corners of L-shaped and U-shaped counter sections of cabinets there will be a considerable amount of space which will be inaccessible by the conventional type of cabinet door. That space can be made available by the use of *carousel shelving*—circular shelves which turn on a central shaft, with bearings top and bottom. The unit is available as a hardware item and must be installed while the counter section is being framed (see Fig. 12–65).

Counter Tops

The top of the counter section is generally made from ⅝- or ¾-in. (16- or 19-mm) plywood or particle board, cut wide enough to allow for a 1-in. (25-mm) overhang at the front. It is fixed in place by glue and by nailing through the top into the edges of panels, top rail, and wall strap.

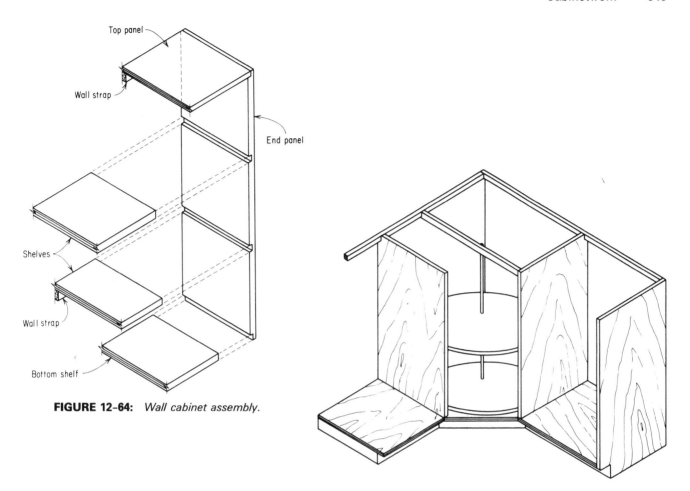

FIGURE 12-64: *Wall cabinet assembly.*

FIGURE 12-65: *Revolving shelving in place.*

Most counter tops are covered with some type of plastic laminate which is attached to the counter surface with *contact cement*. The cement is applied to the back of the laminate and the surface of the counter allowed to dry before the material is set in place. Care must be taken to ensure that the topping is in the proper position before contact is made because adhesion between the two surfaces is instantaneous.

Exposed counter edges are covered first. After the strip is in place, the top edge is dressed flush with the counter top. The top sheet must be cut to fit snugly against the wall and overhang the front edge slightly. Cement is applied to both meeting surfaces and allowed to dry. When the sheet is ready for installation, the dried cement on the counter surface may be covered with paper, but the cement *must be dry* before this is done so that the paper will not stick to the surface. The laminate sheet can then be laid on the paper, adjustments made in position, and the paper removed, allowing the two cement surfaces to contact. Full contact should be ensured by tapping the surface with a wooden mallet or rubber hammer. The slight overhang can then be removed with a coarse file or with a special cutter available to fit a router.

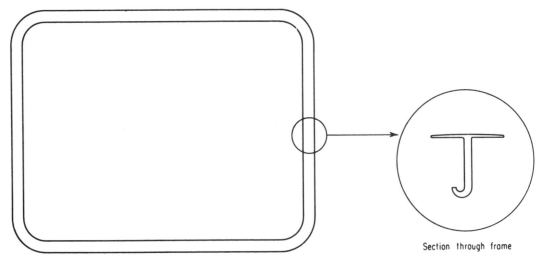

FIGURE 12-66: *Sink frame.*

Section through frame

Sinks

One important item in cabinet building is the installation of a sink, and it is common practice to wait until the plastic laminate top is on before installing it.

The first step is to mark the shape of the sink rim in its proper location on the counter top and cut a hole slightly larger than the sink rim in the top.

Enameled sinks are held in place by a special frame, the same shape as the sink, having a cross section as illustrated in Fig. 12-66. One lip of the frame rests on the counter top and the other on the sink rim, and the web projects down into the opening.

The sink is placed in the opening and temporarily supported and the sink frame fitted over it. To ensure a watertight fit, a light bead of caulking compound should be run all around the edge of the hole before the sink frame is set down. Then, from the underside, clips such as the one shown in Fig. 12-67 are used to secure the sink in place. Metal sinks are held in place as shown in Fig. 12-68.

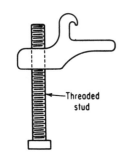

FIGURE 12-67: *Sink frame clip.*

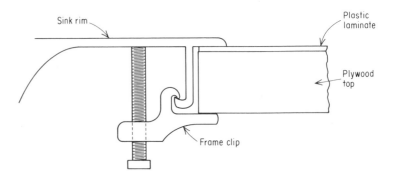

FIGURE 12-68: *Sink in place.*

REVIEW QUESTIONS

12-1. (a) Explain how heat is lost from a building by conduction.

(b) How is it lost by radiation?

12-2. (a) What types of material prevent heat loss by conduction?

(b) What types of material prevent heat loss by radiation?

12-3. List four types of insulation used to prevent heat loss by conduction.

12-4. Illustrate by diagram how batts are installed between studs.

12-5. What is the difference between moisture barrier paper and vapor barrier paper?

12-6. How is gypsum lath reinforced

(a) At internal corners?

(b) At external corners?

12-7. Explain what is meant by a *laminated dry-wall finish.*

12-8. Illustrate four methods of treating joints between sheets of plywood used as interior finish.

12-9. Explain how the pre-expanding of hardboards is carried out.

12-10. (a) List four types of hardwood flooring in common use.

(b) Why is the bottom surface of hardwood flooring strips concave?

(c) What is ½-in. (12-mm) space left between flooring and wall?

(d) What is meant by "blind" nailing?

12-11. What is the basic difference between a mortise lock and a cylindrical lock?

12-12. (a) Where is the strike plate located?

(b) What is the normal height of a door knob from the floor?

(c) What is a hinge *gain?*

12-13. (a) How is the hand of a door determined?

(b) If a door swings from the outside toward you, with the hinges on the left hand, it is a _____ door.

12-14. Name six commonly used *trim* materials.

12-15. What is meant by *coping* with a joint in trim?

12-16. Given the usual dimensions for

(a) Height of toe space under kitchen counter section

(b) Height of counter section above floor

(c) Distance between upper and lower sections of kitchen cabinets

(d) Maximum drawer width

13

ENERGY-EFFICIENT HOUSING

Housing in cooler regions has taken a significant change due to the ever-increasing concern with saving energy. The cost of fuel has risen dramatically in the past few years, resulting in additional cost to homeowners to keep dwellings comfortable. Clearly the construction of houses that are as energy-efficient as possible makes good long-term common sense. These houses are likely to be here for many years, and as costs will likely rise in the future, efficient construction is going to be continually more important.

Initially *typical construction* was upgraded by increasing the thickness of the insulation and being more careful with the placement of the vapor barrier. Insulation of the foundation and the outside of the wall frame followed, resulting in more efficient housing. Recently better design and construction technology has brought forth a new breed of highly efficient houses. These houses are described by such terms as "super-energy-efficient" or "super-insulated." Energy-efficient housing does not dwell exclusively on cutting heat losses. Heat gains are augmented where possible through an effort to increase *passive solar gain*.

The increase of methods used to reduce heat loss should not take place without an understanding of the possible problems associated with moisture in buildings. In a typical house the moisture generated escapes to the outside through flues and chimneys and by seeping out through cracks and holes, carrying water vapor with it. Water vapor also escapes by diffusing through building materials. Moisture escaping through cracks and holes and by diffusing through materials does not disappear without a trace. As air and moisture pass through the structure from the heated interior, air is cooled, thus reducing its ability to hold moisture. This cooling may result in condensation within the walls or the attic of the house (see Fig. 13-1). Water condensing in the wall can cause a loss in insulation effectiveness and increase degradation of materials.

In older houses, a vapor barrier was normally installed to prevent moisture damage. Vapor barriers do reduce the moisture loss through materials but do very little to control condensation associated with air leakage. Conventional houses are so leaky that the inside air tends to be dry, and less condensation will take place. Energy-efficient housing is more vulnerable to condensation problems as it does not benefit from this air movement. Careful control is necessary to effectively seal moisture inside the house. The installation of an air barrier will prevent air leakage into the shell.

Since energy-efficient houses are built to be airtight, the indoor air supply must be regulated. A fresh air supply is needed

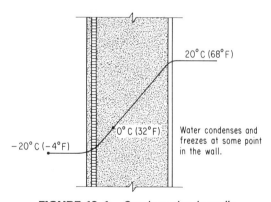

20° C (68° F)

0° C (32° F)

−20° C (−4° F)

Water condenses and freezes at some point in the wall.

FIGURE 13-1: *Condensation in walls.*

to maintain good-quality air. The provision of controlled ventilation is needed to replace stale air. Stale air is exhausted through one vent, and fresh air is drawn through another vent. Heat loss is minimized by maintaining the minimum acceptable rate and by incorporating a heat exchanger to recover heat from the exhaust air (see Fig. 13-2).

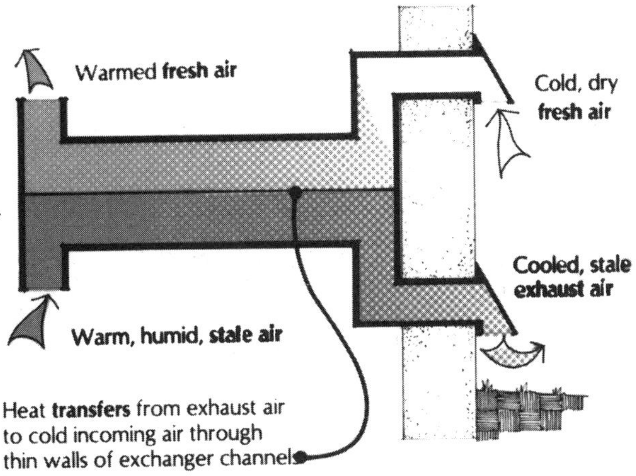

FIGURE 13-2: *Air-to-air heat exchanger. (Courtesy Home & Community Design Branch, Alberta Agriculture)*

FOUNDATIONS

Foundations to be considered in this section will include *slabs-on-grade*, *crawl spaces*, and *basements*. These foundation types can be insulated from the interior of the building or the exterior, each requiring special considerations.

Slabs-on-Grade

In this type of foundation, the concrete slab can be the combined foundation and finish floor (see Fig. 13-3), or the slab can be isolated from the load-bearing foundation wall (see Fig. 13-4).

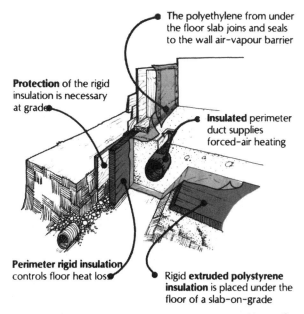

FIGURE 13-3: *Slab-on-grade. (Courtesy Home & Community Design Branch, Alberta Agriculture)*

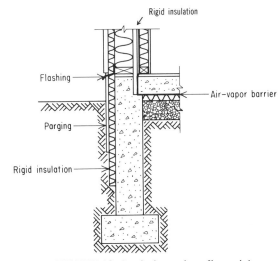

FIGURE 13-4: *Independent floor slab and foundation wall.*

Rigid polystyrene insulation is used below the slab to lower floor heat loss (see Fig. 13-3). Perimeter insulation is also used at the edge of the floor slab to reduce heat loss. Polyethylene is placed under the floor slab and seals to the wall air-vapor barrier (see Fig. 13-3). Protection in the form of parging or preserved plywood should be placed over the insulation around the perimeter. Even though slabs-on-grade should be insulated on the outside of the foundation, insulation is placed on the inside on some occasions (see Fig. 13-5).

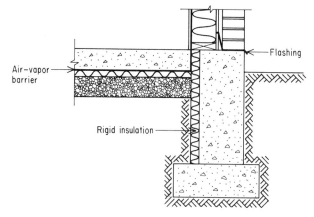

FIGURE 13-5: *Slab-on-grade (interior insulation).*

TABLE 13-1: *Appropriate energy-efficient insulation levels*

	"R" value	"RSI" (metric)
Ceilings	45–60	8–11
Walls	30–40	5–7
Foundation walls (50% or more below grade)	12–25	2–4
Floors over crawl spaces	30–40	5–7
Slabs-on-grade	12–25	2–4
Basement floors	7–10	1.5–2

Crawl Spaces

In a crawl space foundation, the main concern is to provide insulation in the right places and protect the crawl space from moisture buildup. This space is often used for mechanical services, so insulation is needed to keep the temperature above freezing. The main insulation should be provided between the floor joists. "R" (RSI) values should equal or exceed those indicated in Table 13-1. When batt insulation is used, it should be supported from underneath with particle board or with wire mesh of small enough size to prevent entry of rodents (see Fig. 13-6). Foundation insulation should be on the outside of the wall to protect concrete against temperature effects. A moisture barrier, placed over the ground surface, is necessary to keep the space dry. Summer ventilation according to code requirements should be provided. The vents should be provided with dampers that can be closed in winter and opened in summer.

Basements

Preserved wood and concrete are the two main types of basements, though concrete block is popular in some areas. Wood foundation walls can be economical and easy to insulate. The spaces between the studs provide an ideal place for friction-fit batt insulation. An air-vapor barrier covers the insulation on the inside. (see Fig. 13-7). The polyethylene from under the floor is sealed to the wall vapor barrier (see Fig. 13-8). Two types of basement floors used in wood basements are illustrated in Fig. 13-8.

Overhanging the joists and sill creates a smooth exterior joint between the wall sheathing and rigid insulation protection

Note how the floor air-vapour barrier can be **sealed** to the wall layer

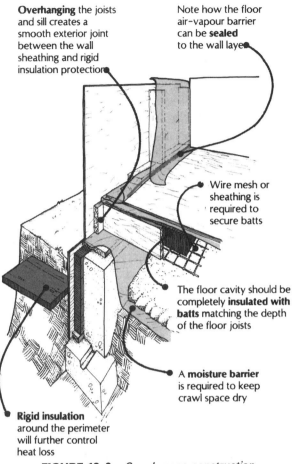

Wire mesh or sheathing is required to secure batts

The floor cavity should be completely **insulated with batts** matching the depth of the floor joists

A **moisture barrier** is required to keep crawl space dry

Rigid insulation around the perimeter will further control heat loss

FIGURE 13-6: *Crawl space construction. (Courtesy Alberta Agriculture)*

Pressure treated plywood on pressure treated studs provides the foundation structure

A 300mm wide (12 in.) **treated plywood "wear strip"** is placed at grade level

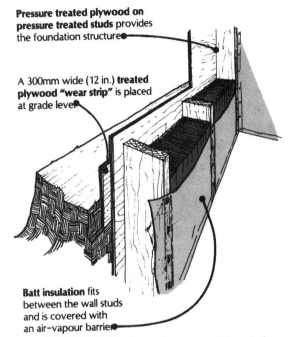

Batt insulation fits between the wall studs and is covered with an air-vapour barrier

FIGURE 13-7: *Insulating a wood foundation. (Courtesy Alberta Agriculture)*

Horizontal insulation keeps footing area frost-free

Note how wall air-vapour barrier is **sealed** to the polyethylene from under the floor

Rigid **extruded polystyrene insulation** can be used under the floor slab

A pressure treated wood floor can be **insulated with batts**

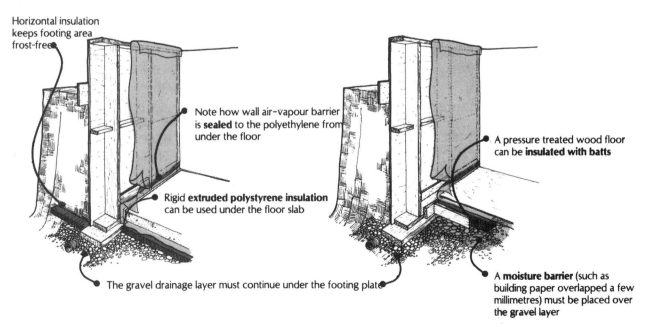

The gravel drainage layer must continue under the footing plate

A **moisture barrier** (such as building paper overlapped a few millimetres) must be placed over the gravel layer

FIGURE 13-8: *Wood foundation floors. (Courtesy Alberta Agriculture)*

The exterior must be **protected** from mechanical damage with stucco, treated plywood or any siding material suitable for earth burial

The polyethylene should be **protected** against punctures with some type of wall finish

Rigid glass fibre or extruded polystyrene **insulation** is placed over concrete or masonry foundation walls

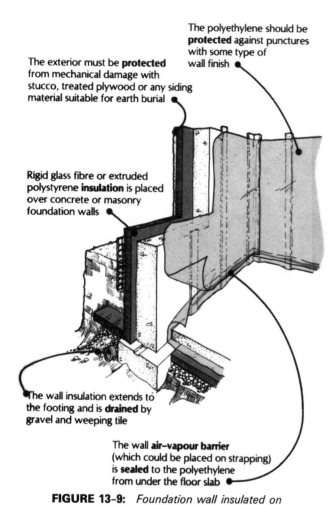

The wall insulation extends to the footing and is **drained** by gravel and weeping tile

The wall **air-vapour barrier** (which could be placed on strapping) is **sealed** to the polyethylene from under the floor slab

FIGURE 13-9: *Foundation wall insulated on outside. (Courtesy Alberta Agriculture)*

A **moisture barrier** is placed on the wall from the grade level to the floor

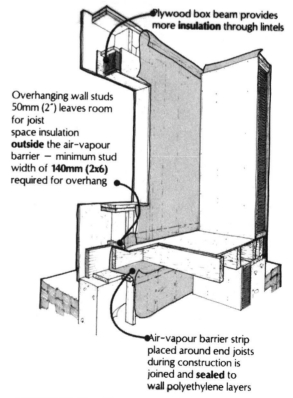

Strapping is used to form an insulation cavity and backing for wall finishing materials

Set the face of the strapping out 150mm (6″) to allow for RSI 3.5 batt insulation (R20)

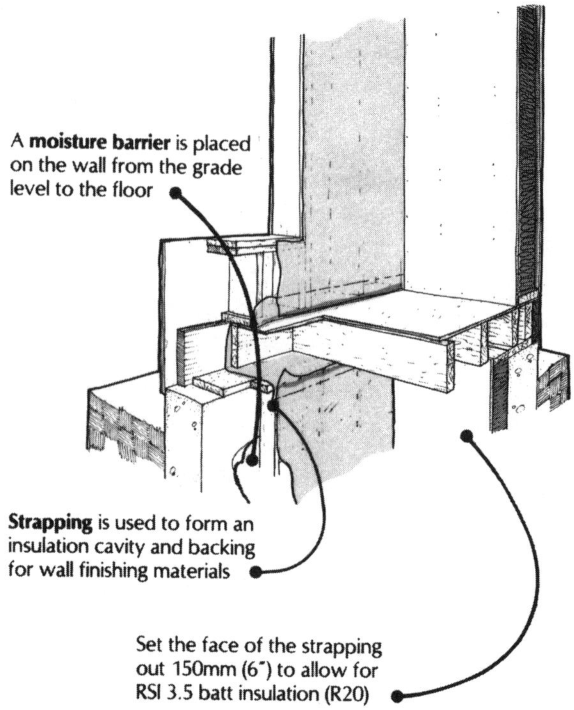

FIGURE 13-10: *Foundation wall insulated on inside.*

Plywood box beam provides more **insulation** through lintels

Overhanging wall studs 50mm (2″) leaves room for joist space insulation **outside** the air-vapour barrier — minimum stud width of **140mm (2x6)** required for overhang

Air-vapour barrier strip placed around end joists during construction is joined and **sealed** to wall polyethylene layers

FIGURE 13-11: *Placement of vapor barrier around header. (Courtesy Alberta Agriculture)*

The best place for insulation in a concrete basement is on the outside. The foundation is then less susceptible to frost damage and leaking. The foundation wall is inside the insulation, and so its large thermal mass acts as a vehicle for heat storage. The exterior insulation can be continued up the wall, making an effective blanket. The polyethylene is placed on the inside of the concrete wall and sealed at the bottom to the poly from under the floor and at the top to the polyethylene around the header. Insulation and a vapor barrier are placed under the floor to provide a warm dry floor (see Fig. 13-9).

Most masonry and many concrete walls will continue to be insulated from the inside, as illustrated in Fig. 13-10. A moisture barrier must be placed against the concrete wall from floor level up to the grade (see Fig. 13-10). This will protect the building materials from moisture migration.

FLOOR FRAME

The floor frame is a major point of heat loss as it is more difficult to insulate and still have a continuous air-vapor barrier at this point (see Fig. 13-11). A strip of polyethylene must be installed around the joist header during framing. The polyethylene must be under the floor frame and over the foundation wall sealed to the basement wall polyethylene. The upper edge will extend under the wall plate and be sealed to the wall vapor barrier. Two layers of insulation are placed outside the header to give it sufficient "R" value. To make a better connection, the bottom wall plate should be narrower to allow room for the insulation to cover the outside (see Fig. 13-12).

Cantilevered floor projections create a special problem. They do provide for additional floor space but create problems in providing a continuous vapor barrier. Figure 13-13 illustrates a method using polystyrene insulation between the floor joists as a vapor barrier. Even with sealing around these pieces, the floor joists still do pierce the vapor barrier.

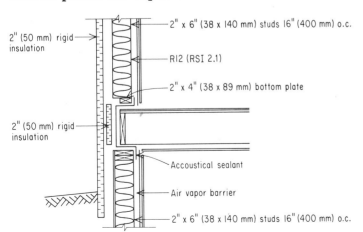

FIGURE 13-12: *Insulating header.*

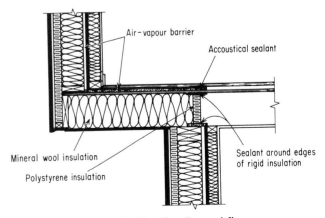

FIGURE 13-13: *Cantilevered floor.*

FIGURE 13-14: *Single stud wall with exterior application of rigid insulation.*

WALL FRAME

Traditional wall construction seriously restricts the level of energy efficiency that can be achieved. A number of variations in design have been developed to increase the efficiency of walls. The typical single-stud walls have been increased in thickness to accommodate additional insulation by using larger studs and more effectively by strapping the inside of the studs (see Fig. 13-14). This strapping allows a place for the installation of electrical wires without interfering with the vapor barrier, as the barrier is placed between the strapping and the studs (see Fig. 13-15). Double-stud walls are very effective in increasing the energy efficiency of walls (see Fig. 13-16).

Single-Stud Walls

The use of single-stud walls is the most common form of construction. The size of the stud has been commonly increased to a 2 × 6 (38 × 140 mm) to be able to increase "R" values (RSI values) in the cavity (see Fig. 13-11). The walls are often offset over the edge of the floor frame to allow for a layer of insulation on the outside of the joist header. In an effort to provide additional insulation and to cut down on the thermal bridges (studs) in the wall, a rigid insulation is added to the outside of the sheathing (see Fig. 13-14). Probably the most effective way of increasing the efficiency of this wall is to strap the inside of the wall with 2 × 2's (38 × 38 mm) at right angles to the wall studs after the insulation and air-vapor barrier have been applied to the exterior wall (see Fig. 13-15). This provide a convenient space for electrical wires and some plumbing vents so the barrier is not punctured. Two-thirds of the insulation value should be outside the vapor barrier.

Double-Stud Walls

Double-stud walls include two frame walls with a space between them. These walls can be built to almost any thickness to achieve very high insulative values. The air-vapor barrier can be easily isolated in a protected position in the wall assembly (see Fig. 13-16). The inside wall is the structural wall and includes lintels, double plates, and sheathing. The vapor barrier is placed under the outside sheathing on the outside of this wall. The outside wall is placed out from the structural wall, and it provides support for the exterior finishing material. Plywood spacers [½ in. (12.5 mm) thick] can be used to position the exterior wall (see Fig. 13-16). The vapor barrier flaps at the top and bottom of the structural wall are stapled to the inside of the plates to limit damage and to provide for easy sealing with polyethylene from the floor frame and the attic. This air-vapor barrier is usually 6 mil (150 μm) and is sealed with an acoustical sealant.

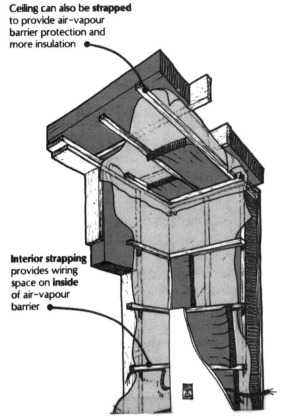

Ceiling can also be **strapped** to provide air-vapour barrier protection and more insulation •

Interior strapping provides wiring space on **inside** of air-vapour barrier •

FIGURE 13-15: *Interior wall strapping on single stud wall. (Courtesy Alberta Agriculture)*

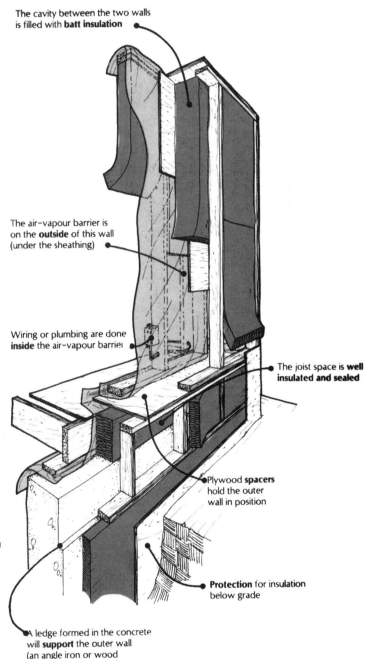

The cavity between the two walls is filled with **batt insulation** •

The air-vapour barrier is on the **outside** of this wall (under the sheathing) •

Wiring or plumbing are done **inside** the air-vapour barrier •

The joist space is **well insulated and sealed**

Plywood **spacers** hold the outer wall in position

Protection for insulation below grade

A ledge formed in the concrete will **support** the outer wall (an angle iron or wood member bolted to the wall can be used)

FIGURE 13-16: *Double wall. (Courtesy Alberta Agriculture)*

FIGURE 13-17: *Polystyrene lintels and panels.*

Modular Wall System

Some manufacturers produce modular energy-efficient systems. One available on the market is a system using solid polystyrene panels. Chemically bonded within the panel are wood studs or structural steel tubing (see Fig. 13–17). The panels are lightweight and easy to handle, allowing for speedy assembly. Special lintels are provided over window and door openings (see Fig. 13–17). Figure 13–18 illustrates the use of polystyrene panels in residential construction.

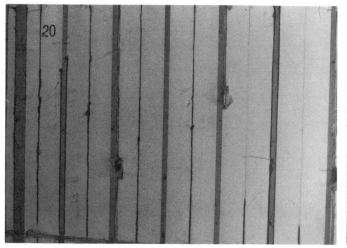

(a)

(b)

FIGURE 13-18: *a) Panels in place.*
b) Sealing edge of panels.
c) Completed modular system.

(c)

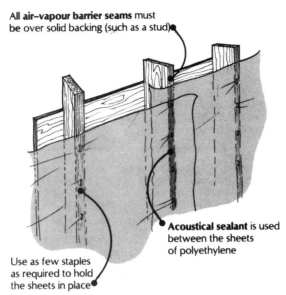

All air–vapour barrier seams must be over solid backing (such as a stud)

Acoustical sealant is used between the sheets of polyethylene

Use as few staples as required to hold the sheets in place

FIGURE 13-19: *Joining air-vapor barrier layers.*

Air-Vapor Barrier

The air-vapor barrier in an energy-efficient house must be installed in such a way as to provide a nearly completely sealed building envelope. It must be sealed with an acoustical sealant to provide a long-lasting seal. An acoustical sealant does not harden or form a skin and is used for sealing polyethylene. The air-vapor barrier controls air leakage and prevents vapor movement. To ensure an effective seal, all joints between polyethylene must be made on a wood backing. The first layer is stapled to the framing member, a continuous bead of sealant is placed, and the second sheet is placed over the first and stapled (see Fig. 13-19).

Special consideration is needed to seal the air-vapor barrier at obstructions. To achieve a continuous envelope at partition walls in single-stud walls, a piece of polyethylene is placed behind the end stud and then sealed to the vapor barrier (see Figs. 5-33 and 13-20). At plumbing pipe openings a plywood backing is installed and sealed to the vapor barrier (see Fig. 13-21). Around

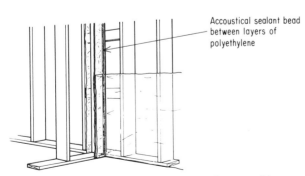

Accoustical sealant bead between layers of polyethylene

FIGURE 13-20: *Sealing air-vapor barrier at partition.*

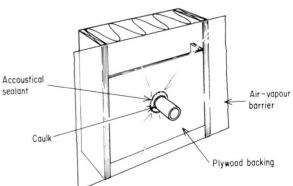

Accoustical sealant

Caulk

Air–vapour barrier

Plywood backing

FIGURE 13-21: *Sealing plumbing pipe.*

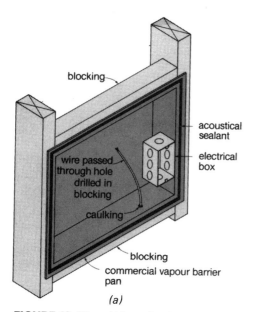

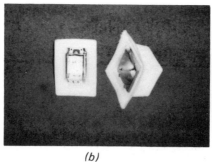

(b)

(c)

FIGURE 13-22: a) Vapor barrier around electrical box. b) and c) Commercial vapor barrier pans.

electrical outlets a *vapor barrier pan* can be used to achieve a seal (see Fig. 13-22). Placing a piece of polyethylene around the box before installation and then sealing to the vapor barrier can be quite effective (see Fig. 13-23).

Windows and doors also need special consideration. To achieve an energy-efficient building, energy-efficient windows and doors must be used. Casement, awning, and hopper windows are good performers. Metal insulated doors are good as they do not warp easily, and so a good seal is maintained. Figure 13-24 illustrates an effective method of achieving a good vapor barrier seal around openings. Double or triple glazing does not provide much of a barrier against heat loss. Some homeowners have added thermal shades to cut heat loss (see Fig. 13-25).

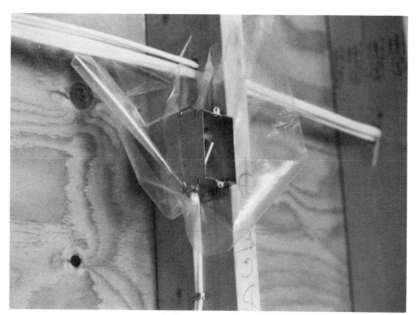

FIGURE 13-23: Vapor barrier behind box.

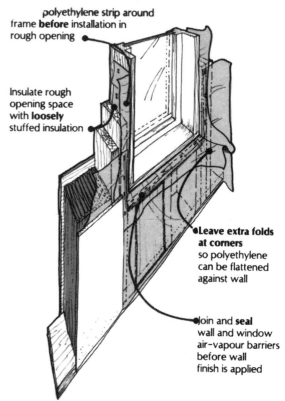

polyethylene strip around frame **before** installation in rough opening

Insulate rough opening space with **loosely** stuffed insulation

•**Leave extra folds at corners** so polyethylene can be flattened against wall

•Join and **seal** wall and window air-vapour barriers before wall finish is applied

FIGURE 13-24: Sealing around a window opening. (Courtesy Alberta Agriculture)

FIGURE 13-25: *Window shades.*

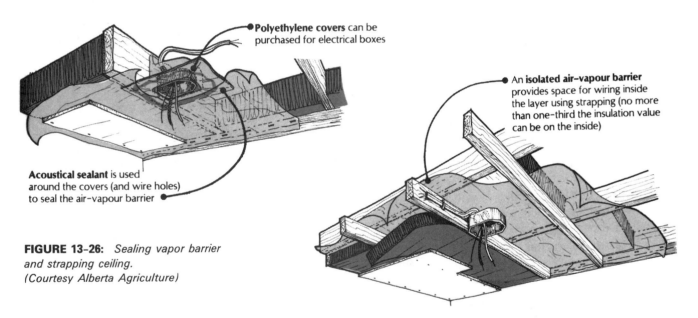

● **Polyethylene covers** can be purchased for electrical boxes

Acoustical sealant is used around the covers (and wire holes) to seal the air-vapour barrier

FIGURE 13-26: *Sealing vapor barrier and strapping ceiling. (Courtesy Alberta Agriculture)*

● An **isolated air-vapour barrier** provides space for wiring inside the layer using strapping (no more than one-third the insulation value can be on the inside)

CEILINGS AND ROOFS

Most houses have sloped roofs with interior ceilings either flat or sloped. A complete, well-sealed air-vapor barrier is essential but, because of light fixtures, chimneys, plumbing vents, and attic access, is difficult to install. The ceiling can be strapped in a manner similar to the method described for walls, thus allowing room for electrical wires (see Fig. 13-26). Polyethylene can be placed above electrical boxes and sealed to the vapor barrier to provide a continuous membrane. Metal firestops around chimneys should be sealed at the ceiling level to stop the passage of air (see Fig. 13-27). Vent stacks can be sealed around the opening they pass

FIGURE 13-27: *Firestop around chimney.*

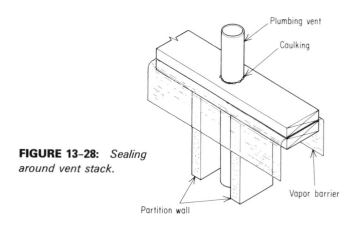

FIGURE 13-28: *Sealing around vent stack.*

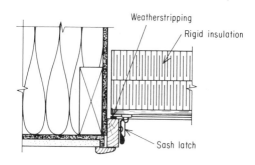

FIGURE 13-29: *Sealing and insulating attic access.*

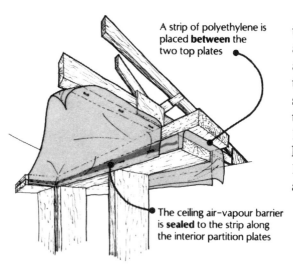

A strip of polyethylene is placed **between** the two top plates

The ceiling air-vapour barrier is **sealed** to the strip along the interior partition plates

FIGURE 13-30: *Seal over partition. (Courtesy Alberta Agriculture)*

through at the wall plate line (see Fig. 13-28). Attic access probably provides the greatest loss of heat and vapor leakage. If the access cannot be eliminated from the ceiling, seal and insulate the hatch as illustrated in Figure 13-29. To provide a continuous seal over partitions, a piece of polyethylene is placed between the top plates and then sealed to the ceiling vapor barrier (see Fig. 13-30). Attic ventilation can be maintained by installing insulation stops between the trusses above the plate line (see Fig. 13-31). Ventilation can be provided in a sloped ceiling by using a wide truss with openings for ventilation (see Fig. 13-32).

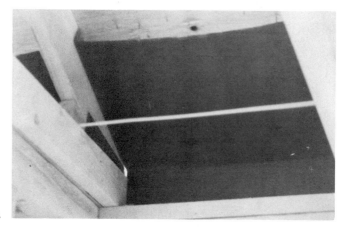

FIGURE 13-31: *Insulation stops.*

FIGURE 13-32: *Sloped truss ceiling.*

PASSIVE SOLAR GAIN

A building designed for passive solar gain catches heat from the sun, stores it, and then releases it slowly to reduce energy costs. Passive systems are designed as part of the building; as such they have little cost and operate by themselves. Active systems, on the other hand, utilize collectors to collect energy and can be expensive to operate (see Fig. 13–33). Passive systems utilize south-facing windows and patio doors to collect energy. Windows on the north side lose heat in the winter due to cold north winds. Windows on the north side should be as small as possible, and large windows should face south.

FIGURE 13-33: *Solar collector panels.*

On sunny days, rooms with southern exposure may become overheated while other areas are still cold. To maintain a more uniform temperature, the heat gained must be stored and released when the temperature drops. This heat gain can be stored in a large thermal mass such as concrete or masonry walls and floors or a rock bin or water storage (see Fig. 13–34). Care must be taken

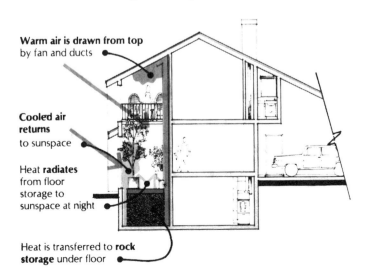

Warm air is drawn from top by fan and ducts

Cooled air returns to sunspace

Heat **radiates** from floor storage to sunspace at night

Heat is transferred to **rock storage** under floor

FIGURE 13-34: *Passive heat storage. (Courtesy Alberta Agriculture)*

FIGURE 13-35: *Sun space.*

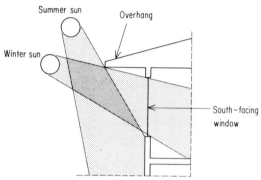

FIGURE 13-36: *Use of overhang to regulate heat gain.*

that windows do not take up too much wall area, making temperature difficult to manage. Sun spaces can be effective, as they are directly heated and will provide indirect heating (see Fig. 13-35). Ventilation must be provided for these spaces so excess heat can be released in hot seasons.

The overhang of the building should be designed to allow full sun into the house during the cold season and provide shade during the hot season (see Fig. 13-36). An adjustable overhang is ideal. A large awning or movable overhang will do this quite effectively. Window shutters or shades can be used to prevent overheating during warmer weather (see Fig. 13-25).

REVIEW QUESTIONS

13-1. Why is it difficult to achieve an effective air-vapor barrier in typical single-stud wall construction?

13-2. Explain how an air-to-air heat exchanger works.

13-3. What should be done with spaces in walls and ceilings that are too small to insulate or to eliminate condensation associated with air leakage?

13-4. What problems can occur with buildings that are sealed so tight that air exchanges are minimal?

13-5. Why is it better to insulate concrete foundations on the outside as opposed to the inside?

13-6. What properties do acoustical sealants have that make them suitable for sealing air-vapor barriers?

13-7. How can the thermal bridges in a single-stud wall be eliminated?

13-8. What percentage of the insulation thickness must be outside the air-vapor barrier to achieve an effective barrier against moisture condensation?

13-9. Explain the principle of passive solar systems.

The basic materials involved in masonry construction are *brick, stone* and *concrete block,* together with the *mortar* used to bind them together. Brick and stone have been primary building materials since time immemorial, while the other made its appearance in more recent times, the first concrete blocks being molded in the 1880s.

Increasing use is being made of all these materials, and no discussions of light construction would be complete without a description of the basic methods of using them in modern buildings.

14

MASONRY CONSTRUCTION

CONCRETE BLOCK CONSTRUCTION

Concrete block is the name given to a type of masonry unit made from a mix of fine aggregate and cement, with or without color added. A very dry mix is used, and the blocks are molded under pressure by a vibrating machine, like that illustrated in Fig. 14-1.

FIGURE 14-1: *Concrete block making machine.*

A great variety of blocks is available, in a wide range of *types, shapes, sizes,* and *surface textures,* each for a specific purpose. Figure 14-2 illustrates some of the common shapes used in building construction.

Concrete block can be used for *basement walls,* for *exterior* and *partition walls,* for *foundation piers,* and for *curtain walls* in steel or reinforced concrete frame buildings. Regardless of the use to which the blocks are to be put, a basic requirement for good results, both from the standpoint of structural ability and good appearance, is good mortar.

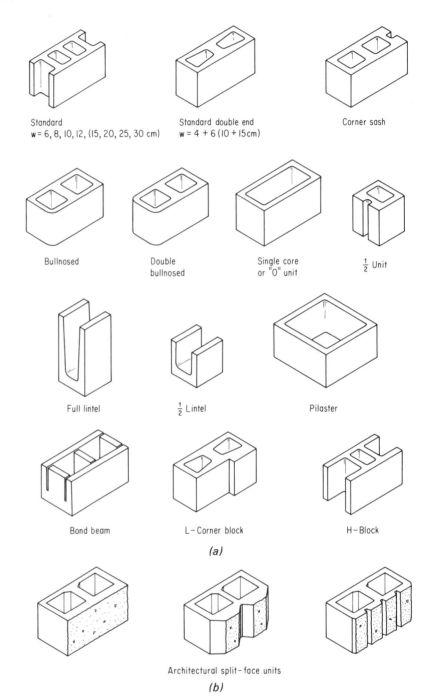

Standard
w = 6, 8, 10, 12, (15, 20, 25, 30 cm)

Standard double end
w = 4 + 6 (10 + 15cm)

Corner sash

Bullnosed

Double
bullnosed

Single core
or "O" unit

½ Unit

Full lintel

½ Lintel

Pilaster

Bond beam

L – Corner block

H – Block

(a)

Architectural split–face units

(b)

FIGURE 14-2: *a) Typical block shapes.
b) Architectural split-face units.*

Mortar

Mortar serves a number of purposes, the main one being to *join the masonry units together* into a strong, well-knit structure. In addition, mortar is required to *produce tight seals between units,* to *bond to steel reinforcement, metal ties, and anchor bolts,* to *provide a bed* which will accommodate variations in the size of units, and to *provide an architectural effect* by the various treatments given to mortar joints in exposed walls.

Masonry mortar is composed of one or more *cementatious materials* (normal portland cement, masonry cement, and hydrated lime), clean, well-graded *masonry sand,* and enough *water* to produce a plastic, workable mixture. In addition, *admixtures* (accelerators, retarders, and water-reducing agents) may be added for some special purpose.

A number of mortar types are recognized, based on strength and composed of varying amounts of cement and hydrated lime, by volume. Table 14-1 indicates the types and the proportions of ingredients in each case.

To obtain good workability and allow the development of the maximum strength possible, mortar ingredients must be thoroughly mixed. Whenever possible, the mixing should be done by machine, except when only a small quantity of mortar is required.

Mixing time should be from 3 to 5 minutes after all ingredients have been added. A shorter mixing time may result in poor-quality mortar, while a longer mixing time may adversely affect the air content of mortars made with air-entraining cements.

If the mortar becomes stiff because of water evaporation, it may be *retempered* by the addition of a little water and thorough remixing. However, if the stiffness is due to partial hydration, the material should be discarded. Mortar should be used within 2 hours after the original mixing if the temperature is above 80 °F (26 °C) or within 3 hours if the temperature is below that point.

Two methods are used for applying mortar to concrete masonry units. One is to apply the mortar to the two long edges only. This is known as *face-shell bedding* (see Fig. 14-3). The other is to apply mortar to the cross webs as well as the face shells, and this is known as *full mortar bedding* (see Fig. 14-4).

FIGURE 14-3: *Face-shell bedding.*

FIGURE 14-4: *Full mortar bedding.*

TABLE 14-1: *Mortar Types by Cement and Lime Proportions*

| | | Parts by volume | | |
Specification	Mortar type	Portland cement	Masonry cement	Hydrated lime or lime putty
	M	1	1	—
		1	—	¼
	S	½	1	—
		1	—	Over ¼ to ½
For plain masonry ASTM C270 CSA A179	N	—	1	—
		1	—	Over ½ to 1¼
	O	—	1	—
		1	—	Over 1¼ to 2½
	K	1	—	Over 2½ to 4
For reinforced masonry C476	PM	1	1	—
	PL	1	—	¼ to ½

(Courtesy Portland Cement Association).

Note: The total aggregate will not be less than two and one quarter or more than three and one half times the total volume of cementitious material.

FIGURE 14-5: *Test lay-up for spacing.*

Laying Blocks

The first step in laying blocks is to locate accurately the positions of the corners of the building on the footings and establish the line of the outside of the wall. This can be done by taking the dimensions from the plans and snapping chalk lines on the footing to indicate the corners and building lines. Then the first course may be laid without mortar to ascertain what spacing is required between blocks (see Fig. 14-5), although, in general, that spacing should be ⅜ in. (10 mm) in order to maintain a 16-in. (400-mm) module, center to center of mortar joints.

Now remove the blocks, spread a full bead of mortar long enough to accommodate at least three blocks, and lay the corner block, making sure that it is *to the line, plumb,* and *level.* Butter the ends of the face shells of the second block, bring it over its final position, and set it down into the mortar bed, while at the same time pressing it against the previously laid block to ensure a tight vertical joint (see Fig. 14-6).

FIGURE 14-6: *Full mortar bed for first course.*

Now butter the ends of several blocks, as shown in Fig. 14-7, so that they can be laid up in quick succession. After several blocks have been laid each way, use a straightedge and level to make sure that the blocks are *aligned,* brought to the *correct level,* and *plumb,* as illustrated in Fig. 14-8.

After the first course is laid, build up the corners as shown in Fig. 14-9. Use a tape to ensure that the bedding joints are maintained at the same thickness for each course, so that the four corners will remain level with one another. Use the straightedge and level frequently to make sure that the corners are plumb and level (see Fig. 14-10).

When the corners have been built up, the walls are completed between them. To do so, a line is run from corner to corner, along the top edge of the course to be laid (see Fig. 14-11). The line is held by a *line holder* (see Fig. 14-12) attached to each corner and adjusted so that the line is at the correct height—level with the top of the block. The line is then drawn as tight as possible to provide a horizontal guide for the blocks in the course.

FIGURE 14-7: *Mortar on block ends.*

FIGURE 14-8. *Block aligning, levelling, and plumbing.*

FIGURE 14-9: *Building up corners.*

FIGURE 14-10: *Plumbing and levelling corners.*

FIGURE 14-11: *Line from corner to corner.*   **FIGURE 14-12:** *Line holder in place.*

FIGURE 14-13: *Mechanical mortar spreader.*

The bed joint mortar is then laid, either by hand or by machine (see Fig. 14-13), and the ends of enough blocks buttered to complete the course (see Fig. 14-3). Set each block carefully and tap it down until it comes to the line (see Fig. 14-11).

The final block in each course is the *closure* block. Butter all the edges of the opening (see Fig. 14-14) and the four vertical edges of the closure block. Set it carefully into place (see Fig. 14-15) and make sure that the mortar is pressed firmly into the joint. Finally, remove any extruded mortar that appears on both the exterior and interior faces of the wall (see Fig. 14-16).

FIGURE 14-14: *Vertical edges buttered.*

FIGURE 14-15: *Closure block set in place.*

FIGURE 14-16: *Removing excess mortar extruded from joint.*

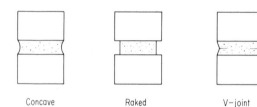

Concave   Raked   V–joint

FIGURE 14-17: *Tooled mortar joints.*

After the mortar is set hard enough that it can just be dented by the thumb nail, the joint may be *tooled*—shaped and compacted. Several types of tooled joints are made, including *concave, V-joint,* and *raked* (see Fig. 14-17), all of which are watertight if properly done. Figure 14-18 illustrates the tooling of a concave joint. Other types of joints are also used but are not as watertight as they are not tooled. Figure 14-19 shows two common styles—*flush* and *extruded.*

FIGURE 14-18: *Concave mortar joint.*

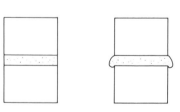

Flush   Extruded

FIGURE 14-19: *Non-tooled joints.*

It is sometimes necessary to use short lengths of block to complete a course. These will usually be cut from a block with a diamond-toothed blade, though, in some cases, the cutting is done by hand. The piece is introduced into the wall in exactly the same way as a whole or half block (see Fig. 14-20).

Plate Anchored to Block Wall

When a wooden plate is to be fastened to the top of a block wall, anchor bolts are used. They should be ½ in. (12 mm) in diameter, 18 in. (450 mm) long, and not more than 4 ft (1.2 m) apart. Lay a piece of metal lath over the cell in which the bolt will be set, two courses below the top of the wall, as shown in Fig. 14-21. When the wall is complete, fill the cell with concrete and set the bolt in place so that at least 2 in. (50 mm) project above the wall.

FIGURE 14-20: *Short piece of block built into wall.*

FIGURE 14-21: *Setting anchor bolts in block wall.*

Control Joints

Movements do occur in masonry walls from various kinds of stresses and if these are not controlled, cracking will occur in step fashion along the mortar joints. To control these movements and eliminate cracking, *control joints* are used. These are continuous vertical joints built into the masonry wall at critical points (see Fig. 14-22). The joint should first be laid up with mortar like any

FIGURE 14-22: *Constructing control joint.*

FIGURE 14-23: *Bond breaker at control joint.*

FIGURE 14-24: *Control joint lateral tie.*

other vertical joint, but the bond between blocks on either side of the joint should be minimized. This may be done by inserting pieces of asphalt-impregnated paper between blocks (see Fig. 14-23). To provide lateral support, metal ties are laid across the joint, in every second horizontal course, as shown in Fig. 14-24. Later, the control joint is raked out about ¾ in. 20 mm deep and the groove filled with caulking compound (see Fig. 14-25).

FIGURE 14-25: *Raking and caulking control joint.*

Intersecting Walls

Two methods are used to join two intersecting walls. One is to butt the cross wall against the inner face of the other, in which case whole and half blocks are required alternately, as the start of each course [see Fig. 14-26(a)]. In the other a *chase* (recess) is built into the main wall, and the blocks of the cross wall are set into it [see Fig. 14-26(b)]. In either case, a control joint is involved at the intersection, and the two walls should be tied together with metal straps or by wire mesh laid across the junction (see Fig. 14-27).

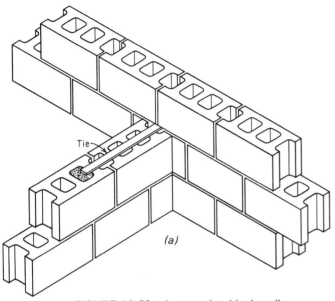

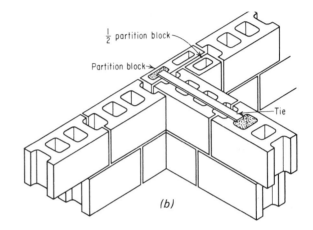

$\frac{1}{2}$ partition block

Partition block

Tie

FIGURE 14-26: *Intersecting block walls.*

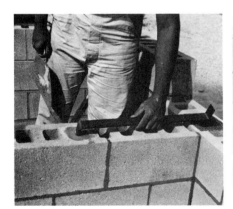

FIGURE 14-27:
Ties for intersecting walls.

Lintels

The wall load over door and window openings in block construction must be supported by a horizontal member which spans the opening and is carried on solid bearing at each end. Such a member is called a *lintel,* and it may be provided by two different methods. One is to use a precast concrete lintel, designed for the span, with its ends resting on the block at each side of the opening (see Fig. 14-28).

The other method involves the use of *lintel blocks,* illustrated in Fig. 14-2. They are laid across the top of the opening and filled with reinforced concrete. Lintel blocks must be supported until the concrete has hardened, and this may be done in one of two ways. One is to set the window frame in place and use it to support the blocks, as illustrated in Fig. 14-29. The other is to provide a temporary support, as shown in Fig. 14-30, which will be removed when the concrete has reached its design strength.

FIGURE 14-28: *Precast lintel in place.*

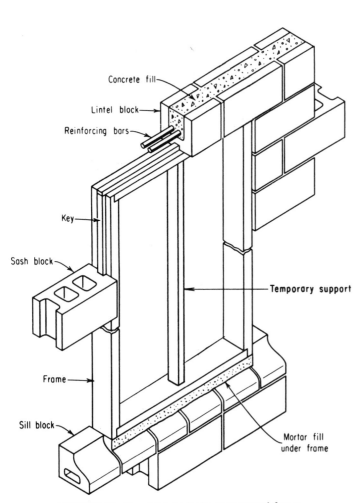

FIGURE 14-29: *Lintel blocks over wood frame.*

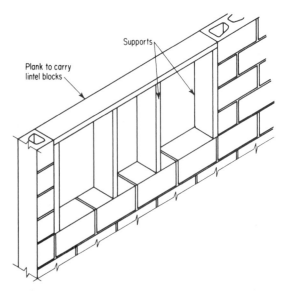

FIGURE 14-30: *Temporary support for lintel block.*

Floors Supported on Block Walls

Wood floors on block walls are usually supported by wooden floor joists resting on a *sill plate* anchored to the top of the wall. The plate is anchored by bolts set in *grout* cast into the block cores (see Fig. 14-21).

Concrete floors may consist of a solid *flat* slab, like that illustrated in Fig. 14-31.

When a flat slab is specified, the outer rim may be formed by a course of 4 in. 100-mm solid blocks, laid on top of the regular blocks and set flush with the outside face of the wall (see Fig. 14-31). The slab form will consist of a plywood deck supported from below on posts or shores.

Pilasters

A pilaster is a type of *column*, incorporated into a block wall for the purpose of providing additional lateral support and/or providing a larger bearing surface for beam ends carried on the wall.

In concrete block construction, the pilaster is formed by means of a *pilaster block*, one of which is illustrated in Fig. 14-2. Pilaster blocks may be filled with reinforced concrete to give them additional strength.

Cutting Concrete Blocks

Concrete blocks are usually made in half sizes as well as full-length units. However, it is sometimes necessary to cut a block to fit a particular location. This can be done in two ways. The block can be scored on both sides with a chisel, a shown in Fig. 14-32(a), and broken cleanly along the score lines. Blocks may also be cut with a masonry saw [Fig. 14-32(b)]. This saw is particularly useful when only a portion of the block is to be cut away.

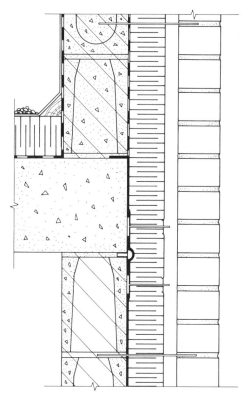

FIGURE 14-31: *Concrete slab and block wall connection.*

(a)   (b)

FIGURE 14-32: *Cutting concrete block.*

FIGURE 14-33: *Precast window sill.*

Window and Door Frames in Block Walls

A number of methods are used to set wood window and door frames into a concrete block wall. One method is to set the frame on *sill blocks,* as illustrated in Fig. 14-29. In place of sill blocks, a precast concrete sill may be installed at the bottom of the opening (see Fig. 14-33) and the frame set on it. A wood sill may be used for the same purpose, as illustrated in Fig. 11-58(d). Frames with no slope on their sill may be set directly on the blocks at the bottom of the opening.

Wood frames may be held in place by a *key* which is fastened to the outside of the frame and fits into the notches in the sash block framing the sides of the opening.

Metal straps, with one end nailed to the side jambs of the frame and the other laid in the mortar joints between the blocks along the sides of the opening, may also be used to anchor frames.

Steel window and door frames are commonly used in block construction (see Fig. 14-34). Flanges on their inner and outer edges form rebates into which standard width blocks will fit (see Fig. 14-35).

Cavity Walls

A cavity wall is constructed using two thicknesses (wythes) of block, separated by a continuous airspace (see Fig. 14-36) and tied together by wire ties embedded in the mortar joints in every second course.

When required, the space between the wythes can be filled with loose fill or rigid insulation for added protection against heat loss through the wall. If loose fill insulation is to be introduced, it is good practice to keep the inner faces of the walls free from

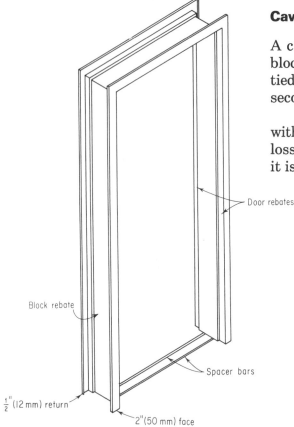

Door rebates

Block rebate

Spacer bars

$\frac{1}{2}$" (12 mm) return

2" (50 mm) face

FIGURE 14-34: *Steel door frame for block wall.*

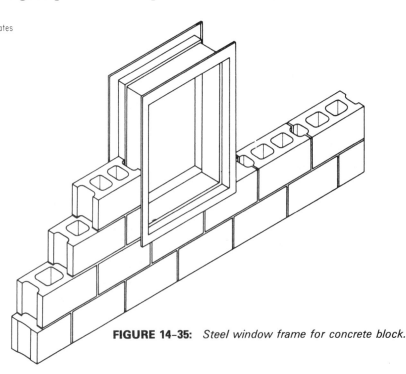

FIGURE 14-35: *Steel window frame for concrete block.*

FIGURE 14-36:
Cavity wall with cross ties.

protruding ridges of mortar, which could later interfere with the flow of loose fill insulation into the space. This can be done by the use of a *clean-out bar*—a strip of board slightly narrower than the width of the space between wythes. It has a light wire attached to each end and rests on a row of wire ties in the cavity. When the two following courses have been laid, the bar is raised by means of wires, bringing with it any mortar which has extruded from the joints or has dropped into the cavity (see Fig. 14-37).

When rigid insulation is specified, it is placed in the cavity as the blocks are being laid.

FIGURE 14-37: *Cavity clean-out bar.*

Reinforced Block Walls

Block walls may be reinforced either vertically or horizontally, and either single-wythe or cavity walls may be reinforced.

To reinforce single-wythe walls vertically, the reinforcing bars are introduced into the cores at the specified spacing, and those cores are filled with a high-strength concrete grout, with a relatively high slump of from 8 to 10 in. (200–250 mm). Two-core blocks are preferred to three-core because of the ease in placing reinforcement and grout, and special shaped units have been developed for use in single-wythe, reinforced block construction. They are *H-blocks,* illustrated in Fig. 14-2. The advantage of this unit is that it can be laid around vertical steel, rather than having to be threaded down over the rods.

Reinforcement is placed in the cavity of cavity wall construction, and the entire space between wythes is filled with grout.

Horizontal reinforcing is accomplished by making one complete course of block into a reinforced concrete beam at a specified level in the wall. Such a beam is known as a *bond beam.*

Several methods are used to construct a bond beam, one of which is to use bond beam blocks, illustrated in Fig. 14-2. A complete course of these is laid in a full mortar bed, with the corner blocks being mitered to allow continuous reinforcing around the corners. Reinforcing bars are placed in the block channels (see Fig. 14-38), which are subsequently filled with concrete. Further courses of block may then be added in the usual way above the bond beam.

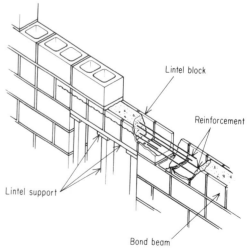

FIGURE 14-38: *Bond beam.*

Lintel block

Reinforcement

Lintel support

Bond beam

(a)

(b)

(c)

FIGURE 14-39: *Parging block wall.*

A bond beam may also be made by forming and casting an all-concrete beam the same width as the blocks and any desired height.

Watertight Block Walls

To ensure that block walls below grade will be watertight, they must be parged and sealed. Parging consists of applying two ¼-in. (6-mm) coats of plaster, using the same mortar that was used for laying the blocks (see Fig. 14-39). Dampen the wall before applying the plaster in order to get a better bond. The first coat should extend from 6 in. (150 mm) above the grade line down to the footing. When it is partially set up, roughen with a wire brush and then allow it to harden for at least 24 hours. Before the second coat is applied, the wall should be dampened again and the plaster kept damp for 48 hours.

In poorly drained soils, the plaster should be covered with two coats of an asphalt waterproofing, brushed on (see Fig. 14-40).

In heavy, wet soils, the wall may be further protected by laying a line of drainage tile around the outside of the footing to prevent a buildup of moisture in the area. Such a technique is illustrated in Fig. 14-41. The joints in the tile should be covered with the strips of asphalt-impregnated building paper and the tile covered with about 12 in. (300 mm) of coarse gravel before the backfilling is done.

FIGURE 14-40: *Waterproofing below grade.*

FIGURE 14-41: *Laying drainage tile.*

Protection of Block

Concrete blocks require some protection both *before* and *after* laying. Blocks should be laid *dry* and must therefore be protected from wetting by rain or snow by being kept in an enclosed space or by some type of cover, such as that illustrated in Fig. 14-42.

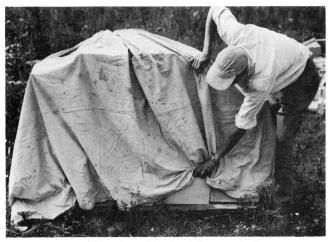

FIGURE 14-42: *Block protected from rain.*

FIGURE 14-43: *Block structure protected from drying out.*

Block should also be kept from getting too cold just prior to laying and may require a heated enclosure for that purpose.

After blocks are laid, the mortar should be protected from drying out too quickly in hot weather, and this may be accomplished by covering newly laid block with a tarpaulin, dampened if necessary (see Fig. 14-43).

In cold weather, block structures should be kept at a reasonable temperature [50 °F (10 °C) or better] for several days until the mortar has had an opportunity to harden properly. This may be done by some type of space heater.

BRICK CONSTRUCTION

The basic ingredient is *clay,* finely ground, mixed with water, molded, and burned in a kiln to form *brick*—one of the oldest building materials known to man.

Sizes and shapes of units have changed considerably over the years, and there is still some variation from one area to another. But as a result of consultation and cooperation among planners, designers, manufacturers, and governmental authorities, a great deal has been accomplished in the way of standardization of brick sizes and in the application of the principle of *modular coordination* to the manufacture of brick.

Brick Shapes

The best known shape is the *common brick,* universally recognized as a brick shape with a generally accepted set of dimensions. In addition, a number of other shapes have been developed, some regionally, some for a special purpose, and some for reasons of economy. Included in these are *Metro, Monarch, Giant, Bullnose, Cant, Radial, Slice,* and *Firebrick* (see Fig. 14-44). Each shape has a range of widths in which it is normally produced.

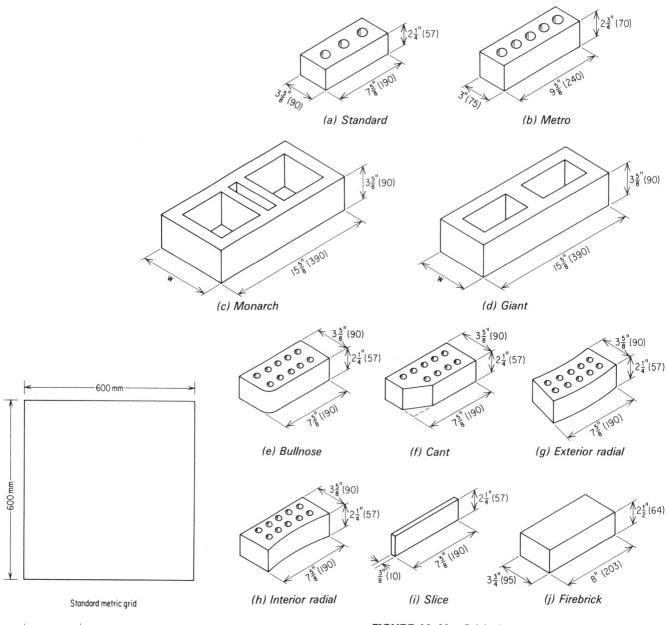

FIGURE 14-44: *Brick shapes.*

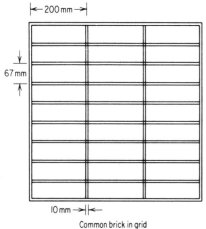

FIGURE 14-45: *Planning grid.*

Brick Sizes

It has been customary to designate unit masonry products by their *nominal* dimensions (sizes by which they are named), but their actual or *modular* dimensions are such that, measured center to center of mortar joints of specified thickness, the dimension will conform to a standard planning module.

In the Imperial System, modular coordination has resulted in the manufacture of brick of such sizes that when they are laid up three courses will equal 8 in. center to center of mortar joints. Another modular brick is made 3⅝ by 5⅝ by 15⅝ in., and many building codes permit its use in making single-wythe walls (see Fig. 14-44—Giant).

Under the SI metric system, the recognized planning modules are 100, 300, and 600 mm, with the internationally recognized *planning grid* being a square 600 ×600 mm. Under this system the recognized module for brick is 100 mm.

The nominal metric size for common brick is 100 × 67 × 200 mm, which means that, using standard 10-mm mortar joints, the actual size of common brick will be 90 × 57 × 190 mm. Thus one brick length, two brick widths, or three brick thicknesses will equal 200 mm—two modules. Three common brick modular lengths or nine thicknesses fit into the 600 × 600 mm planning grid (see Fig. 14-45).

Brick Terminology

Course types. Brick may be laid with four different exposures. One is with the long edge horizontal. This is called a *stretcher.* Another is with the end horizontal—a *header.* A third is with the long edge vertical—a *soldier,* and the fourth, with the end vertical, is a *rowlock* (see Fig. 14-46).

Brick bond. The word *bond* used in connection with brick may have several different meanings, depending on the context. The method by which the units are tied together in a wall, either by overlapping or by the use of ties, is called the *structural bond.* The pattern formed by the units in the exposed face of the wall is called the *pattern;* the adhesion of mortar to the brick is known as the *mortar bond.*

A great number of patterns are in use, some of the more common ones being running, common, Flemish, English, Dutch, and stack. Figure 14-47 illustrates these.

Stretcher   Header

Soldier   Rowlock

FIGURE 14-46: *Types of brick course.*

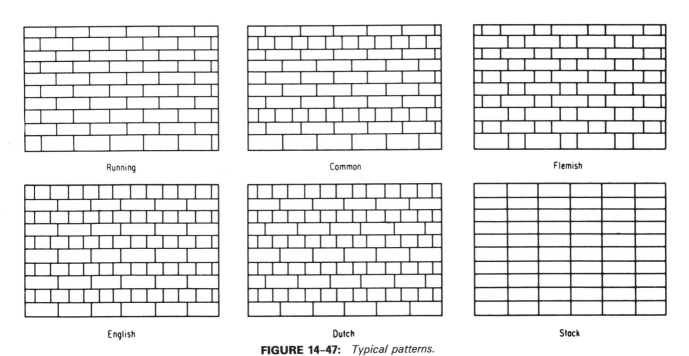

Running   Common   Flemish

English   Dutch   Stack

FIGURE 14-47: *Typical patterns.*

Types of Brick Construction

Brick may be used in walls in a number of ways in light construction. One is to use it as a *veneer* over a wood frame, concrete block, or concrete backup wall. Another is to build a *conventional* brick wall, which will be at least 3⅝ in. (190 mm) thick. Another is to build *cavity walls,* with brick for both exterior and interior wythes. In buildings having a skeleton frame of wood, steel, or reinforced concrete, brick may be used as a *curtain wall* to close in the spaces between framing members.

Brick Veneer

Over wood frame. A single wythe of brick, 3⅝ in. (90 mm) thick, is often used to face a sheathed wall, framed in wood. Both the building frame and the brick wythe must be supported on the foundation wall (see Fig. 14–48), and a 1-in. (25-mm) space should be left between brick and sheathing.

The brick must be anchored to the frame by noncorrosive metal straps, not less than 22 gauge (0.76 mm) thick and 1 in. (25 mm) wide, spaced in accordance with Table 14–2.

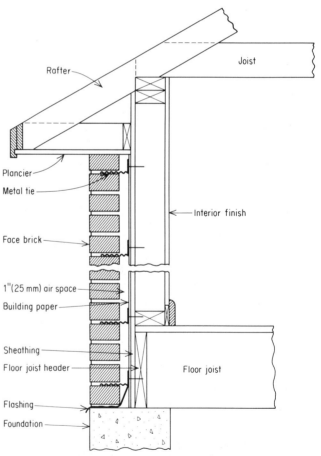

FIGURE 14–48: *Typical wall section, brick veneer on frame construction.*

TABLE 14-2: *Brick veneer tie spacing [in. (mm)]*

Horizontal spacing	Maximum vertical spacing
16 (400)	24 (600)
24 (600)	20 (500)
31 (800)	16 (400)

Over concrete block. Four, six-, or eight-inch (100-, 150-, or 200-mm) block may be used as a backup wall for a 4-in. (100-mm) brick veneer face, depending on the height of wall. This veneer should be separated from the backup by an unfilled space of at least 1 in. (25 mm) (see Fig. 14-49). When rigid insulation is used on the outside of the block backup, the space must be increased in size so the 1-in. (25-mm) space exists between the insulation and the facing (see Fig. 14-50). The facing wythe has to carry its own weight but otherwise is not considered to contribute to the vertical or lateral load resistance of the wall. The backup is designed to resist vertical or lateral loads. The space serves as a drainage gap between the veneer and the backup wall, and any forces on the facing are transmitted across the gap to the backup by ties.

Tying to the backup wall can be accomplished by using corrugated ties embedded in the mortar joints with the tie spacing regulated by Table 14-2 or by using continuous joint reinforcing in every second or third row of the block backup (see Fig. 14-51).

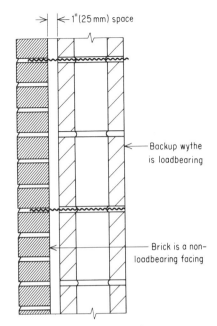

FIGURE 14-49: *Brick veneer.*

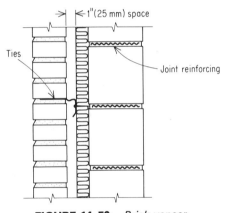

FIGURE 14-50: *Brick veneer with cavity insulated.*

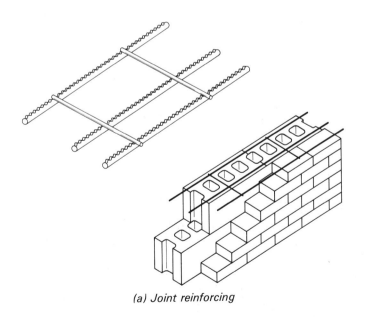

(a) Joint reinforcing

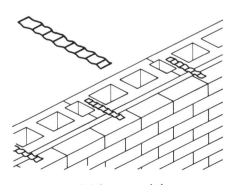

(b) Corrugated ties

FIGURE 14-51: *Joint reinforcing and ties.*

The brick veneer over openings in walls must be supported by a *lintel,* usually a steel angle with its ends supported on the brick on either side of the opening. The maximum allowable opening span depends on the size of lintel used, and Table 14–3 gives the maximum span allowed for various lintel angle sizes using both 3- and 3⅝-in. (75- and 90-mm) brick veneer.

TABLE 14–3: *Maximum opening span allowed for various lintel angle sizes*

Angle size [in. (mm)]	3-in. (75-mm) brick	3½-in. (90-mm) brick
3½ × 3 × ¼	8 ft, 4 in.	·—
(90 × 75 × 6)	(2.55 m)	—
3½ × 3½ × ¼	8 ft, 6 in.	8 ft, 1 in.
(90 × 90 × 6)	(2.59 m)	(2.47 m)
3⅞ × 3½ × ¼	9 ft, 2 in.	8 ft, 9 in.
(100 × 90 × 6)	(2.79 m)	(2.66 m)
4⅞ × 3½ × ⁵⁄₁₆	11 ft, 5 in.	10 ft, 10 in.
(125 × 90 × 8)	(3.47 m)	(3.31 m)
4⅞ × 3½ × ⅜	11 ft, 11 in.	11 ft, 5 in.
(125 × 90 × 10)	(3.64 m)	(3.48 m)

Conventional Brick Wall

The conventional brick wall is made by using headers and stretchers in various combinations or patterns, as illustrated in Fig. 14–47, and will be at least 7⅝ in. (190 mm) in thickness. Walls of this type may be used as the load-bearing walls in buildings of two stories or more. In such cases, the exterior walls of the bottom stories of two-story buildings and all the walls of three-story buildings must not be less than 7⅝ in. (190 mm) thick.

Single-Wythe Brick Walls

Solid brick walls in one-story buildings and the top story of two-story buildings may be constructed of 5½-in. (140-mm) solid units, provided that the wall is not over 9 ft, 2 in. (2.80 m) high at the eaves and not more than 15 ft, 1 in. (4.60 m) high at the peaks of the gable ends. Solid units or grouted hollow units are normally used for this type of construction.

On the inside surface, the wall may be *furred out,* and interior finish is applied to this furring. This furring allows for application of a vapor barrier, makes for easy installation of electrical facilities, and provides space for the introduction of insulation. However, it is possible to plaster directly to the inside face of the brick, or face brick may be used, in which case no other finish is required.

An 8 in. (200-mm) foundation wall is adequate, but the details of construction will depend on the type of floor being used. Figure 14–52 illustrates two typical methods of construction using a single-wythe wall.

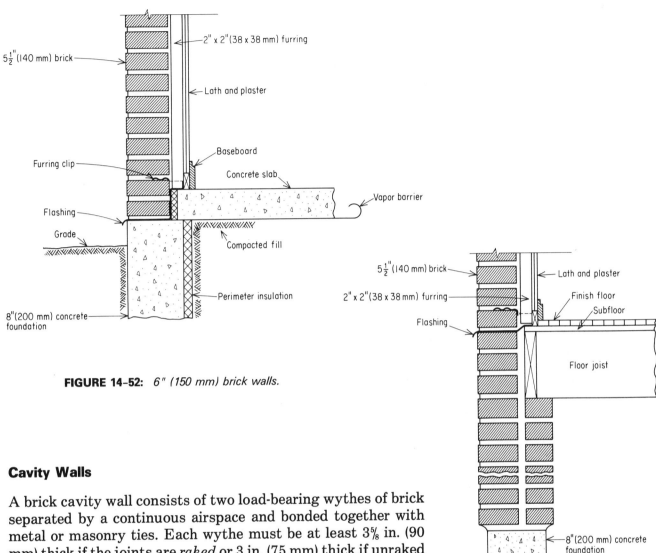

FIGURE 14-52: *6" (150 mm) brick walls.*

Cavity Walls

A brick cavity wall consists of two load-bearing wythes of brick separated by a continuous airspace and bonded together with metal or masonry ties. Each wythe must be at least 3⅝ in. (90 mm) thick if the joints are *raked* or 3 in. (75 mm) thick if unraked joints are used. The airspace must not be less than 2 in. (50 mm) or more than 3 in. (75 mm) wide if metal ties are used and not less than 3 in. (75 mm) or more than 4 in. (100 mm) wide if masonry ties are used.

The minimum thickness of cavity walls above the foundation must not be less than 9-¹⁄₁₆ in. (230 mm) for the top 11 ft, 9 in. (3.6 m) and not less than 11½ in. (290 mm) for the remaining lower portion when 3⅝ in.- (190-mm-) wide units are used to a maximum height of 23 ft, 7 in. (7.2 m). Where a cavity wall is non-load-bearing, the total thickness of wythes and cavities shall not be less than 9¹⁄₁₆ in. (230 mm).

The ties commonly used to provide the structural bond between the two wythes of a cavity wall are the continuous type and are spaced vertically every 16–24 in. (400–600 mm).

Flashing is provided at the bottom of cavity walls to direct any moisture which collects within the wall toward the outside wythe (see Fig. 14-53), and *weepholes* in the outside wythe allow that moisture to drain to the outside (see Fig. 14-54).

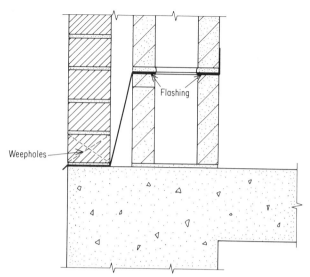

FIGURE 14-53: *Flashing at base of cavity wall.*

FIGURE 14-54: *Weephole in brick cavity wall.*

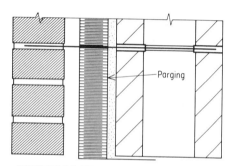

FIGURE 14-55: *Cavity wall with rigid insulation.*

As was the case in concrete block construction, the cavity in a brick cavity wall may be filled with insulation. In such a case, the outer face of the inner wythe should be sealed with an asphalt waterproof coating. A more common method is to use rigid insulation in the space cemented to the interior wythes, allowing an airspace between the back of the exterior wythe and the insulation so there is no migration of moisture to the inside (see Fig. 14-55).

Curtain Walls

Brick curtain walls used in skeleton frame construction may be *laid up in place* or *prefabricated* in a plant, transported to the site, and set into the opening as a unit (see Fig. 14-56). Once in place, the units are fastened to the frame and to one another by bolting or welding (see Fig. 14-57).

FIGURE 14-56: *Prefabricated brick panels ready for installation.*

FIGURE 14-57: *Welding section of prefabricated brick wall together.*

Laying Brick

One of the primary requisites for a good brick structure is good mortar. To be able to fulfill its purpose, mortar must possess a number of important qualities, both in the *plastic* stage and after it has hardened.

In the plastic stage, the important qualities are *workability, water retentivity,* and a *consistent rate of hardening.* Hardened mortar must have *good bond, durability,* good *compressive strength,* and *good appearance.*

Good workability is the result of a combination of factors, including the *quality of aggregate* used, the use of a small quantity of *hydrated lime or lime putty* (see Table 14–1), the use of *masonry cement,* the *amount of water* used, proper *mixing facilities,* and the ability of the mortar to *retain water.*

Mortar with good workability should *slip readily on the trowel, spread easily* on the masonry unit, *adhere to vertical surfaces,* and *extrude readily* from joints as the unit is being placed, without dropping. The consistency must be such that the unit can be properly bedded but its weight and the weight of following courses will not cause further extrusion of the mortar.

The ability of a mortar to retain water—its *water retentivity*—is very important because loss of water will result in premature stiffening of the mortar, which will prevent it from achieving a good bond and watertight joints.

The causes of poor water retentivity in mortar are likely to be *poorly graded aggregate, oversized aggregate, too short a mixing time,* or the *wrong type of cement.* The addition of an air-entraining agent and increased mixing may also improve the water retentivity.

To obtain good masonry construction, it is essential that all mortar joints be completely filled as the bricks are being laid. Failure to do so will result in voids through which water can penetrate. Not only will water pass through the wall to mar the interior finish, but water in the wall may dissolve salts from the brick and then deposit them on the surface as *efflorescence* when it returns to the outside and evaporates. In addition, water in the wall may freeze and cause deterioration in the wall itself.

Mortar for the bed joint should be spread thickly, with a shallow furrow down the center of the bed (see Fig. 14–58). There will then be enough excess mortar in the bed to fill the furrow and allow some mortar to be extruded at the joint when the bricks are bedded to the line. The bed mortar should be spread over only a few bricks at a time so that water will not evaporate before the bricks are laid and thus result in poor adhesion. Figure 14–59 illustrates mortar which has good adhesion qualities.

Care must be taken that all vertical joints in both stretcher and header courses are completely filled with mortar. To obtain a full head joint in a stretcher course, apply plenty of mortar to the end of the brick being placed, so that when it is set, mortar

FIGURE 14-58: *Furrowed bed joints.*

FIGURE 14-59: *Mortar with good adhesion.*

FIGURE 14-60: *Full head joint.*

will be extruded at the top of the head joint (see Fig. 14-60). To ensure full vertical cross joints in header courses, spread mortar over the entire side of the header brick to be placed, as illustrated in Fig. 14-61. When the brick is set, mortar should be extruded from the face and the top of the joint (see Fig. 14-61).

Closures in both stretcher and header courses need careful attention. Mortar should be spotted on the sides or ends of both bricks already in place, and both sides or ends of the brick to be placed should be well buttered. Then set the closure brick without disturbing those already in place (see Fig. 14-62).

Tooling the mortar joint compacts the mortar, making it more dense, and helps to seal any fine cracks between brick and mortar (see Fig. 14-63). A number of styles of joint are used, similar to those used in concrete block (see Figs. 14-17 and 19).

FIGURE 14-61:
Buttering and setting header brick.

FIGURE 14-62:
Setting closure brick.

FIGURE 14-63: *Before and after tooling joint.*

(a)

(b)

FIGURE 14-64: *a) Coursed stone ashlar. b) Random stone ashlar.*

STONE CONSTRUCTION

In modern light construction, stone is used almost entirely as a veneer—an exterior facing over a wood frame or unit masonry structural wall or as an interior decorative material used for fireplaces, mantels, feature walls, and finish floors.

Stone for this purpose is available in two forms. One is a small stone block, commonly known as *ashlar,* usually 2, 3, or 4 in. (50, 75, or 100 mm) thick, with regular or irregular face dimensions. Stones will usually not exceed 2 ft (600 mm) in length, while the height will vary from 4 to 12 in. (100–300 mm). The other is a thin, flat slab, from ½ to 1 in. (12-25 mm) thick, with either regular or random face dimensions.

Ashlar veneer is applied in the same way as brick veneer. The stone must rest on the foundation (see Fig. 14–48) and is bonded to the wall with metal ties, with one end nailed to a wood frame backup wall or laid in the mortar joints of a unit masonry wall. If the stones are cut to specific dimensions, they can be laid with regular course lines, and the result is known as *coursed ashlar.* But if the face dimensions are irregular, the result will be *random ashlar,* illustrated in Fig. 14-64.

Thin slabs of stone—slate, galena, or argillite—are fixed to the wall in a bed of mortar (see Chapter 11).

REVIEW QUESTIONS

14-1. Outline the primary purpose of mortar used in masonry construction.

14-2. What are the basic ingredients of a good mortar?

14-3. Under what circumstances is it not necessary to use hydrated lime in mortar?

14-4. Explain what is meant by *retempering* mortar.

14-5. What is the purpose of building up the corners of block walls first?

14-6. List four types of mortar joints that may be used in masonry work.

14-7. Describe how a *control joint* is made in block masonry.

14-8. What is the purpose of a control joint?

14-9. Describe two methods of bonding two intersecting block walls together.

14-10. What is the purpose of a *lintel* in masonry construction?

14-11. Give two reasons for using pilasters in masonry construction.

14-12. Outline two advantages of cavity walls in masonry construction.

14-13. What is the purpose of a *bond beam* in a block wall?

14-14. Describe two methods of fastening interior finishing material to the inside surface of a brick wall.

INDEX